Numerical Simulation in Biomechanics and Biomedical Engineering

Numerical Simulation in Biomechanics and Biomedical Engineering

Editor

Mauro Malvè

MDPI • Basel • Beijing • Wuhan • Barcelona • Belgrade • Manchester • Tokyo • Cluj • Tianjin

Editor
Mauro Malvè
Universidad Pública de Navarra
Spain

Editorial Office
MDPI
St. Alban-Anlage 66
4052 Basel, Switzerland

This is a reprint of articles from the Special Issue published online in the open access journal *Mathematics* (ISSN 2227-7390) (available at: http://www.mdpi.com).

For citation purposes, cite each article independently as indicated on the article page online and as indicated below:

LastName, A.A.; LastName, B.B.; LastName, C.C. Article Title. *Journal Name* **Year**, *Volume Number*, Page Range.

ISBN 978-3-0365-2211-1 (Hbk)
ISBN 978-3-0365-2212-8 (PDF)

Contents

About the Editor

Mauro Malvè is currently Associate Professor of Structural Mechanics at the Public University of Navarra located in Pamplona (Spain). He belongs to the Engineering Department of this institution. He moved to Pamplona after his Postdoctoral Fellowship at the Bioengineering Research Networking (CIBER-BBN) of the University of Zaragoza (Spain). He took the degree of Aeronautical Engineering at the Politecnico di Milano (Italy). After a brief experience in the industry, he did his PhD in Mechanical Engineering, with focus on the computational cardiovascular mechanics, at the University of Karlsruhe (Germany). Nowadays, his research work is focused on the computational fluid dynamics and on the fluid–structure interaction analysis applied to human and veterinary medicine. In particular, he has extensive experience in the numerical simulation of the respiration mechanics and of cardiovascular fluid dynamics. His interests also include the design and simulation of cardiovascular stents and respiratory prosthesis. He is currently a member of the Bioengineering Research Networking (CIBER-BBN) and of the European Society of Biomechanics. He maintains collaborations with several Institutions in Spain and around the world such as the University of Zaragoza (Spain); the University of Saskatchewan (Canada); the École Vétérinaire Maisons Alfort, where he performed three research stayings in the recent years; and the University of Grenobles-Alpes (France). He has published over 40 scientific publications indexed with the Journal of Citations Report, and has presented his research work in over 60 national and international conferences.

Preface to "Numerical Simulation in Biomechanics and Biomedical Engineering"

In the last decades, the improvement of the computational technology has allowed the introduction of advanced numerical models and high-performance simulations in several fields of engineering. In particular, biomedical engineering, which can be a bridge discipline between medicine and engineering, and combines the knowledge of several aspects of both fields, has received great attention from the scientific community for its direct relation to human health. In a more general meaning, biomedical engineering also includes the study of the processes related to nature and animals. Specific applications can be found in the understanding of human pathologies and diseases; in the advancement of the medical health care; and in the improvement of the diagnosis, therapies, medical devices, and clinical outcomes, among other aspects. However, biomedical engineering should theoretically also help to reduce the number of tests in animals, and should also contribute to the improvement of their health care. More recent applications can be found in the analysis of biological problems, such as the cells' culture and motility, and the microfluidic and diffusion processes.

Numerical methods and computer simulation have been widely used to help the biomedical engineering for providing computational models able to reproduce many aspects associated to the human medicine and to the biology. Considerable research has been obtained with the improvement of the computer performances that allows for the increase of more and more complexity in such in silico modeling. Despite the extensive investigation in this field and the large improvement in computer technology, the complex mechanism of different biological problems and related pathologies has been not fully understood. This is partially due to the difficulties to reproduce, with the necessary accuracy, the complexity of certain phenomena and the overall limitations of the computational and experimental modeling.

This e-book presents a collection of several examples of application of the numerical modeling to complex problems in the field of biomechanics and biomedical engineering. Some of the fields included in the book are tissue engineering, computational biofluid dynamics, structural analysis of muscle skeletal system and bone tissue, design and analysis of medical devices, 3D printing technique for the biomedical engineering, analytical and numerical solution of blood flow, and analysis of topological data.

The Editor thanks the contribution, effort, and dedication of the authors to describe and show by means of their papers some of the application of the mathematics by means of the numerical models to the biomedical engineering. Their recognized expertise in the mentioned fields of the biomechanics and biomedical engineering have contributed to the scientific quality of this book that will certainly be appreciated by the readers.

Mauro Malvè

Editor

Review

Wall Shear Stress Topological Skeleton Analysis in Cardiovascular Flows: Methods and Applications

Valentina Mazzi, Umberto Morbiducci, Karol Calò, Giuseppe De Nisco, Maurizio Lodi Rizzini, Elena Torta, Giuseppe Carlo Alp Caridi, Claudio Chiastra and Diego Gallo *

PoliTo^BIO^Med Lab, Department of Mechanical and Aerospace Engineering, Politecnico di Torino, Corso Duca degli Abruzzi 24, 10129 Turin, Italy; valentina.mazzi@polito.it (V.M.); umberto.morbiducci@polito.it (U.M.); karol.calo@polito.it (K.C.); giuseppe.denisco@polito.it (G.D.N.); maurizio.lodirizzini@polito.it (M.L.R.); elena.torta@polito.it (E.T.); giuseppe.caridi@polito.it (G.C.A.C.); claudio.chiastra@polito.it (C.C.)
* Correspondence: diego.gallo@polito.it; Tel.: +39-011-0906574

Abstract: A marked interest has recently emerged regarding the analysis of the wall shear stress (WSS) vector field topological skeleton in cardiovascular flows. Based on dynamical system theory, the WSS topological skeleton is composed of fixed points, i.e., focal points where WSS locally vanishes, and unstable/stable manifolds, consisting of contraction/expansion regions linking fixed points. Such an interest arises from its ability to reflect the presence of near-wall hemodynamic features associated with the onset and progression of vascular diseases. Over the years, Lagrangian-based and Eulerian-based post-processing techniques have been proposed aiming at identifying the topological skeleton features of the WSS. Here, the theoretical and methodological bases supporting the Lagrangian- and Eulerian-based methods currently used in the literature are reported and discussed, highlighting their application to cardiovascular flows. The final aim is to promote the use of WSS topological skeleton analysis in hemodynamic applications and to encourage its application in future mechanobiology studies in order to increase the chance of elucidating the mechanistic links between blood flow disturbances, vascular disease, and clinical observations.

Keywords: fixed points; manifolds; divergence; hemodynamics; computational fluid dynamics

Citation: Mazzi, V.; Morbiducci, U.; Calò, K.; De Nisco, G.; Lodi Rizzini, M.; Torta, E.; Caridi, G.C.A.; Chiastra, C.; Gallo, D. Wall Shear Stress Topological Skeleton Analysis in Cardiovascular Flows: Methods and Applications. *Mathematics* **2021**, *9*, 720. https://doi.org/10.3390/math9070720

Academic Editor: Mauro Malvè

Received: 24 February 2021
Accepted: 22 March 2021
Published: 26 March 2021

Publisher's Note: MDPI stays neutral with regard to jurisdictional claims in published maps and institutional affiliations.

1. Introduction

Recent advances in medical imaging, modeling, and computational fluid dynamics (CFD) have allowed the modeling of local hemodynamics in realistic, personalized cardiovascular models, fostering understanding of the association between local hemodynamics and the initiation and progression of vascular disease, and in a wider perspective, contributing to the translation of computational methods in real-world clinical settings to complement clinical information.

It has long been recognized that hemodynamic factors regulate several aspects of vascular pathophysiology [1,2]. Wall shear stress (WSS), the frictional force per unit area exerted by streaming blood on the endothelium, has been identified as a major biomechanical factor involved in vascular homeostasis. In fact, WSS is sensed through several vascular mechanosensors and biochemical machineries that regulate the expression of genes coding for extra- and intra-cellular proteins, playing a relevant role in the development, growth, remodeling, and maintenance of the vascular system [3,4]. In this scenario, a multitude of WSS-based descriptors of the near-wall hemodynamics has been proposed over the years to provide potential indicators of flow disturbances associated with aggravating biological events. In particular, regions at the luminal surface presenting with low [5] and oscillatory [6] WSS have been identified as localizing factors of vascular disease [3,6]. However, the complex hemodynamic milieu the endothelium is exposed to can be only

1

partially characterized by low and oscillatory WSS [7,8], as confirmed by a large body of literature reporting poor-to-moderate (and sometimes, contradictory) associations between low and oscillatory WSS with respect to vascular disease, e.g., [7–13]. This indicates a limited current understanding of the mechanistic link between WSS and vascular disease that hampers the use of WSS not only as a biomarker of vascular disease but also as a predictor of its progression within a clinical context [14].

Stimulated by the need to improve the understanding of the link between altered hemodynamics and clinical observations, the topological skeleton of the WSS vector field at the luminal surface of an artery is receiving increasing interest [15–20]. Based on dynamical system theory, the WSS topological skeleton is composed of a collection of fixed points, i.e., focal points where WSS locally vanishes, and unstable/stable manifolds, consisting of contraction/expansion regions linking fixed points. Such an interest arises from the ability of WSS topological skeleton features to reflect cardiovascular flow features like flow stagnation, separation and recirculation that are known to be promoting factors for vascular disease [2,17]. Very recent studies have demonstrated that the WSS topological skeleton governs the near-wall biochemical transport in arteries [15,16,18,21], which plays a fundamental role in, e.g., the initiation of atherosclerosis and thrombogenesis [22,23]. In addition, evidence of a direct association between WSS topological skeleton features and markers of vascular diseases from real-world clinical data have recently emerged [20,24].

In the present study, we report and discuss the theoretical background of Lagrangian- and Eulerian-based methods currently applied to the analysis of the WSS topological skeleton. Based on the recent promising findings highlighting a link between WSS topological skeleton features and markers of vascular disease [17–21,24], the aim of this study is to encourage the application of WSS topological skeleton analysis to cardiovascular flows as an ad hoc instrument that is potentially able to further elucidate the mechanistic link between WSS and vascular pathophysiology.

2. Topological Skeleton of a Vector Field

Topological features of a vector field have been largely studied in the context of dynamical systems theory. A dynamical system is defined as a set of n differential equations:

$$\dot{x}(t) = u(x,t);$$
$$x(t_0) = x_0, \tag{1}$$

where $t \in \mathbb{R}^+$ is the time, $x_0 \in \mathbb{R}^n$ the initial position at time point t_0, i.e., $x_0 = x(t_0)$, and $u(x,t)$ the velocity field. Given the initial condition $x_0 \in \mathbb{R}^n$, a unique solution of Equation (1) exists, called trajectory, given by:

$$x(t) = x(t_0) + \int_{t_0}^{t} u(x(s),s)\, ds. \tag{2}$$

Associated with the dynamical system defined in Equation (1), the so-called flow map can be defined as follows:

$$\Phi_{t_0}^{t} : x_0 \rightarrow x(t), \tag{3}$$

providing the expression of all the system trajectories at time t. In general, the topological skeleton of the vector field u is recognized to provide the organizing structures of the system itself.

In steady-state conditions (i.e., when vector field $u(x,t)$ in Equation (1) does not explicitly depend on time), the topological skeleton of a vector field consists of a collection of fixed points (Figure 1A) and the associated stable and unstable manifolds connecting them (Figure 1B). A fixed point (or critical point) is a point $x_{fp} \in \mathbb{R}^n$ where the vector field locally vanishes. The nature of fixed points can be stable or unstable. A stable fixed point is characterized by a sink configuration, and it attracts the nearby trajectories, while an unstable fixed point is characterized by a source configuration, and it repels the nearby trajectories (Figure 1A).

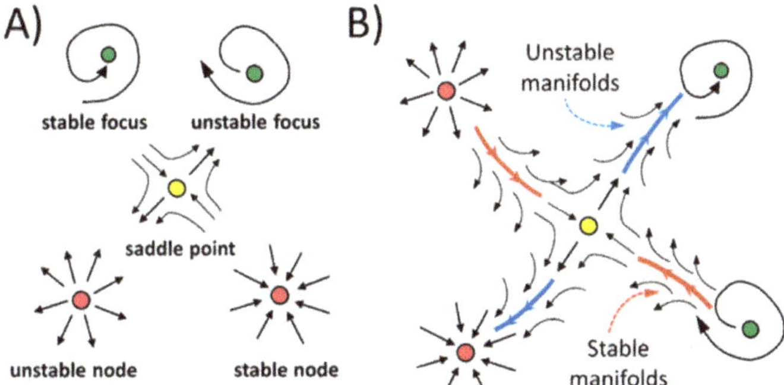

Figure 1. (**A**) Possible configurations for a fixed point of a vector field. (**B**) Explanatory sketch of the stable/unstable manifolds connecting fixed points.

A fixed point can be classified as a saddle point, node, or focus (Figure 1A): (1) a saddle point is a point attracting and repelling nearby trajectories along different directions (i.e., where the streamlines of the vector field intersect themselves); (2) a stable/unstable node is characterized by converging/diverging streamlines; (3) a focus is characterized by spiraling trajectories, and it can be attracting or repelling.

Technically, the exact location of fixed points in a domain of interest can be identified by computing the Poincaré index [25], a topological invariant index quantifying how many times a vector field rotates in the neighborhood of a point. For the sake of simplicity, we consider the dynamical system in Equation (1) under steady-state conditions and lying in a 2D space, i.e., $u(x) = (X(x), Y(x))$, with $x \in \mathbb{R}^2$. An explanatory example of how to calculate the Poincaré index can then be provided. Let $x_{fp} \in \mathbb{R}^2$ be an isolated fixed point of u with a neighborhood N such that there are no other fixed points in N than x_{fp}, and let γ be a closed curve inscribing N. Then, the Poincaré index $\mathcal{I}(\gamma, u)$ of the curve γ relative to u is the number of the positive field rotations while traveling along γ in a positive direction:

$$\mathcal{I}(\gamma, u) = \frac{1}{2\pi} \int_\Gamma d\theta = \frac{1}{2\pi} \int_\Gamma d\, arctan\left(\frac{Y}{X}\right), \tag{4}$$

where θ is the vector field rotation angle. The Poincaré index is equal to -1 at saddle point locations (Figure 1A), 1 at node or focus locations (Figure 1A), and 0 otherwise. The algorithm for computing the Poincaré index for a 3D vector field defined on unstructured triangle meshes is extensively described elsewhere [19].

The Poincaré index allows identifying fixed point locations, but it does not provide information about the fixed points nature. Therefore, a criterion to distinguish between a node or a focus and between the attractive or repelling nature of a fixed point is needed. In light of this, the vector field u around the fixed point x_{fp} can be expressed by linearization as:

$$u(x) = u\left(x_{fp}\right) + J\left(x_{fp}\right)\left(x - x_{fp}\right), \tag{5}$$

where J is the Jacobian matrix of u. The classification of fixed points can be thus performed by computing the eigenvalues of the Jacobian matrix J, as summarized in Table 1. In detail, two real eigenvalues with different signs identify a saddle point. Two real eigenvalues with the same sign identify nodes characterized as attracting or repelling (i.e., stable or unstable, respectively) according to their sign (negative or positive, respectively). Complex conjugate eigenvalues identify a stable or unstable focus according to the sign of the real part (negative or positive, respectively).

Table 1. Classification of fixed points based on the eigenvalues of the Jacobian matrix.

λ	Fixed Point
$\lambda_1 < 0 < \lambda_2$	Saddle point
$\lambda_1, \lambda_2 > 0$	Unstable node
$\lambda_1, \lambda_2 < 0$	Stable node
$\lambda_{1,2} = \alpha \pm \beta i$	Unstable focus
$\lambda_{1,2} = -\alpha \pm \beta i$	Stable focus

The stable and unstable manifolds (or critical lines) associated with a fixed point x_{fp} are composed of all initial conditions $x_0 \in \mathbb{R}^n$ such that the trajectories initiated at these points x_0 approach the fixed point x_{fp} asymptotically. By construction, stable and unstable manifolds act as separatrices of the vector field, portioning regions of different behavior and dynamics. In detail, an unstable manifold attracts nearby trajectories, as opposed to the stable manifold, which repels nearby trajectories (Figure 1B). In mathematical terms, an unstable manifold W^u associated with the generic fixed point x_{fp} is defined as follows:

$$W^u\left(x_{fp}\right) = \left\{ x_0 \in \mathbb{R}^n \ : \ \Phi_{t_0}^t(x_0) \to x_{fp} \text{ as } t \to +\infty \right\}, \tag{6}$$

while a stable manifold W^s can be expressed as:

$$W^s\left(x_{fp}\right) = \left\{ x_0 \in \mathbb{R}^n \ : \ \Phi_{t_0}^t(x_0) \to x_{fp} \text{ as } t \to -\infty \right\}. \tag{7}$$

In general, two different perspectives have been proposed to identify manifolds of a vector field, namely the Lagrangian and Eulerian perspectives. The Lagrangian perspective considers individual particles, tracking their motion along their paths as they are advected by the flow field. By contrast, the Eulerian perspective considers the properties of the vector field under analysis at each fixed location in space and time. In the following sections, a brief theoretical background is reported for a better understanding of the theory supporting the Lagrangian and Eulerian approaches for the analysis of vector field topology, with particular emphasis on their application to cardiovascular flows.

3. Lagrangian Approach

3.1. Lagrangian Coherent Structures

When the vector field $u(x, t)$ in Equation (1) is time-dependent, solutions can be complex and chaotic, making the interpretation of the topological skeleton made of W^u, W^s and x_{fp} difficult. The need to robustly define intrinsic structures governing fluid/mass transport under unsteady-state conditions has led to the development of the concept of coherent structures (CS). Technically, CS are defined as emergent patterns, influencing the transport of tracers/mass in time-dependent flows [26]. In this context, Lagrangian Coherent Structures (LCS) are coherent structures identified by applying methods based on a Lagrangian approach. The theoretical bases of LCS lie in methods of nonlinear dynamics, chaos theory, and fluid dynamics.

From a mathematical perspective and in relation to fluid mechanics, LCS can be defined as material surfaces in the flow field that are dominant in attracting or repelling neighboring fluid elements over a defined time interval [27,28]. These material surfaces are able to localize where the flow field experiences the largest and the smallest stretching [29]. In detail, material surfaces in the flow field attracting trajectories more strongly than any other nearby material surface are referred to as attracting LCS. Oppositely, material surfaces repelling trajectories more strongly than any other nearby material surface are referred to as repelling LCS.

The detection and visualization of LCS is usually performed by applying two different Lagrangian-based approaches, namely (1) Lagrangian particle tracking and (2) the computation of the finite-time Lyapunov exponent (FTLE). Both approaches are based on particle path information derived from the post-processing of velocity data obtained by CFD simulation or by in vivo (e.g., phase contrast magnetic resonance imaging (MRI)) and in vitro (e.g., particle image velocimetry) measurements. The workflow of the Lagrangian-based approaches to visualize LCS is sketched in Figure 2.

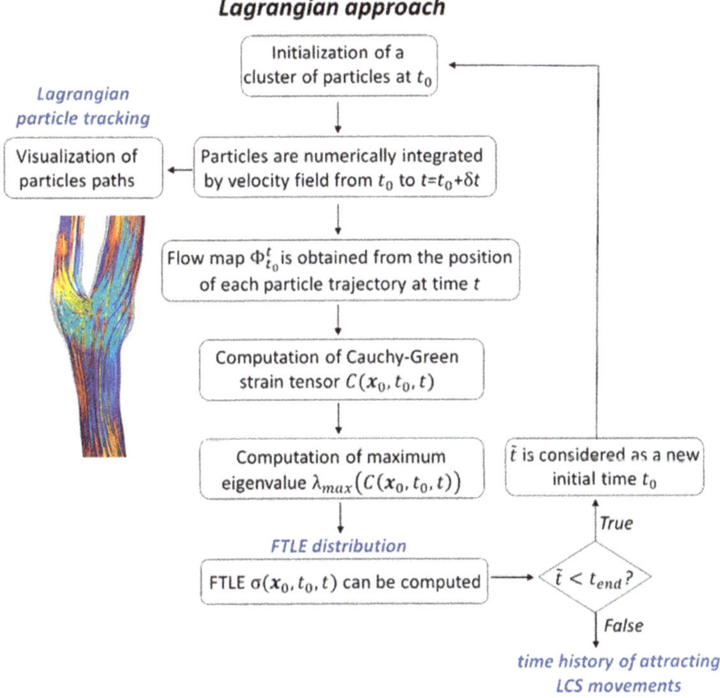

Figure 2. Workflow of the Lagrangian-based approaches to visualize attracting Lagrangian coherent structures (LCS) starting from a cluster of particles at time t_0 over the domain of interest. The same procedure applies to repelling LCS by considering reversing time. FTLE: finite time Lyapunov exponent.

The Lagrangian particle tracking is performed by seeding the domain of interest with tracer particles and by visualizing their motion (Figure 2). The aim of this approach is to reveal coherent features revealing how the flow under analysis is organized. From a mathematical perspective, the position of a tracer particle is governed by the differential equation reported in Equation (1). To obtain the position of such a particle at a desired time t, Equation (1) is numerically integrated from t_0 to t. The direct integration of tracer particles allows for an in-depth understanding of how tracers are transported through the domain of interest. In detail, attracting LCS will be generally distinguishable, since tracer particles are attracted to and along these surfaces (Figure 3). Analogously, repelling LCS will be distinguishable from the advection of tracer particles by reversing time (Figure 3). Attracting LCS are traced out with forward time integration of particles, while repelling LCS are traced out with backward time integration of particles.

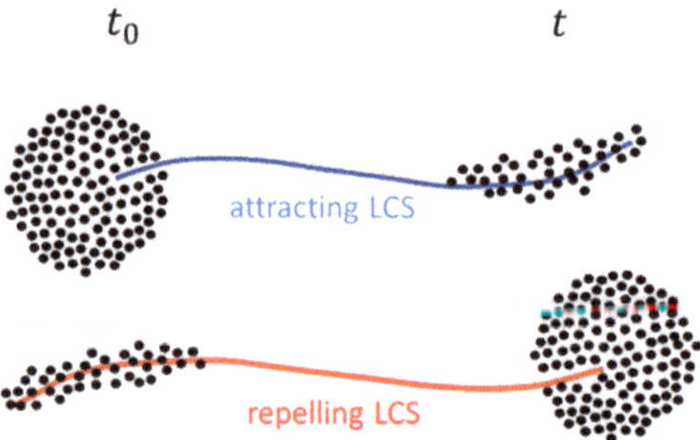

Figure 3. Explanatory sketch of attracting and repelling LCS over time interval $[t_0, t]$. A sphere of tracer particles released at time t_0 will spread out along the attracting LCS (time t). The opposite occurs for a repelling LCS.

Lagrangian particle tracking represents a Lagrangian-based technique aiming at overcoming issues related to standard approaches used for topological skeleton extraction of vector fields with unsteady-state conditions. However, the resulting tracer particle motion complexity could obscure the interpretation of the vector field topology. For this motivation, the second approach consists of the computation of the FTLE (Figure 2). Based on theory, a LCS can be defined as the material surface locally maximizing the FTLE [27,30], the Lyapunov exponent being a measure of the sensitivity to the initial position of a dynamical system. Technically, the finite time Lyapunov exponent $\sigma(x_0, t_0, t)$ [27,28,31–35] is defined as:

$$\sigma(x_0, t_0, t) = \frac{1}{|t - t_0|} \ln \sqrt{\lambda_{max}(C(x_0, t_0, t))}, \tag{8}$$

where $\lambda_{max}(C(x_0, t_0, t))$ is the maximum eigenvalue of the right Cauchy–Green strain tensor $C(x_0, t_0, t)$:

$$C(x_0, t_0, t) = \nabla\Phi_{t_0}^t(x_0)^T \nabla\Phi_{t_0}^t(x_0), \tag{9}$$

where $\nabla\Phi_{t_0}^t(x_0)^T$ denotes the transpose of the gradient of the flow map in Equation (3). From a physical point of view, $C(x_0, t_0, t)$ in Equation (9) represents the material deformation of infinitesimal volume elements of fluid, and it is a symmetric and positive-definite matrix. Roughly speaking, the FTLE σ defined in Equation (8) measures the rate of separation of initially close vector field trajectories. Let δ_0 be a small distance between two material points at time t_0, as depicted in Figure 4 (Panel A). It can be demonstrated [26] that the separation δ_t after the time interval $|t - t_0|$ satisfies the inequality:

$$||\delta_t|| \leq e^{\sigma(x_0, t_0, t)|t - t_0|}||\delta_0||, \tag{10}$$

where equality holds if the initial distance δ_0 is aligned with the eigenvector of $C(x_0, t_0, t)$ associated with λ_{max}.

The algorithm for LCS identification based on FTLE computation starts with the initialization of a cluster of massless elemental particles at time t_0 over the domain of interest (Figure 2). Then, particles are numerically integrated by the field in Equation (1) from t_0 to t, and their trajectories are calculated. The flow map $\Phi_{t_0}^t$ (Equation (3)) is obtained from the final position of each particle trajectory at time t in the domain, and subsequently its gradient $\nabla\Phi_{t_0}^t(x_0)$ can be computed. For a structured grid like the one shown in Figure 4 (panel B), $\nabla\Phi_{t_0}^t(x_0)$ can be calculated by finite differencing, e.g., using central differencing as follows:

$$\nabla \Phi_{t_0}^{t}(x_0) \approx \begin{bmatrix} \dfrac{x_{(i+1)jk}(t)-x_{(i-1)jk}(t)}{x_{(i+1)jk}(t_0)-x_{(i-1)jk}(t_0)} & \dfrac{x_{i(j+1)k}(t)-x_{i(j-1)k}(t)}{y_{i(j+1)k}(t_0)-y_{i(j-1)k}(t_0)} & \dfrac{x_{ij(k+1)}(t)-x_{ij(k-1)}(t)}{z_{ij(k+1)}(t_0)-z_{ij(k-1)}(t_0)} \\[1.2em] \dfrac{y_{(i+1)jk}(t)-y_{(i-1)jk}(t)}{x_{(i+1)jk}(t_0)-x_{(i-1)jk}(t_0)} & \dfrac{y_{i(j+1)k}(t)-y_{i(j-1)k}(t)}{y_{i(j+1)k}(t_0)-y_{i(j-1)k}(t_0)} & \dfrac{y_{ij(k+1)}(t)-y_{ij(k-1)}(t)}{z_{ij(k+1)}(t_0)-z_{ij(k-1)}(t_0)} \\[1.2em] \dfrac{z_{(i+1)jk}(t)-z_{(i-1)jk}(t)}{x_{(i+1)jk}(t_0)-x_{(i-1)jk}(t_0)} & \dfrac{z_{i(j+1)k}(t)-z_{i(j-1)k}(t)}{y_{i(j+1)k}(t_0)-y_{i(j-1)k}(t_0)} & \dfrac{z_{ij(k+1)}(t)-z_{ij(k-1)}(t)}{z_{ij(k+1)}(t_0)-z_{ij(k-1)}(t_0)} \end{bmatrix}. \tag{11}$$

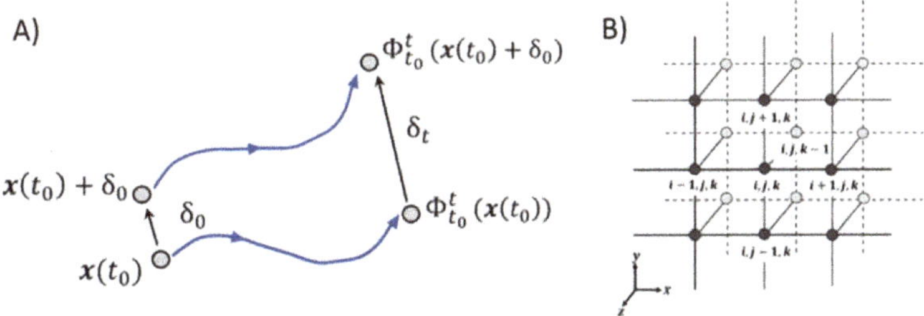

Figure 4. (**A**) Explanatory sketch illustrating the separation of nearby particles due to the flow map $\Phi_{t_0}^{t}$, during time interval $|t - t_0|$. (**B**) Nodal indexing of elemental cells in a 3D-structured mesh. Indices i, j, k represent the positions along the x, y, z directions, respectively.

Once the flow map gradient is obtained, the Cauchy–Green strain tensor $C(x_0, t_0, t)$ can be computed according to Equation (9).

Finally, the maximum eigenvalue $\lambda_{max}(C(x_0, t_0, t))$ and the FTLE $\sigma(x_0, t_0, t)$ can be computed according to Equation (8) (Figure 2). The obtained $\sigma(x_0, t_0, t)$ value for each particle is assigned to the particle position at time t_0. This procedure is repeated, varying the time t_0 (e.g., within the cardiac cycle in cardiovascular applications) and aiming at providing the time series of FTLE values and ultimately the time history of LCS movements (Figure 2). Positive integration times reveal repelling LCS in the FTLE field, while negative integration times reveal attracting LCS in the FTLE field.

In general, the computation of the spatial variation of the FTLE field requires the vector field to be interpolated in both time and space, and high-order integration and interpolation schemes are needed to ensure accuracy of results. Furthermore, the mesh used to compute the FTLE distribution over the domain of interest usually needs to be more resolved than the computational mesh for a more robust detection of LCS.

3.2. LCS Application to Intravascular Flows

Lagrangian-based approaches have been largely applied to identifying LCS in intravascular flows. Indeed, Lagrangian particle tracking has been massively applied to explore the complexity of intravascular flows, e.g., to provide a measure of stasis in idealized computational bifurcation models [36], or to study vortices generation and their potential role in thrombogenesis in idealized aneurysm models [37,38]. Several studies have applied particle tracking to identify flow disturbances in, e.g., carotid bifurcation models, contributing to providing a deeper understanding of the hemodynamics-driven processes underlying atherosclerosis onset/progression [39–42]. Moreover, particle tracking has been used to study the hepatic perfusion in the Fontan circulation [43,44], identify the optimal left ventricular assist device cannula outflow configurations [45], obtain a deeper understanding of the dynamics of embolic particles within arteries [46], and detect peculiar intravascular helical flow patterns in the aorta from in vivo, 4D-flow MRI data [47,48].

Regarding the FTLE-based analysis of the flow field, its extension to intravascular flows is relatively recent, motivated by the fact that LCS are determined by blood flow structures associated to adverse vascular events including flow stagnation, separation, and recirculation. Among the main contributions, here we mention that Shadden and Taylor [32] used LCS to quantify the extent of flow stagnation to determine where flow separated and to understand how flow was partitioned to downstream vasculature in computational hemodynamic models of large vessels. LCS have been proposed as a powerful method to capture vortex transport in blood flow. In this regard, Arzani and Shadden [33] used LCS to characterize the hemodynamics in abdominal aortic aneurysm (AAA) models, suggesting that AAA intravascular flow topology is dictated by systolic vortex propagation through the abnormal vessel. Arzani et al. [49] computed FTLE fields and associated LCS to capture a large coherent vortex in AAA computational models. Furthermore, LCS have been applied to identify left ventricle (LV) blood flow features during heart filling. In detail, Gharib et al. [50] used LCS to demonstrate the existence of a link between the vortex ring formation inside the LV and the ejection fraction. Charonko et al. [51] quantified the vortex ring volume by computing LCS from in vivo LV phase contrast MRI data of healthy and diseased patients. Töger et al. [52] extracted LCS from in vivo LV phase contrast MRI data to measure the vortex ring volume during LV rapid filling. The identification of attracting and repelling LCS from LV Doppler-echocardiography data was adopted as a criterion to discriminate between healthy and diseased patients [53]. Other studies applied LCS to characterize the flow field through heart valves. In particular, LCS were extracted to delineate the boundaries of the outflow jet downstream of aortic valves and used as a measure of the severity of the valve's stenosis [54,55]. In a very recent study [56], FTLE-based LCS detection on computational hemodynamics models of aortic bicuspid and mechanical heart valves was used to study mass transport processes that might be related to valve disease. The analysis of the fluid dynamics in the neighborhood of blood clots was another effective application of LCS to hemodynamics [57]. In addition, FTLE-based analysis was adopted to highlight the hemodynamic impact of flow diverter stents in the treatment of intracranial aneurysms [58,59].

We refer the interested reader to reference [31] for a broader, detailed overview of Lagrangian methods used in post-processing of velocity data in cardiovascular flows.

3.3. LCS Application to Near-Wall Flow Features

Recently, in the study of cardiovascular flows, the concept of LCS has been extended to analyze the near-wall flow topology, i.e., the topology of the flow field close to the luminal surface of arteries. The rationale is in the well-established involvement of near-wall mass transport in most of the processes concurring to determine vascular pathophysiology [5]: in the near-wall region, blood flow regulates the local biotransport processes and imparts mechanical shear stress on the endothelium (i.e., the WSS), which in turn regulates important developmental, homeostatic, and adaptive mechanisms in arteries, as well as susceptibility to and progression of atherosclerosis [1].

Based on theory, it has been demonstrated [60] that the WSS vector field can be scaled to provide a first-order approximation for the near-wall blood flow velocity vector field as follows:

$$u_\pi = \frac{\tau \delta n}{\mu} + \mathrm{O}\left(\delta n^2\right), \tag{12}$$

where $u_\pi \in \mathbb{R}^3$ is the near-wall velocity, $\tau \in \mathbb{R}^3$ represents the WSS vector field, μ is the dynamic viscosity, and δn is the distance from the wall where the velocity is evaluated. By construction, the vector field in Equation (12) is defined on the luminal surface of the vessel, and it represents the near-wall velocity, as the velocity is zero on the surface itself due to the no-slip condition. The LCS underlying theory described in Section 3.1 can be extended to analyze the near-wall flow topology by using the expression of near-wall velocity u_π (given by Equation (12)) in Equation (1). Such near-wall Lagrangian structures, computed from the WSS vector field, are referred to as WSS LCS [15].

Computationally, WSS LCS can be identified on the luminal surface of the vessel by numerically integrating a high number of luminal surface tracer particles, applying the procedure described in the first part of Section 3.1. In detail, attracting and repelling WSS LCS can be traced out with forward and backward time integration of surface tracer particles based on the near-wall blood flow velocity (Equation (12)), respectively.

The recent interest in WSS LCS from the cardiovascular fluid mechanics research community was driven by WSS LCS ability to highlight blood flow features associated with vascular disease initiation and progression, like flow stagnation, separation, recirculation, flow impingement, and the interaction of vortex structures with the vascular wall [19,61,62] These blood flow features have been classified as "aggravating flow events", as they trigger biological processes leading to the development or progression of vascular disease [2,17]. An example of attracting WSS LCS on the luminal surface of a patient-specific computational hemodynamic model of carotid bifurcation is presented in Figure 5. Details on the carotid bifurcation hemodynamic modeling are reported elsewhere [9,14,20,63]. In this specific case, luminal surface tracer particles (Figure 5A) are numerically integrated in forward time. The resulting LCS is located at the carotid bulb, a region characterized by flow disturbances (slow, recirculating blood flow) promoting atherosclerosis [2,4]. In detail, the attracting WSS LCS provides the boundary at the luminal surface of the slow vortex structure formed inside the carotid bulb (Figure 5C, where the recirculation region is highlighted visualizing the streamlines of the cycle-average velocity vector field).

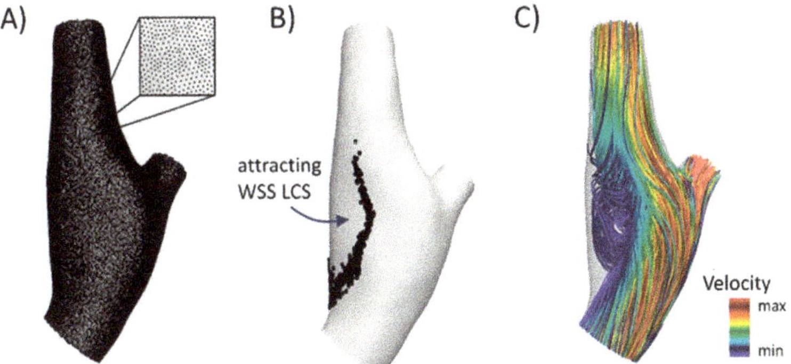

Figure 5. (**A**) Initial tracer particle position on the luminal surface of a carotid bifurcation model. (**B**) Attracting wall shear stress Lagrangian coherent structures (WSS LCS) traced out from forward time integration of WSS trajectories. (**C**) Streamlines of the cycle-average velocity vector field, colored by cycle-average velocity magnitude.

In addition, the shear forces exerted by the streaming blood flow in the near-wall region on the endothelium affect biotransport processes, i.e., the transport of biochemicals through the subendothelial layer [22]. Biotransport is of paramount importance in many cardiovascular processes, including the initiation of atherosclerosis and thrombogenesis [23]. In general, cardiovascular mass transport is investigated in silico by coupling the governing equations of motion, the Navier–Stokes equations, with the advection–diffusion equation, given by:

$$\frac{dC}{dt} + \boldsymbol{u} \cdot \nabla C - D\nabla^2 C = 0, \tag{13}$$

where C is a non-dimensional concentration of the species transported in the domain, $\boldsymbol{u}$ is the fluid velocity vector, and D is the mass diffusion coefficient. However, high computational costs are associated with the class of numerical simulations used to accurately solve the near-wall transport and blood-wall transfer [64,65], making this approach expensive in hemodynamics applications. To overcome this limitation, and based on the well-established role that WSS plays in conditioning the permeability of the endothelium and the near-wall mass transport process, recent studies [15,18] have brilliantly demonstrated that WSS LCS can be used as a template for near-wall mass transport. This allows reduction of the computational effort needed to solve the full transport problem, represented by Equation (13) [15]. In particular, it has been demonstrated that attracting WSS

LCS attract biochemicals, leading to high near-wall concentration in their neighborhood, whereas repelling WSS LCS have been shown to act as near-wall transport barriers [15,17].

In the context of cardiovascular flows, it has been recently demonstrated that attracting/repelling WSS LCS on the luminal surface of an artery match the unstable/stable manifolds of the cycle-average WSS vector field [15,18], defined as:

$$\overline{\tau} = \frac{1}{T}\int_0^T \tau(x,t)dt, \tag{14}$$

where τ is the instantaneous local WSS value and T is the time duration of the cardiac cycle. Technically, the first step in the topological analysis of cycle-average WSS at the luminal surface of a vessel is the identification of WSS fixed points. The exact position of WSS fixed points can be identified by computing, e.g., the Poincaré index, as explained in Section 2. Then, the cycle-average WSS field around a fixed point x_{fp}, according to Equation (5), by linearization can be expressed as:

$$\overline{\tau}(x) = \overline{\tau}\left(x_{fp}\right) + J\left(x_{fp}\right)\left(x - x_{fp}\right), \tag{15}$$

where J is the Jacobian matrix of $\overline{\tau}$ (see Equation (14)). The identified WSS fixed points can be classified according to their nature (i.e., saddle, node, or focus, Figure 1A) by analyzing the eigenvalues of the Jacobian matrix J of $\overline{\tau}$ (Table 1), as described in Section 2. Note that the WSS vector field is embedded in a three-dimensional space even if it lies in a two-dimensional space (the luminal surface of a vessel). To perform a two-dimensional analysis, two strategies are possible. In the first strategy, a projection of the vector field into two orthogonal directions (hence, in a two-dimensional space) is needed. In the second one, avoiding the projection of the vector field (and thus reducing the computational steps), a three-dimensional analysis is performed, thus obtaining three eigenvalues of the Jacobian matrix, with one of them having a value close to zero. Then, the eigenvalue-based analysis for the WSS fixed points classification considers only the two eigenvalues different from zero.

Saddle-type fixed points are of particular interest, since typically a stable or unstable manifold starts from a saddle point and vanishes into a source or sink, respectively, as depicted in Figure 1B. Saddle point locations (where the Poincaré index is -1 and the eigenvalues are real with different signs) are perturbed along the positive eigenvector of J in two opposite directions, obtaining two initial conditions [18,61]. Unstable manifolds can be traced out by numerically integrating $\overline{\tau}$ from these initial conditions in forward time until trajectories reach a stable fixed point (sink configuration) or leave the domain. Similarly, stable manifolds are delineated by integrating $\overline{\tau}$ in backward time starting from the perturbation of saddle point locations along the negative eigenvector of J until trajectories reach an unstable fixed point (source configuration) or leave the domain.

An example of unstable manifolds of cycle-average WSS on the luminal surface of a patient-specific computational hemodynamic model of carotid bifurcation is presented in Figure 6. Details on carotid bifurcation hemodynamic modeling are reported elsewhere [9,14,20,63]. WSS fixed points were identified by computing the Poincaré index (Figure 6A), and subsequently, unstable manifolds were traced out by applying Runge-Kutta 4-5 numerical integration schemes (Figure 6B). By visual inspection of Figure 6C, it can be appreciated that cycle-average WSS unstable manifolds co-localize with attracting WSS LCS, confirming the capability of the latter to identify critical lines of the WSS field.

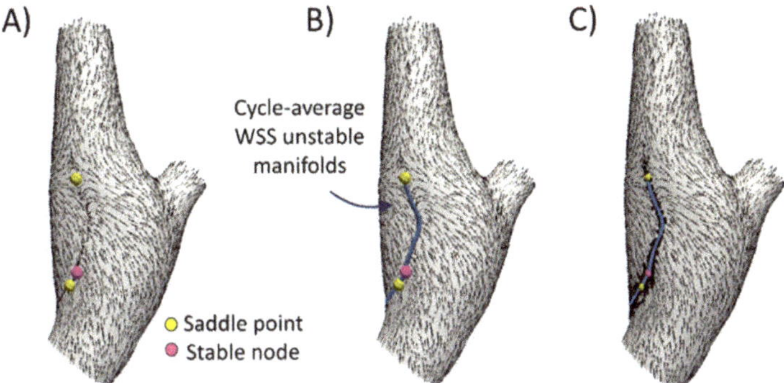

Figure 6. (**A**) Cycle-average WSS fixed points on the luminal surface of a carotid bifurcation model. (**B**) Unstable manifolds of cycle-average WSS (blue lines) traced out by integrating cycle-average WSS vectors, starting from saddle point positions. (**C**) Cycle-average WSS unstable manifolds superimposed on the attracting WSS LCS. WSS vectors are normalized for visualization.

The analysis of cycle-average WSS fixed points and manifolds has been applied to analyze cardiovascular flows. Arzani et al. [18] used WSS LCS from stable and unstable manifolds of cycle-average WSS on patient-specific computational hemodynamics models of AAAs, carotid arteries, cerebral aneurysms, and coronary aneurysms to characterize near-wall flow topology and biochemical transport. Farghadan et al. [16] used WSS topology and magnitude analysis to predict surface concentration patterns in cardiovascular transport problems by computing WSS LCS from manifolds of cycle-average WSS in image-based coronary and carotid artery models. Mahmoudi et al. [21] studied the near-wall transport of some of the prominent biochemicals contributing to the initiation and progression of atherosclerosis in computational hemodynamic models of the coronary artery, highlighting the strength of cycle-average WSS LCS as a template for luminal surface concentration and flux patterns of biochemicals transported with blood.

Summarizing, the Lagrangian approach for identifying near-wall topological features is schematized in Figure 7, where the link between attracting/repelling WSS LCS with unstable/stable cycle-average WSS manifolds, respectively, is highlighted. In addition, Figure 7 presents a brief summary of the link between Lagrangian-based near-wall flow topology and mass transport. For a more in-depth analysis, the interested reader can refer to recent literature [15,17,18,21] where the link between WSS LCS, cycle-average WSS manifolds, and biochemical transport in cardiovascular flows is unambiguously documented.

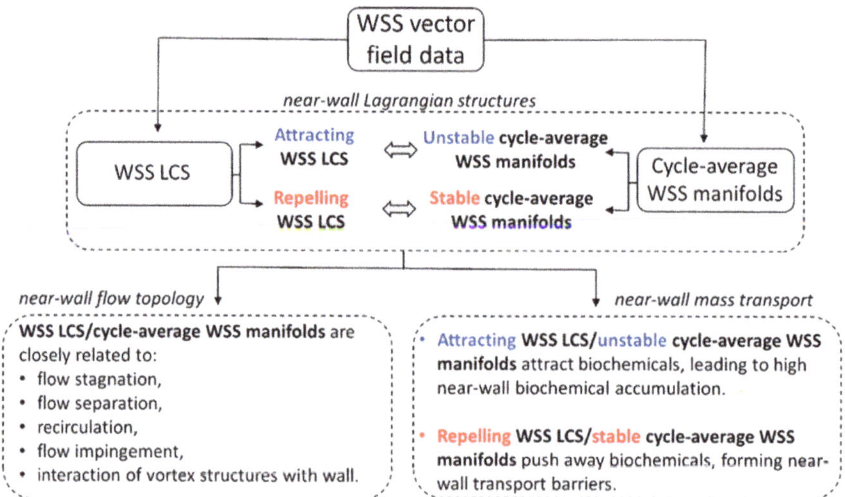

Figure 7. Identification and significance of the near-wall Lagrangian structures. The link between WSS LCS and cycle-average WSS manifolds and their role in near-wall flow topology and near-wall mass transport is highlighted.

4. Eulerian Approach

4.1. Volume Contraction Theory

From a Eulerian perspective, the volume contraction theory provides a simple alternative way to analyze the behavior of a dynamical system. Contrarily to Lagrangian-based approaches, the Eulerian perspective considers vector field properties at each point in space and time. The here-presented volume contraction theory, based on fluid mechanics and differential geometry, is focused on the temporal change of an elemental volume (of fluid, for the case of interest) in the phase space of a dynamical system (fluid flow, for the case of interest). Let $V(t)$ be an arbitrary volume in the phase space of the dynamical system defined in Equation (1). Let $S(t)$ be a closed surface enclosing the volume $V(t)$, i.e., such that $S(t) = \delta V(t)$. $S(t)$ evolves during the time interval dt resulting in a contraction or expansion of the volume, as depicted in Figure 8. The rate of volume variation, which we will call volume contraction rate in the following, can be expressed as follows as a consequence of the application of the Gauss theorem:

$$\frac{dV(t)}{dt} = \int\int_{S} \boldsymbol{u} \cdot \boldsymbol{n} \, dS = \int\int\int_{V} \nabla \cdot \boldsymbol{u} \, dV, \tag{16}$$

where $\boldsymbol{u}$ is the vector field defined in Equation (1) and $\boldsymbol{n}$ is the unit normal to the surface S (Figure 8). Shrinking the near-wall volume V to a point, it can be written:

$$\lim_{V \to 0} \frac{1}{V} \frac{dV(t)}{dt} = \lim_{V \to 0} \frac{1}{V} \int\int\int_{V} \nabla \cdot \boldsymbol{u} \, dV = (\nabla \cdot \boldsymbol{u}). \tag{17}$$

Equation (17) clearly shows that in the limit as V approaches zero, the local value of vector $\boldsymbol{u}$ divergence is equal to its total flux per unit volume.

In general, in non-conservative dynamical systems, the volume of phase space is not preserved, as it can contract or expand. Thus, trajectories tend toward a lower-dimensional subset of phase space. From Equation (17), the volume contraction rate $\Lambda(x,t)$ of a n-dimensional system, representing the rate of separation of infinitesimal close trajectories, can be obtained as:

$$\Lambda(\boldsymbol{x},t) = \nabla \cdot \boldsymbol{u}(\boldsymbol{x},t) = tr\, J(\boldsymbol{u}) = \sum_{i=1}^{n} \lambda_i, \tag{18}$$

where $tr\,J(u)$ is the trace of the Jacobian matrix of vector field u and λ_i are its eigenvalues. Physically, the Jacobian matrix describes how a small change at a starting point x_0 propagates to the final point of the flow map $\Phi_{t_0}^t(x_0)$ of Equation (3). In this sense, Equation (18) represents the sum of the Lyapunov exponents of Equation (8).

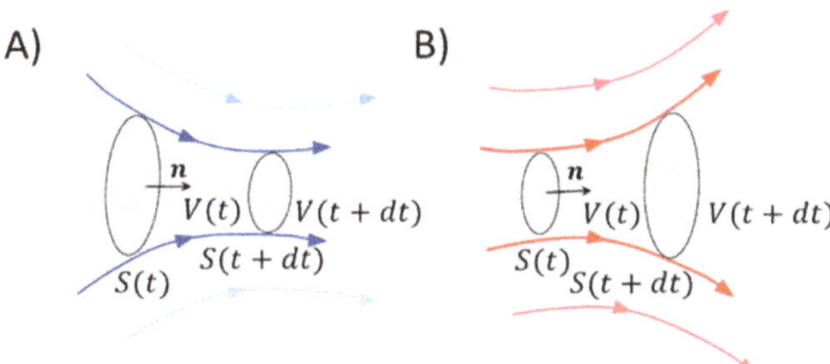

Figure 8. Explanatory sketch of (**A**) volume contraction and (**B**) volume expansion in the phase space of a dynamical system.

4.2. Eulerian-Based Approach for WSS Topological Skeleton Identification

It has been recently demonstrated [19] that the application of the volume contraction theory to cardiovascular flows allows the analysis of the WSS topological skeleton on the luminal surface of a vessel through the direct calculation of the WSS divergence. Briefly, considering the expression of the near-wall blood flow velocity vector u_π given in Equation (12) and substituting it in Equation (17), it follows that:

$$\lim_{V \to 0} \frac{1}{V} \frac{dV(t)}{dt} = \lim_{V \to 0} \frac{\delta n}{V \mu} \int \int \int_V \nabla \cdot \tau \, dV = (\nabla \cdot \tau). \tag{19}$$

Based on Equation (19), the WSS divergence gives practical information about the WSS topological skeleton. Note that in general, the WSS vector field defined at the luminal surface of a vessel is not conservative, even in the case of incompressible flows.

Contextualizing the physical meaning of Equation (19) to the study of the phenomena at the interface between blood flow and vessel wall, it can be observed that as the divergence represents the volume density of the outward flux of a vector field from an infinitesimal volume around a given point:

- a local positive value of the divergence of the WSS field at the luminal surface means that locally shear forces exert an expansion action on the endothelium;
- a local negative value of the divergence of the WSS field at the luminal surface means that locally shear forces exert a contraction action on the endothelium.

In general, the application of the volume contraction theory to the analysis of a dynamical system faces one limitation in cases where the distance between two neighboring trajectories increases/decreases in spite of a negative/positive value of divergence, respectively. As WSS divergence depends by construction upon the algebraic summation of the magnitude of the single gradients of WSS vector components, in some cases, it might fail to properly identify WSS expansion/contraction configuration patterns. In fact, these regions describe specific directional arrangements of the vectors, but both variations in magnitude and in direction are accounted for in the divergence. To overcome this limitation, which could markedly affect the application of the Eulerian-based approach to study WSS manifolds in cardiovascular flows, the use of the divergence of the normalized WSS vector field has been recently proposed [19]:

$$DIV_W = \nabla \cdot \tau_u = \nabla \cdot \left(\frac{\tau}{|\tau|} \right), \tag{20}$$

where τ_u is the WSS unit vector. Equation (20) can be used to encase the connections between fixed points, i.e., manifolds, identify basins of attraction, and subdivide the domain into different vector field behaviors. Then, in the light of the above and as depicted in Figure 9, luminal surface regions characterized by negative values of DIV_W are referred to as contraction regions and approximate unstable manifolds. Similarly, regions at the luminal surface characterized by positive values of DIV_W are referred to as expansion regions and approximate stable manifolds (Figure 9).

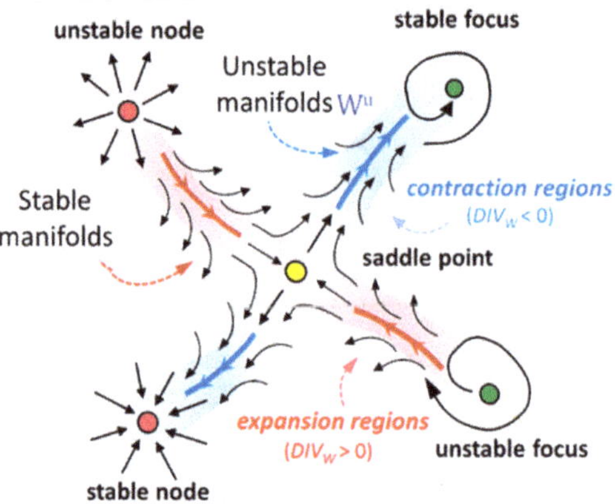

Figure 9. Explanatory sketch of the topological skeleton of a vector field. Contraction/expansion regions, colored in blue/red, respectively, approximating unstable/stable manifolds, are highlighted.

To complete the Eulerian-based WSS topological skeleton analysis, once WSS manifolds have been identified using DIV_W, the WSS fixed point location can be carried out using the Poincaré index, as in the Lagrangian-based analysis (as described in Section 2). Then, the eigenvalues of the Jacobian matrix of the WSS vector field can be used to distinguish between a node or a focus and between the attractive or repelling nature of a fixed point, as described in Section 2 in general terms (Table 1) and in Section 3.3 for the specific case of a WSS vector field defined on the luminal surface of a vessel.

An example of Eulerian-based topological skeleton analysis of the cycle-average WSS field on the luminal surface of a patient-specific computational hemodynamic model of carotid bifurcation is presented in Figure 10A. Details on carotid bifurcation hemodynamic modeling are reported elsewhere [9,14,20,63]. WSS fixed points were identified and classified by computing the Poincaré index and eigenvalues of the Jacobian matrix, respectively, whereas contraction/expansion regions were identified by computing the divergence of the normalized cycle-average WSS vector field. By visual inspection of Figure 10B, it can be noted that cycle-average WSS contraction regions co-localize with cycle-average WSS unstable manifolds, traced out by integrating cycle-average WSS starting from saddle point positions, thus confirming the capability of the contraction regions to encase WSS manifolds.

The Eulerian-based approach to analyze the WSS topological skeleton can be easily implemented. It requires only single snapshots of the WSS vector field, and the post-processing algorithms, based on a robust theory, are easily reproduced. This approach does not need the Lagrangian surface transport computation, as required for Lagrangian-based and integrated trajectory-based methods, thus reducing computational effort. Furthermore, it is characterized by modularity in the sense that the method can be applied only for the purpose of fixed point identification and/or classification or only for contraction/expansion region identification.

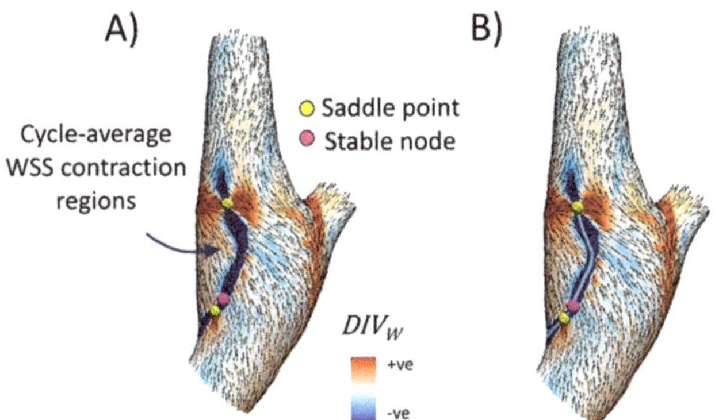

Figure 10. (**A**) Cycle-average WSS topological skeleton on the luminal surface of a carotid bifurcation model using the Eulerian-based approach. Cycle-average WSS fixed points and contraction/expansion regions (blue/red regions, respectively) are computed simultaneously. (**B**) Cycle-average WSS topological skeleton superimposed on the cycle-average WSS unstable manifolds, traced out by integrating cycle-average WSS vectors starting from saddle point positions. Vectors are normalized for visualization.

4.3. Application of the Eulerian-Based Method for WSS Topological Skeleton Analysis to Cardiovascular Flows

The described Eulerian-based method to identify the WSS topological skeleton on the luminal surface of an artery can be easily applied (1) to cycle average WSS vectors (defined in Equation (14) in Section 3.3) and (2) to instantaneous WSS vectors along the cardiac cycle.

The cycle-average WSS topological skeleton highlights blood flow features associated with vascular disease development, and it is strongly related to arterial near-wall mass transport. In detail, on the one hand, contraction/expansion regions of cycle-average WSS vectors, because of their capability to encase unstable/stable cycle-average WSS manifolds, can be used to identify biochemical concentration patterns at the arterial luminal surface. On the other hand, the instantaneous WSS topological skeleton allows analyzing the unsteady nature of WSS fixed points and contraction/expansion regions. In detail, instantaneous WSS fixed points may have a potential impact on the endothelial cells (ECs) function. By definition, a WSS fixed point represents a focal point on the luminal surface of a vessel where WSS vanishes, and low WSS is a biomechanical factor involved in vascular dysfunction. In light of this, quantitative descriptors of WSS fixed points residence times along the cardiac cycle were proposed [17,19], aiming at characterizing their unsteady nature. In detail, a WSS fixed point residence time, that for each surface element measures the accumulated amount of time that WSS fixed points spend inside that element, weighted by the sum of the absolute values of the eigenvalues of the instantaneous WSS Jacobian matrix, was proposed elsewhere [17]. More recently, a different formulation for quantifying WSS fixed points was proposed [19] where the local residence time of WSS fixed points were weighted by the absolute value of the sum of the eigenvalues of the WSS Jacobian matrix (i.e., according to Equation (18), the absolute value of the WSS divergence, representing the strength of the contraction/expansion of the WSS around the fixed point). In mathematical terms:

$$RT\nabla_{x_{fp}}(e) = \frac{\overline{A}}{A_e} \frac{1}{T} \int_0^T \mathbb{I}_e\left(x_{fp}, t\right) |(\nabla \cdot \boldsymbol{\tau})_e| dt, \tag{21}$$

where x_{fp} is the location of a WSS fixed point at time $t \in [0, T]$, T is the cardiac cycle duration, e is the generic triangular element of the superficial mesh of area A_e, $\overline{A}$ is the average surface area of all triangular elements of the superficial mesh, $\mathbb{I}_e$ is the indicator function (equal to 1 if $x_{fp} \in e$, 0 otherwise) and $(\nabla \cdot \boldsymbol{\tau})_e$ is the instantaneous WSS divergence value, representing the local strength of the contraction/expansion of the WSS around the considered fixed point. Roughly speaking, Equation (21) allows quantifying the fraction of the cardiac cycle for which a generic triangle mesh surface element e on the vessel luminal surface hosted as a fixed point, weighting the residence time by the strength of the local contraction/expansion action.

Furthermore, the strength and the nature of WSS contraction and expansion action on the ECs lining the luminal surface, as identified by WSS contraction/expansion regions, is expected to have biological consequences linked to vascular pathophysiology. In particular, the exposure to high variability of WSS contraction and expansion action may mechanically induce a recurring variation in EC stimulation along the cardiac cycle, with consequent widening cell–cell junctions and associated increased endothelium permeability and EC dysfunction and apoptosis [7,66]. The amount of variation in the WSS contraction/expansion action exerted at the luminal surface of a vessel along the cardiac cycle can be quantified using the quantity topological shear variation index ($TSVI$) [24]:

$$TSVI = \left\{ \frac{1}{T} \int_0^T \left[DIV_W - \overline{DIV_W} \right]^2 dt \right\}^{1/2}. \tag{22}$$

Equation (22) allows localizing regions on the vessel luminal surface exposed to large variations in WSS contraction/expansion action exerted by the flowing blood along the cardiac cycle.

An example of the distribution of WSS fixed point weighted residence time (Equation (21)) and the topological shear variation index (Equation (22)) on the luminal surface of a patient-specific computational hemodynamic model of carotid bifurcation is presented in Figure 11. Details on the carotid bifurcation hemodynamic modeling are reported elsewhere [9,14,20,63]. From Figure 11, it emerged that the highest $RT\nabla_{x_{fp}}(e)$ values and highest variation in the contraction/expansion action exerted by the WSS along the cardiac cycle were mainly located at the carotid bulb and around the bifurcation apex.

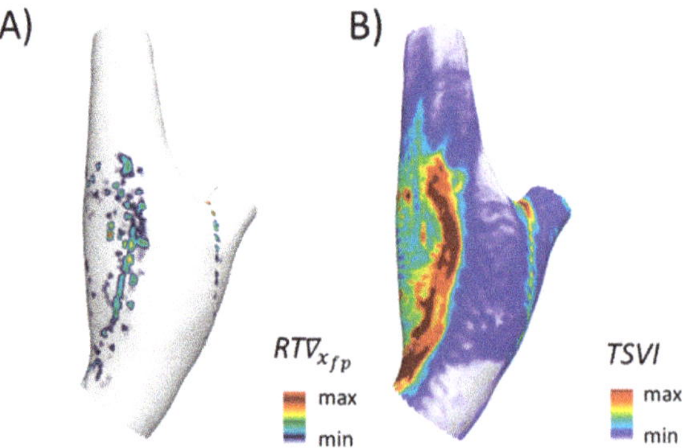

Figure 11. (**A**) Distribution of WSS fixed-point weighted residence $RT\nabla_{x_{fp}}(e)$ and (**B**) Topological shear variation index ($TSVI$) on a carotid bifurcation model.

Interestingly, very recent studies highlighted a link between WSS contraction/expansion variability along the cardiac cycle and aggravating biological events at the arterial wall. In particular, De Nisco et al. [24] applied the Eulerian-based approach for the analysis of the WSS topological skeleton for personalized computational hemodynamic models of ascending thoracic aorta aneurysm (ATAA) and healthy aorta, reporting that: (1) the different spatiotemporal heterogeneity characterizing the ATAA and healthy hemodynamics markedly reflect on their WSS topological skeleton features; (2) a link emerged between the variability of the contraction/expansion action exerted by WSS on the endothelium (as quantified by the *TSVI*) along the cardiac cycle and ATAA wall stiffness. Morbiducci et al. [20] demonstrated in a longitudinal study integrating clinical data with computational hemodynamics that WSS topological skeleton features quantified by the *TSVI* independently predicted long-term restenosis after carotid bifurcation endarterectomy.

5. Future Directions

The translation into clinical settings of the WSS topological skeleton is hampered by several barriers that add up to those affecting in general the translation of computational hemodynamics and the derived knowledge, as discussed elsewhere [67]. Specifically, as a first step, the analysis of the topological skeleton needs to be distilled into intuitive, clinically relevant criteria. To this aim, only semi-quantitative results are obtained from the definition of fixed points and stable/unstable manifolds, consisting of contraction/expansion regions. However, quantitative results can be obtained by focusing on specific features by using ad-hoc topological skeleton descriptors, such as the fixed point weighted residence time $RT\nabla_{x_{fp}}(e)$ (Equation (21)) or the topological shear variation index, *TSVI* (Equation (22)). Then, the definition of clinically relevant criteria based on the WSS topological descriptors require cut-off values for an effective translation into the clinic. These cut-off values need to be accurately defined and tested in terms of performance including accuracy, sensitivity, specificity, and positive predictive value, among others. Therefore, the determination of cut-off values requires adequate statistical power, obtained usually through multiple prospective, randomized trials. Moreover, the endpoint to be predicted should be clearly defined, as different endpoints correspond to different cut-off values.

In the perspective of an effective translation into the clinic of quantitative topological skeleton features, in a previous study [20], we proved that exposure to high values of both descriptors $RT\nabla_{x_{fp}}(e)$ and *TSVI* was correlated with intima-media thickness (a marker of vascular disease) at 60 month follow-ups in carotid bifurcations after carotid endarterectomy. To determine the cut-off values of the descriptors, a pooled distribution for each descriptor was calculated from 46 models of healthy carotid bifurcation. The 80th percentile of those distributions was then used. This approach allowed definition of the cut-off values for abnormally high values of $RT\nabla_{x_{fp}}(e)$ or the *TSVI*.

It is evident that cut-off values are specific to the vascular region and to the predicted endpoint and therefore cannot be extrapolated to other conditions. In the future, the continuous improvements in imaging and data acquisition, the increasing availability of computational power, and the development of more and more efficient and robust methodologies for blood flow modeling are expected to accelerate the translation into the clinic of the analysis of the WSS topological skeleton. Our paper aims to give the methodological basis to tackle these future efforts.

6. Conclusions

The need for the identification of hemodynamic coherent structures in blood vessels is dictated by the so-called hemodynamic risk hypothesis, suggesting a major role of flow disturbances in vascular pathophysiology [2]. The action of fluid forces on the endothelial mechanosensors and biochemical machinery has been historically explained in terms of WSS [3,4]. However, only moderate (and sometimes contradictory) associations between vascular disease and WSS-based descriptors have emerged to date, motivating a more in-depth analysis of the fluid near-wall transport phenomena. In this sense, the capability of the WSS topological skeleton to capture features reflecting cardiovascular flow complexity [17–20] and having a direct link to adverse vascular biological events has recently attracted a strong research interest. In this regard, recent studies have demonstrated that the cycle-average WSS topological skeleton governs the near-wall biochemical transport in arteries [15,16,18], a process linked to, e.g., endothelium-mediated vasoregulation, thrombosis, and atherosclerosis [23]. Furthermore, evidence about the role of WSS topological skeleton features in vascular pathophysiology emerged from very recent studies suggesting a direct link between WSS topological skeleton features and, e.g., aortic wall stiffness [24] and late restenosis in endarterectomized carotid arteries [20].

Motivated by the need to characterize more precisely the WSS phenotype(s) linked to aggravating biological events, here we provided an overview of the theoretical and methodological basis for analyzing the WSS topological skeleton in cardiovascular flows. In detail, the present study is intended to: (1) promote the application of WSS topological skeleton analysis to cardiovascular flows, aiming at elucidating the role that peculiar WSS features play in vascular pathophysiology; (2) facilitate the reproducibility and comparability of results from WSS topological skeleton analyses among different studies; (3) confirm its potential as a tool for increasing the chance of elucidating the mechanistic link between flow disturbance and clinical outcomes when applied to real-world clinical data.

Here, both WSS topological skeleton Lagrangian- and Eulerian-based methods currently adopted in the literature are presented. Lagrangian-based approaches start from the processing of Eulerian data, which represent the typical outputs of current in vivo (e.g., phase contrast MRI), in vitro (e.g., particle image velocimetry), and computational methods used for the investigation of cardiovascular flows. On the one hand, Lagrangian approaches are particularly useful for revealing the global organization of the vector field and characterizing its evolution over time, making the relevant features easy to detect by visual inspection, as they offer effective three-dimensional (or even four-dimensional, i.e., including time) visualizations. On the other hand, Lagrangian techniques rely on the numerical integration of particle trajectories, requiring sufficiently resolved data in both time and space, thus, in principle, making such methods computationally expensive and time consuming [29]. Moreover, adopting a Lagrangian approach may result in a poor control over the zone of investigation, which is determined by particle motion and accumulation. For this reason, it can also be difficult to get a complete picture of the flow at specific time instants. Furthermore, the influence of particle distribution and of particle seeding schemes on quantities of interest is poorly investigated.

In contrast, Eulerian-based approaches usually simplify the data analysis workflow, as they can be directly applied to the output given by the main current techniques used for the investigation of cardiovascular flows (e.g., phase contrast MRI, CFD data). Moreover, they usually have a simpler definition, making their implementation easy and characterized by a reduced computational cost. More importantly, they can give a picture of the entire vector field. However, the inherent unsteady nature of the hemodynamic vector fields (e.g., velocity, WSS) can make the characterization of the dynamic evolution of the vector field features difficult with Eulerian-based approaches.

In conclusion, the theoretical background of the advanced methods of analysis of the WSS presented here and the recent findings related to their application to cardiovascular flows support their use to further elucidate the cause–effect relationships at the basis of the links between local hemodynamics and vascular disease. Based on the reported evidence about the physiological significance of the WSS topological skeleton in cardiovascular flows, its application in future studies, including longitudinal data, biological mechanism, and mechanobiology studies, is strongly encouraged and warranted.

Author Contributions: Conceptualization, V.M., U.M., and D.G.; data curation, V.M. and D.G.; data processing: V.M. and D.G.; data interpretation: V.M., U.M., and D.G.; writing—original draft preparation, V.M., U.M., K.C., G.D.N., M.L.R., E.T., G.C.A.C., C.C., D.G.; writing—review and editing, U.M., C.C., and D.G.; supervision, U.M. and D.G. All authors have read and agreed to the published version of the manuscript.

Funding: This research was funded by MIUR FISR—FISR2019_03221 CECOMES.

Acknowledgments: The authors are thankful to David A. Steinman for providing the patient-specific computational hemodynamic model of carotid bifurcation.

Conflicts of Interest: The authors declare no conflict of interest.

References

1. Kwak, B.R.; Bäck, M.; Bochaton-Piallat, M.-L.; Caligiuri, G.; Daemen, M.J.A.P.; Davies, P.F.; Hoefer, I.E.; Holvoet, P.; Jo, H.; Krams, R.; et al. Biomechanical factors in atherosclerosis: Mechanisms and clinical implications. *Eur. Heart J.* **2014**, *35*, 3013–3020. [CrossRef] [PubMed]
2. Morbiducci, U.; Kok, A.M.; Kwak, B.R.; Stone, P.H.; Steinman, D.A.; Wentzel, J.J. Atherosclerosis at arterial bifurcations: Evidence for the role of haemodynamics and geometry. *Thromb. Haemost.* **2016**, *115*, 484–492. [CrossRef]
3. Malek, A.M.; Alper, S.L.; Izumo, S. Hemodynamic shear stress and its role in atherosclerosis. *J. Am. Med. Assoc.* **1999**, *282*, 2035–2042. [CrossRef]
4. Zarins, C.K.; Giddens, D.P.; Bharadvaj, B.K.; Sottiurai, V.S.; Mabon, R.F.; Glagov, S. Carotid bifurcation atherosclerosis. Quantitative correlation of plaque localization with flow velocity profiles and wall shear stress. *Circ. Res.* **1983**, *53*, 502–514. [CrossRef]
5. Caro, C.G.; Fitz-Gerald, J.M.; Schroter, R.C. Atheroma and arterial wall shear. Observation, correlation and proposal of a shear dependent mass transfer mechanism for atherogenesis. *Proc. R. Soc. Lond. Ser. B Biol. Sci.* **1971**, *177*, 109–159. [CrossRef]
6. Ku, D.N.; Giddens, D.P.; Zarins, C.K.; Glagov, S. Pulsatile flow and atherosclerosis in the human carotid bifurcation. Positive correlation between plaque location and low oscillating shear stress. *Arteriosclerosis* **1985**, *5*, 293–302. [CrossRef]
7. Wang, C.; Baker, B.M.; Chen, C.S.; Schwartz, M.A. Endothelial cell sensing of flow direction. *Arterioscler. Thromb. Vasc. Biol.* **2013**, *33*, 2130–2136. [CrossRef]
8. Peiffer, V.; Sherwin, S.J.; Weinberg, P.D. Does low and oscillatory wall shear stress correlate spatially with early atherosclerosis? A systematic review. *Cardiovasc. Res.* **2013**, *99*, 242–250. [CrossRef]
9. Gallo, D.; Bijari, P.B.; Morbiducci, U.; Qiao, Y.; Xie, Y.J.; Etesami, M.; Habets, D.; Lakatta, E.G.; Wasserman, B.A.; Steinman, D.A. Segment-specific associations between local haemodynamic and imaging markers of early atherosclerosis at the carotid artery: An in vivo human study. *J. R. Soc. Interface* **2018**, *15*. [CrossRef]
10. Hoogendoorn, A.; Kok, A.M.; Hartman, E.M.J.; de Nisco, G.; Casadonte, L.; Chiastra, C.; Coenen, A.; Korteland, S.A.; Van der Heiden, K.; Gijsen, F.J.H.; et al. Multidirectional wall shear stress promotes advanced coronary plaque development: Comparing five shear stress metrics. *Cardiovasc. Res.* **2020**, *116*, 1136–1146. [CrossRef]
11. Kok, A.M.; Molony, D.S.; Timmins, L.H.; Ko, Y.-A.; Boersma, E.; Eshtehardi, P.; Wentzel, J.J.; Samady, H. The influence of multidirectional shear stress on plaque progression and composition changes in human coronary arteries. *EuroIntervention* **2019**, *15*, 692–699. [CrossRef]
12. Timmins, L.H.; Molony, D.S.; Eshtehardi, P.; McDaniel, M.C.; Oshinski, J.N.; Giddens, D.P.; Samady, H. Oscillatory wall shear stress is a dominant flow characteristic affecting lesion progression patterns and plaque vulnerability in patients with coronary artery disease. *J. R. Soc. Interface* **2017**, *14*. [CrossRef] [PubMed]
13. Colombo, M.; He, Y.; Corti, A.; Gallo, D.; Casarin, S.; Rozowsky, J.M.; Migliavacca, F.; Berceli, S.; Chiastra, C. Baseline local hemodynamics as predictor of lumen remodeling at 1-year follow-up in stented superficial femoral arteries. *Sci. Rep.* **2021**, *11*, 1613. [CrossRef]
14. Gallo, D.; Steinman, D.A.; Morbiducci, U. Insights into the co-localization of magnitude-based versus direction-based indicators of disturbed shear at the carotid bifurcation. *J. Biomech.* **2016**, *49*, 2413–2419. [CrossRef]
15. Arzani, A.; Gambaruto, A.M.; Chen, G.; Shadden, S.C. Lagrangian wall shear stress structures and near-wall transport in high-Schmidt-number aneurysmal flows. *J. Fluid Mech.* **2016**, *790*, 158–172. [CrossRef]
16. Farghadan, A.; Arzani, A. The combined effect of wall shear stress topology and magnitude on cardiovascular mass transport. *Int. J. Heat Mass Transf.* **2019**, *131*, 252–260. [CrossRef]
17. Arzani, A.; Shadden, S.C. Wall shear stress fixed points in cardiovascular fluid mechanics. *J. Biomech.* **2018**, *73*, 145–152. [CrossRef]
18. Arzani, A.; Gambaruto, A.M.; Chen, G.; Shadden, S.C. Wall shear stress exposure time: A Lagrangian measure of near-wall stagnation and concentration in cardiovascular flows. *Biomech. Model. Mechanobiol.* **2017**, *16*, 787–803. [CrossRef]
19. Mazzi, V.; Gallo, D.; Calò, K.; Najafi, M.; Khan, M.O.; De Nisco, G.; Steinman, D.A.; Morbiducci, U. A Eulerian method to analyze wall shear stress fixed points and manifolds in cardiovascular flows. *Biomech. Model. Mechanobiol.* **2020**, *19*, 1403–1423. [CrossRef]
20. Morbiducci, U.; Mazzi, V.; Domanin, M.; De Nisco, G.; Vergara, C.; Steinman, D.A.; Gallo, D. Wall Shear Stress Topological Skeleton Independently Predicts Long-Term Restenosis After Carotid Bifurcation Endarterectomy. *Ann. Biomed. Eng.* **2020**, *48*, 2936–2949. [CrossRef]
21. Mahmoudi, M.; Farghadan, A.; McConnell, D.; Barker, A.J.; Wentzel, J.J.; Budoff, M.J.; Arzani, A. The Story of Wall Shear Stress in Coronary Artery Atherosclerosis: Biochemical Transport and Mechanotransduction. *J. Biomech. Eng.* **2020**. [CrossRef] [PubMed]
22. Ethier, C.R. Computational modeling of mass transfer and links to atherosclerosis. *Ann. Biomed. Eng.* **2002**, *30*, 461–471. [CrossRef]
23. Tarbell, J.M. Mass transport in arteries and the localization of atherosclerosis. *Annu. Rev. Biomed. Eng.* **2003**, *5*, 79–118. [CrossRef]
24. De Nisco, G.; Tasso, P.; Calò, K.; Mazzi, V.; Gallo, D.; Condemi, F.; Farzaneh, S.; Avril, S.; Morbiducci, U. Deciphering ascending thoracic aortic aneurysm hemodynamics in relation to biomechanical properties. *Med. Eng. Phys.* **2020**, *82*, 119–129. [CrossRef]
25. Garth, C.; Tricoche, X.; Scheuermann, G. Tracking of vector field singularities in unstructured 3D time-dependent datasets. In Proceedings of the IEEE Visualization 2004, Austin, TX, USA, 10–15 October 2004; pp. 329–336.
26. Shadden, S.C.; Lekien, F.; Marsden, J.E. Definition and properties of Lagrangian coherent structures from finite-time Lyapunov exponents in two-dimensional aperiodic flows. *Phys. D Nonlinear Phenom.* **2005**, *212*, 271–304. [CrossRef]
27. Haller, G. Distinguished material surfaces and coherent structures in three-dimensional fluid flows. *Phys. D Nonlinear Phenom.* **2001**, *149*, 248–277. [CrossRef]

28. Haller, G. Lagrangian coherent structures. *Annu. Rev. Fluid Mech.* **2015**, *47*, 137–162. [CrossRef]
29. Nolan, P.J.; Serra, M.; Ross, S.D. Finite-time Lyapunov exponents in the instantaneous limit and material transport. *Nonlinear Dyn.* **2020**, *100*, 3825–3852. [CrossRef]
30. Lekien, F.; Shadden, S.C.; Marsden, J.E. Lagrangian coherent structures in n-dimensional systems. *J. Math. Phys.* **2007**, *48*, 1–19. [CrossRef]
31. Shadden, S.C.; Arzani, A. Lagrangian postprocessing of computational hemodynamics. *Ann. Biomed. Eng.* **2015**, *43*, 41–58. [CrossRef] [PubMed]
32. Shadden, S.C.; Taylor, C.A. Characterization of coherent structures in the cardiovascular system. *Ann. Biomed. Eng.* **2008**, *36*, 1152–1162. [CrossRef]
33. Arzani, A.; Shadden, S.C. Characterization of the transport topology in patient-specific abdominal aortic aneurysm models. *Phys. Fluids (1994).* **2012**, *24*, 81901. [CrossRef]
34. Green, M.A.; Rowley, C.W.; Haller, G. Detection of Lagrangian coherent structures in three-dimensional turbulence. *J. Fluid Mech.* **2007**, *572*, 111–120. [CrossRef]
35. Peacock, T.; Haller, G. Lagrangian coherent structures: The hidden skeleton of fluid flows. *Phys. Today* **2013**, *66*, 41–47. [CrossRef]
36. Ehrlich, L.W.; Friedman, M.H. Particle paths and stasis in unsteady flow through a bifurcation. *J. Biomech.* **1977**, *10*, 561–568. [CrossRef]
37. Perktold, K. On the paths of fluid particles in an axisymmetrical aneurysm. *J. Biomech.* **1987**, *20*, 311–317. [CrossRef]
38. Perktold, K.; Peter, R.; Resch, M. Pulsatile non-Newtonian blood flow simulation through a bifurcation with an aneurysm. *Biorheology* **1989**, *26*, 1011–1030. [CrossRef] [PubMed]
39. Perktold, K.; Hilbert, D. Numerical simulation of pulsatile flow in a carotid bifurcation model. *J. Biomed. Eng.* **1986**, *8*, 193–199. [CrossRef]
40. Steinman, D.A. Simulated pathline visualization of computed periodic blood flow patterns. *J. Biomech.* **2000**, *33*, 623–628. [CrossRef]
41. Tambasco, M.; Steinman, D.A. On assessing the quality of particle tracking through computational fluid dynamic models. *J. Biomech. Eng.* **2002**, *124*, 166–175. [CrossRef]
42. Tambasco, M.; Steinman, D.A. Path-dependent hemodynamics of the stenosed carotid bifurcation. *Ann. Biomed. Eng.* **2003**, *31*, 1054–1065. [CrossRef]
43. Yang, W.; Feinstein, J.A.; Shadden, S.C.; Vignon-Clementel, I.E.; Marsden, A.L. Optimization of a Y-graft design for improved hepatic flow distribution in the fontan circulation. *J. Biomech. Eng.* **2013**, *135*, 11002. [CrossRef]
44. Yang, W.; Vignon-Clementel, I.E.; Troianowski, G.; Reddy, V.M.; Feinstein, J.A.; Marsden, A.L. Hepatic blood flow distribution and performance in conventional and novel Y-graft Fontan geometries: A case series computational fluid dynamics study. *J. Thorac. Cardiovasc. Surg.* **2012**, *143*, 1086–1097. [CrossRef]
45. Amili, O.; MacIver, R.; Coletti, F. Magnetic Resonance Imaging Based Flow Field and Lagrangian Particle Tracking From a Left Ventricular Assist Device. *J. Biomech. Eng.* **2019**, *142*. [CrossRef] [PubMed]
46. Mukherjee, D.; Padilla, J.; Shadden, S.C. Numerical investigation of fluid–particle interactions for embolic stroke. *Theor. Comput. Fluid Dyn.* **2016**, *30*, 23–39. [CrossRef]
47. Morbiducci, U.; Ponzini, R.; Rizzo, G.; Cadioli, M.; Esposito, A.; De Cobelli, F.; Del Maschio, A.; Montevecchi, F.M.; Redaelli, A. In vivo quantification of helical blood flow in human aorta by time-resolved three-dimensional cine phase contrast magnetic resonance imaging. *Ann. Biomed. Eng.* **2009**, *37*, 516–531. [CrossRef]
48. Morbiducci, U.; Ponzini, R.; Rizzo, G.; Cadioli, M.; Esposito, A.; Montevecchi, F.M.; Redaelli, A. Mechanistic insight into the physiological relevance of helical blood flow in the human aorta: An in vivo study. *Biomech. Model. Mechanobiol.* **2011**, *10*, 339–355. [CrossRef] [PubMed]
49. Arzani, A.; Les, A.S.; Dalman, R.L.; Shadden, S.C. Effect of exercise on patient specific abdominal aortic aneurysm flow topology and mixing. *Int. J. Numer. Method Biomed. Eng.* **2014**, *30*, 280–295. [CrossRef]
50. Gharib, M.; Rambod, E.; Kheradvar, A.; Sahn, D.J.; Dabiri, J.O. Optimal vortex formation as an index of cardiac health. *Proc. Natl. Acad. Sci. USA* **2006**, *103*, 6305–6308. [CrossRef] [PubMed]
51. Charonko, J.J.; Kumar, R.; Stewart, K.; Little, W.C.; Vlachos, P.P. Vortices formed on the mitral valve tips aid normal left ventricular filling. *Ann. Biomed. Eng.* **2013**, *41*, 1049–1061. [CrossRef]
52. Töger, J.; Kanski, M.; Carlsson, M.; Kovács, S.J.; Söderlind, G.; Arheden, H.; Heiberg, E. Vortex ring formation in the left ventricle of the heart: Analysis by 4D flow MRI and Lagrangian coherent structures. *Ann. Biomed. Eng.* **2012**, *40*, 2652–2662. [CrossRef]
53. Hendabadi, S.; Bermejo, J.; Benito, Y.; Yotti, R.; Fernández-Avilés, F.; del Álamo, J.C.; Shadden, S.C. Topology of blood transport in the human left ventricle by novel processing of Doppler echocardiography. *Ann. Biomed. Eng.* **2013**, *41*, 2603–2616. [CrossRef]
54. Astorino, M.; Hamers, J.; Shadden, S.C.; Gerbeau, J.-F. A robust and efficient valve model based on resistive immersed surfaces. *Int. J. Numer. Method Biomed. Eng.* **2012**, *28*, 937–959. [CrossRef]
55. Shadden, S.C.; Astorino, M.; Gerbeau, J.-F. Computational analysis of an aortic valve jet with Lagrangian coherent structures. *Chaos* **2010**, *20*, 17512. [CrossRef]
56. Sadrabadi, M.S.; Hedayat, M.; Borazjani, I.; Arzani, A. Fluid-structure coupled biotransport processes in aortic valve disease. *J. Biomech.* **2021**, 110239. [CrossRef]

57. Xu, Z.; Chen, N.; Shadden, S.; Marsden, J.; Kamocka, M.; Rosen, E.; Alber, M.S. Study of blood flow impact on growth of thrombi using a multiscale model. *Soft Matter* **2009**, *5*, 769–779. [CrossRef]
58. Mutlu, O.; Olcay, A.B.; Bilgin, C.; Hakyemez, B. Evaluating the Effectiveness of 2 Different Flow Diverter Stents Based on the Stagnation Region Formation in an Aneurysm Sac Using Lagrangian Coherent Structure. *World Neurosurg.* **2019**, *127*, e727–e737. [CrossRef]
59. Mutlu, O.; Olcay, A.B.; Bilgin, C.; Hakyemez, B. Evaluating the Effect of the Number of Wire of Flow Diverter Stents on the Nonstagnated Region Formation in an Aneurysm Sac Using Lagrangian Coherent Structure and Hyperbolic Time Analysis. *World Neurosurg.* **2020**, *133*, e666–e682. [CrossRef]
60. Gambaruto, A.M.; Doorly, D.J.; Yamaguchi, T. Wall shear stress and near-wall convective transport: Comparisons with vascular remodelling in a peripheral graft anastomosis. *J. Comput. Phys.* **2010**, *229*, 5339–5356. [CrossRef]
61. Gambaruto, A.M.; João, A.J. Computers & Fluids Flow structures in cerebral aneurysms. *Comput. Fluids* **2012**, *65*, 56–65. [CrossRef]
62. Goodarzi Ardakani, V.; Tu, X.; Gambaruto, A.M.; Velho, I.; Tiago, J.; Sequeira, A.; Pereira, R. Near-Wall Flow in Cerebral Aneurysms. *Fluids* **2019**, *4*, 89. [CrossRef]
63. Calò, K.; Gallo, D.; Steinman, D.A.; Mazzi, V.; Scarsoglio, S.; Ridolfi, L.; Morbiducci, U. Spatiotemporal Hemodynamic Complexity in Carotid Arteries: An Integrated Computational Hemodynamics and Complex Networks-Based Approach. *IEEE Trans. Biomed. Eng.* **2020**, *67*, 1841–1853. [CrossRef]
64. Tada, S. Numerical study of oxygen transport in a carotid bifurcation. *Phys. Med. Biol.* **2010**, *55*, 3993–4010. [CrossRef] [PubMed]
65. De Nisco, G.; Zhang, P.; Calò, K.; Liu, X.; Ponzini, R.; Bignardi, C.; Rizzo, G.; Deng, X.; Gallo, D.; Morbiducci, U. What is needed to make low-density lipoprotein transport in human aorta computational models suitable to explore links to atherosclerosis? Impact of initial and inflow boundary conditions. *J. Biomech.* **2018**, *68*, 33–42. [CrossRef] [PubMed]
66. Chiu, J.J.; Chien, S. Effects of disturbed flow on vascular endothelium: Pathophysiological basis and clinical perspectives. *Physiol. Rev.* **2011**, *91*, 327–387. [CrossRef]
67. Morris, P.D.; Narracott, A.; von Tengg-Kobligk, H.; Silva Soto, D.A.; Hsiao, S.; Lungu, A.; Evans, P.; Bressloff, N.W.; Lawford, P.V.; Hose, D.R.; et al. Computational fluid dynamics modelling in cardiovascular medicine. *Heart* **2016**, *102*, 18–28. [CrossRef] [PubMed]

 *mathematics*

Article

Finite Element Analysis of Custom Shoulder Implants Provides Accurate Prediction of Initial Stability

Jonathan Pitocchi [1,2,3,*], Mariska Wesseling [1], Gerrit Harry van Lenthe [2] and María Angeles Pérez [3]

[1] Materialise NV, 3001 Leuven, Belgium; mariska.wesseling@materialise.be
[2] Biomechanics Section, KU Leuven, 3001 Leuven, Belgium; harry.vanlenthe@kuleuven.be
[3] Multiscale in Mechanical and Biological Engineering, Instituto de Investigación en Ingeniería de Aragón (I3A), Instituto de Investigación Sanitaria Aragón (IIS Aragón), University of Zaragoza, 50018 Zaragoza, Spain; angeles@unizar.es
*   Correspondence: jonathan.pitocchi@materialise.be

Received: 26 May 2020; Accepted: 3 July 2020; Published: 6 July 2020

Abstract: Custom reverse shoulder implants represent a valuable solution for patients with large bone defects. Since each implant has unique patient-specific features, finite element (FE) analysis has the potential to guide the design process by virtually comparing the stability of multiple configurations without the need of a mechanical test. The aim of this study was to develop an automated virtual bench test to evaluate the initial stability of custom shoulder implants during the design phase, by simulating a fixation experiment as defined by ASTM F2028-14. Three-dimensional (3D) FE models were generated to simulate the stability test and the predictions were compared to experimental measurements. Good agreement was found between the baseplate displacement measured experimentally and determined from the FE analysis (Spearman's rank test, $p < 0.05$, correlation coefficient $\rho s = 0.81$). Interface micromotion analysis predicted good initial fixation (micromotion <150 µm, commonly used as bone ingrowth threshold). In conclusion, the finite element model presented in this study was able to replicate the mechanical condition of a standard test for a custom shoulder implants.

Keywords: finite element analysis; shoulder implant stability; implant design; reverse shoulder arthroplasty; micromotion

---

1. Introduction

Since its introduction in the late 1980s, reverse shoulder arthroplasty (RSA) has become a standard treatment for patients with rotator cuff arthropathy. More recently, surgeons have expanded its application to fracture care, rheumatoid arthritis, and even failed prior surgery replacements, further increasing the number of surgeries [1,2]. In many cases, the presence of considerable bone loss at the glenoid side, due to degenerative arthritis or secondary to revision surgeries, may complicate baseplate implantation. This limits the treatment options and jeopardizes the clinical outcomes, as insufficient bone stock can lead to suboptimal component fixation and therefore early implant failure.

Different methods have been described to address glenoid defects, depending on the bone loss severity [3]. Eccentric reaming can be performed in case of moderate bone loss, while bone grafting is more suitable for large defects. However, the results of bone grafting are controversial since not all studies have reported satisfactory outcomes [4]. More recently, custom implants have been introduced as an alternative treatment. Together with patient-specific preoperative planning and implant design, custom implants allow for proper joint positioning and fixation of the component in the remaining native bone [5,6].

In order to avoid aseptic loosening of the glenoid component, a stable bone–implant interface is necessary, in which only small relative movements are allowed. Fixation screws are used to provide initial mechanical stability (primary fixation) which subsequently can lead to biological fixation by bone ingrowth (secondary fixation). To enable bone ingrowth, custom implants have a porous titanium structure (spray-coated or 3D printed) [7,8]. However, micro-motion at the bone–implant interface above 150 μm has been shown to inhibit this mechanism and lead to an unstable fibrous tissue layer between the metallic porous layer and the host bone [9]. Therefore, implant design should be optimized to minimize micromotion at the time of initial fixation, thus leading to a stable bone–implant interface and to a better osseointegration.

For patient-specific shoulder implants, the enormous design space, which allows the glenoid component to be adapted to the patient anatomy, represents a challenge to the evaluation of the mechanical stability. While mechanical tests can be performed extensively to assess the stability of standard implants [10–12], for custom implants with a unique design for each patient, it is not practical to use mechanical testing to verify the stability. Alternatively, Finite Element (FE) analysis has been widely used to evaluate the influence of different implant configurations on the initial fixation of an implant [13–19].

Chae et al. analyzed the bone–implant interface micromotion of an inferiorly tilted glenoid component virtually implanted in a scapula model and found that the tilted fixation compromised initial mechanical stability [17]. Suarez et al. investigated how a different type and number of screws impacted the initial stability of a cementless glenoid component, reporting higher interface micromotions when the same implant was tested in poor quality bone [14], even when more physiological loads (e.g., from musculoskeletal model) were applied [18]. Elwell et al. [19] reported similar results, showing that the use of only two fixation screws could amplify the negative effect of baseplate lateralization, thus jeopardizing implant stability and worsening its functional outcome. Hopkins et al. examined multiple standard designs with different screw angle inclination, concluding that increasing the screw inclination enhanced stability more than using longer and thicker screws [15]. Other studies explored instead the effect of the prosthesis repositioning (using different glenosphere sizes or bone grafting) and found that a lateralization of 10 mm was mechanically acceptable for osseointegration [13,16].

However, the effect of different loading directions, which in case of a custom implant cannot be neglected due to the asymmetry of the design shape, was never systematically investigated. It is evident that, since the main parameters (number and type of screws, baseplate dimensions, etc.) are unique for each custom implant, FE analysis has the potential to guide the design process by virtually comparing multiple designs without the need of a mechanical test.

Therefore, the aim of this study is to develop an automated workflow to evaluate the initial stability of custom shoulder implants during the design phase, by simulating a fixation experiment based on ASTM F2028-14 [20]. To our knowledge, this is the first study to automate, evaluate and validate a full in silico modeling of the ASTM F2028-14 for a custom-made prosthesis. Moreover, the FE model can be used to predict the relative motion at the bone–implant interface, which cannot be quantified by the current mechanical tests.

2. Materials and Methods

A custom reverse shoulder implant was designed and 3D printed to comply with ASTM standards [20]. To evaluate the preclinical stability of the implant, displacement of the glenoid baseplate was measured in response to axial and shear loading, after insertion in a bone substitute. The experimental baseplate displacement was compared to the model estimation to validate the virtual bench test. A more detailed explanation regarding the mechanical test and the in silico model is presented in the following sections.

2.1. Experimental Set-Up

The ASTM F2028-14 [20] is a standard method commonly used for assessing the risk of glenoid loosening in shoulder implants. The test protocol includes three subsequent steps: (1) an initial static analysis to measure the baseplate displacement, (2) a fatigue phase in which the implant is cyclically rotated around an axis loaded with a compressive axial force, and (3) an additional static phase to measure the glenoid fixation, similarly to step 1.

The custom implant was inserted into a 20 pcf (pounds per cubic foot) polyurethane block (Sawbones Europe AB, Sweden), which is normally used as substitute of glenoid bone in mechanical tests [21]. Two locking and two nonlocking (compression) screws were used to fix the implant to the artificial bone (Figure 1a). Compression screws are able to close the gap at the bone–implant interface, by pressing the metal component towards the bone. For this reason, nonlocking screws were inserted first, followed by the locking screws, which instead lock the implant in place thanks to the threaded head mating the threaded holes of the implant.

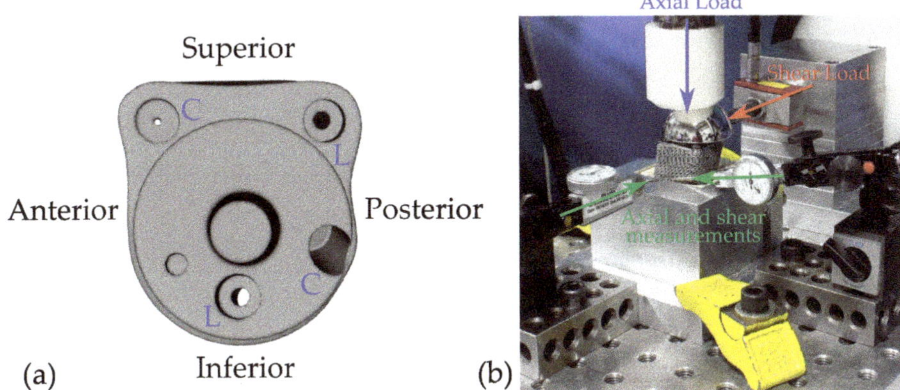

Figure 1. Left (**a**), top view of the custom implant with the four main directions: anterior, posterior, superior and inferior. Four screws were used to fix the implant: two locking (L) and two nonlocking (compression, C). Right (**b**), experimental set-up with a shear load (red arrow) applied inferiorly via a horizontal loading fixture. Axial load was applied through the glenosphere (blue arrow). Axial and shear components of the baseplate displacement were measured superiorly with two dial indicators (green arrows).

An axial compressive load of 430 N was applied perpendicular to the glenoid plane by a flat polyacetal load applicator. An additional shear load of 350 N was applied parallel to the baseplate via a horizontal loading fixture (Figure 1b). Shear and axial forces were defined in a worst-case loading scenario, being respectively 42% and 51% of body weight (assumed to be 86 kg) [20].

Contrary to standard baseplates, which normally have a symmetric round shape, custom implants can present an asymmetric design, consequently the shear load was applied along the four main directions of the implant: anterior, posterior, superior and inferior (Figure 1a). Dial indicators (MTS System, USA) were placed to measure the displacement of the baseplate. For each loading direction, both axial and shear baseplate displacements were measured, resulting in a total of eight measurements. Each measurement was performed three times and averaged value was obtained. The test was repeated for six identical samples under the same conditions.

2.2. Generation of Finite Element Models

An automated workflow was developed to set-up FE simulations of a virtual bench test. To obtain a virtual bench test that can be run multiple times by the design engineers to support possible design decisions and adaptations, the computational time of the simulation needs to be limited. For this reason, the finite element model was created to simulate only the static step of the experimental test, without considering the fatigue aspect, similarly to the work of Virani et al. [13].

2.2.1. Bone and Implant Models

The geometry files (STL) of the implant were imported into the design software 3-matic (v 14.0, Materialise N.V., Leuven, Belgium), that includes a Python scripting interface to automate processes (Figure 2). The bone substitute, which had to match the nonflat contact surface at the interface with the implant, and the loading box were created through a series of Boolean operations.

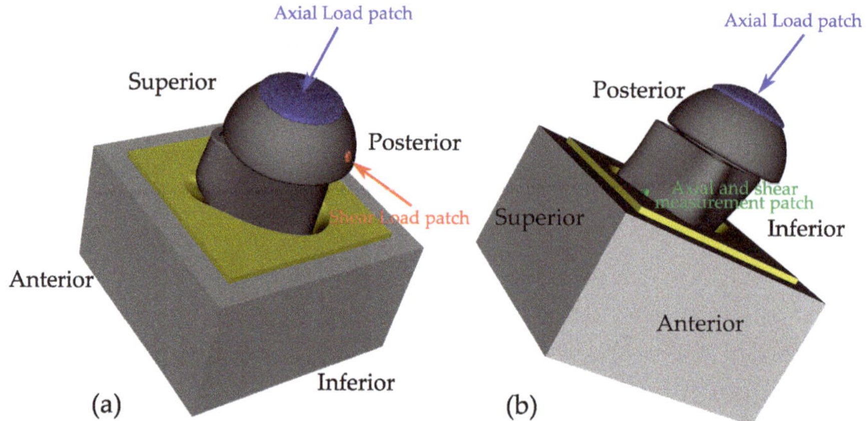

Figure 2. Left (**a**), isometric view of the finite element (FE) model with a shear load applied inferiorly. In blue the patch defined for the application of the axial load, in red the shear load patch. Right (**b**), superior view of the FE model. In green the measurement patch defined to calculate the baseplate displacement.

The 3D FE models were meshed with tetrahedral C3D4 elements. For the loading box, a coarse mesh was used, with element edge lengths ranging from 2 to 4 mm. The bone block was meshed with nonuniform elements, using a more fine mesh at the interface. A mesh convergence study was performed upfront by evaluating the impact of different mesh size on the interface micromotion. Ultimately, an average element edge length of 0.5 mm at the baseplate–bone interface was considered as the converged mesh. Nonmanifold nodes were created at the bone–implant interface, to facilitate the micromotion calculation and the convergence of the contact analysis. Due to this operation the elements nodes in the contact surface were shared between implant and bone. The implant was meshed with an average edge length of 0.5, for a total of approximately 630,000 elements, consistent with the dimensions of the prosthetic components and necessary to capture the complexity of the custom design. Ultimately, the glenosphere was meshed with an average element size of 0.5 mm. The meshing process of the screws is described in Section 2.2.3.

All components were modeled with linear elastic material properties, which is an assumption commonly made under these experimental conditions [22]. The loading box and baseplate were assigned with a Young's modulus of 110,000 MPa and a Poisson's ratio, ν, of 0.3 (corresponding to Titanium Ti-6Al-4V, [23]). The porous structure of the baseplate, mainly consisting of 3D printed Titanium, was modelled as a solid part and characterized by a lower stiffness. A Young's modulus equal to 2000 MPa and a Poisson's ratio of 0.3 were used, consistently with the values reported in the literature for titanium porous scaffolds [24]. The glenosphere was modeled using cobalt-chromium-molybdenum material properties (E = 220,000 MPa, ν = 0.3, [25]). The material properties of the foam block, representative of human glenoid trabecular bone, were taken as reference for the bone substitute (E = 200 MPa, ν = 0.3, [26]).

Contact surfaces were tied or were modelled as a hard contact with friction, depending on the interaction of the component. The interface between glenosphere and baseplate, and loading box and bone block, were considered completely tied, with no relative motion. Coulomb friction contact was implemented at the bone–implant interface. In the literature, values ranging from 0.5 to 0.7 are reported for the friction coefficient between bone and porous metal [13,14,22,27], thus an average friction coefficient of 0.6 was selected for the presented model.

2.2.2. Screw Model

In order to assess the impact of different screw types (compression and locking) on fixation, particular attention was paid to the screw modeling. A recent study showed that an excessive simplification of the screw shaft model has an impact on the micromotion in RSA implant design analysis [22]. Hence, the validity of the simplification assumptions has always to be evaluated against experimental measurements, aiming for a trade-off between acceptable computation times and prediction accuracy.

Screws were modeled following a previously described approach [28]. This approach uses structural elements for the connection to the bone, which avoids the need of meshing screw holes and the associated computational cost related to additional contact analysis (Figure 3a). A script was implemented in Python 3.7 to automate the modeling process and include the screws in the Abaqus input file. As output of the design planning phase, five screw parameters could be extracted: position (head coordinates), length, direction, outer diameter and root diameter.

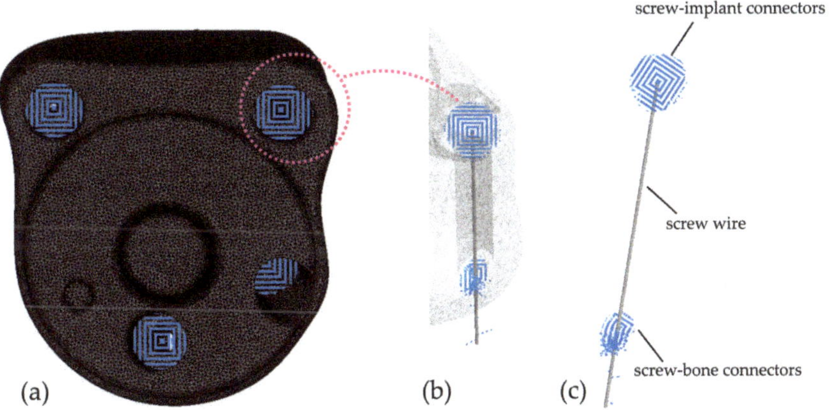

Figure 3. Left (**a**), top view of the model and the four screws. In blue the connectors between screw head and implant. Right (**b**), detail of one screw (implant transparent). Right (**c**), the generated screw model.

Each screw was modeled as a wire connecting the head point (input parameter) to the endpoint (obtained with the length and direction vector) and penetrating the elements of the bone (Figure 3b,c). All the nodes of the bone elements lying around the wire and at a maximum distance equal to the outer screw radius were connected perpendicular to the wire with rigid connector elements. The screw head was fixed to the implant in a similar way, by connecting the node representing the head with the nodes within the baseplate holes. To mesh the screw wire, beam elements (B32, three-node) with a circular cross section equal to the root radius where used, imposing as nodes the calculated intersection points between wire and connector elements. Since titanium screws were used, a Young's modulus of 110,000 MPa and a Poisson's ratio of 0.3 were assigned as material properties.

To differentiate the mechanical behavior between locking and compression screws, additional assumptions were made. To model the loose connection between the unthreaded head of a compression screw and the implant, the stiffness of the first 2 mm of the screw shaft was set to 200 MPa, a value equal to the elastic modulus of the bone substitute [14].

Moreover, nonlocking screws provide an initial compression that constrains the implant towards the bone. The impact of this aspect on FE analysis was already examined in literature, demonstrating that the inclusion of preload in the model is a key parameter when investigating interface micromotion [29]. For this reason, preload was explicitly modeled using the pretension section of Abaqus at the intersection of the screwed and nonscrewed portion of the shaft, similarly to the study of Virani et al. [13]. For the current model, the input values of the insertion force were estimated based on experimental data [30]. Briefly, a custom-made load sensor was built to measure the compression force generated by the screw head. Screws with different lengths were inserted into synthetic bone blocks (Sawbones; Malmö, Sweden) of 20 pcf and the force was acquired until failure of the bone substitute. This resulted in a maximum compression of 370 N and 420 N for the two screws used in the loosening test. Since those values were measured at failure loads, the pretensions in Abaqus were set to 260 N and 300 N, by taking 70% of the force to failure [14].

2.2.3. Boundary Conditions and Simulation Steps

Boundary and loading conditions mimicking the experimental set-up were applied. Specifically, the bottom and side faces of the rectangular metal box were fully constrained in all the directions. The axial load of 430 N was applied perpendicular to the glenoid plane through the glenosphere. A patch (10 mm radius) was defined on top of the glenosphere cup surface and all nodes lying inside were selected to apply the load (Figure 2). For the shear force, a patch of 1 mm radius was defined on the inferior side of the cup, as to simulate the horizontal load fixture (Figure 2a).

To estimate the baseplate displacement, measurement patches of nodes (1 mm radius, representative of the dial indicator tip) were also automatically defined on the baseplate surface, using the known direction vector of the load (Figure 2b). For example, when the shear load was imposed inferiorly (Figure 1a), the measurement patch was defined superiorly, centered at the intersection point between the load direction vector and the edge of the implant surface.

All analyses were performed in Abaqus/Standard 6.14 (Dassault Systèmes, Waltham, MA, USA). To solve the nonlinear equilibrium equations the Newton's method was used [31]. A three-step analysis was implemented to mimic the experimental set-up and take into account the implemented surgical technique, which consists of inserting the compression screws first followed by the locking screws: in the first step, screw pretension was modeled (see Section 2.2.3), in the second step shear load was applied, followed by the axial load in the third step.

The end of the first step was considered as the initial state for the displacement analysis, similarly to the experimental set-up (pretension of the compression screws already present before the application of the loads). Consequently, the final baseplate displacement, used for model validation, was defined as the difference in the average displacement of the patch nodes between the second and third step.

2.3. Statistical Analyses and Sensitivity Study

Predicted implant stability values were calculated as the average of the displacements for the nodes lying in the measurement patch, as defined in Section 2.2.3. Both the shear and axial components of the displacements were taken into account. A Spearman's rank order correlation test was used for comparing the consistency of results between the experimental and in silico analysis, with a significance level set to 0.05. Correlation coefficients whose magnitude were lower than 0.7, between 0.7 and 0.9 and higher than 0.9, indicated respectively a moderate, high and very high correlation [32].

Besides the baseplate displacement, shear and axial micromotion at the bone–implant interface were calculated using the FE method. These micromotions comprised the displacement values for all nodes on the contact surface. Since nonmanifold nodes were created at the bone–implant interface, micromotion was defined as the relative motion between the corresponding nodes after application of the loads. In particular, for each contact node on the implant surface, micromotion U_P was calculated as:

$$U_P = R_P - R_B,\tag{1}$$

where R_P and R_B are the vector positions of the node on the prosthesis (p) and its corresponding one on the bone surface (B), respectively. Shear (U_t) and axial (U_n) micromotion were then calculated by projecting the total micromotion on the corresponding loading direction vectors, as follows:

$$U_t = UP \cdot \hat{t}\tag{2}$$

$$U_n = UP \cdot \hat{n}\tag{3}$$

where $\hat{t}$ and $\hat{n}$ respectively represent the unit vector of the directions along which shear and axial load were applied. The total relative micromotion between glenoid baseplate and bone, is further referred to as peak micromotion [33] and was visualized as a color map on the back of the prosthesis.

To evaluate the impact of changes in the model parameters on the FE output interface micromotion, a sensitivity analysis was performed. In particular, changes in the bone substitute material properties, the friction coefficient and the screw preload were investigated. A summary of these numerical tests is presented in Table 1. Each parameter was modified independently, for a total of 24 simulations (six for each loading condition).

Table 1. Parameter variation for the sensitivity analysis.

Parameter	Baseline Value	Sensitivity Values
Elastic Modulus Bone	200 MPa	150 MPa, 553 MPa
Coefficient of Friction (CoF)	0.6	0.5, 0.7
Screw pre-load	260 N, 300 N	±20%

For the stiffness of the bone surrogate, the Young's modulus was modified to mimic the properties of 15 pcf (osteoporotic bone) and 30 pcf foam blocks, corresponding to 150 MPa and 553 MPa respectively [16,26].

The Coulomb's coefficient was adapted to simulate local changes at the bone–implant interface by imposing values of 0.5 and 0.7, which are representative of the friction ranges found in literature.

Finally, a change in the preload of the compression screws was applied, modifying by ±20% the baseline pretension value.

A paired t-test was used to compare the peak micromotion of the baseline model with each sensitivity model, with a significance level set to 0.01, following a Bonferroni correction of the alpha value ($\alpha = 0.05$, n = 6: $\alpha/n \approx 0.01$).

3. Results

FE results for the baseplate displacement were within the variability of the experimental measurements for all loading directions (Figure 4). The smallest displacements were found when the shear load was applied inferiorly to the baseplate. The Spearman's rank order test revealed a statistically significant ($p < 0.05$) high correlation ($\rho s = 0.81$) between the experimental results and FE results.

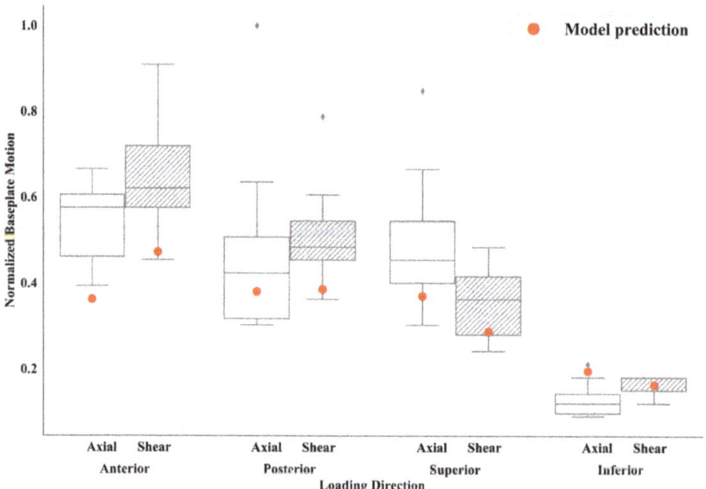

Figure 4. Baseplate displacement measured experimentally (boxplot) and determined from the model (red dots). For the FE analysis, predicted values were calculated as the average of the displacements for the nodes lying in the measurement patch, as defined in Section 2.2.3. Data were normalized to the largest micromotion measured in any of the tests. For each of the four main implant directions, both axial and shear displacements were measured. Gray points represent outliers in the measurements.

The maximum interface micromotion was found for the anterior shear load (Figure 5). For all the loading directions, the median peak micromotion was lower than 50 μm. A 95th percentile of 141 μm, 80 μm, 73 μm and 25 μm was reported for the anterior, posterior, superior and inferior loading respectively. When looking at the axial and shear components, the median shear micromotion was always higher than the axial. For none of the loading directions, micromotion above 150 μm was reported (Figure 6).

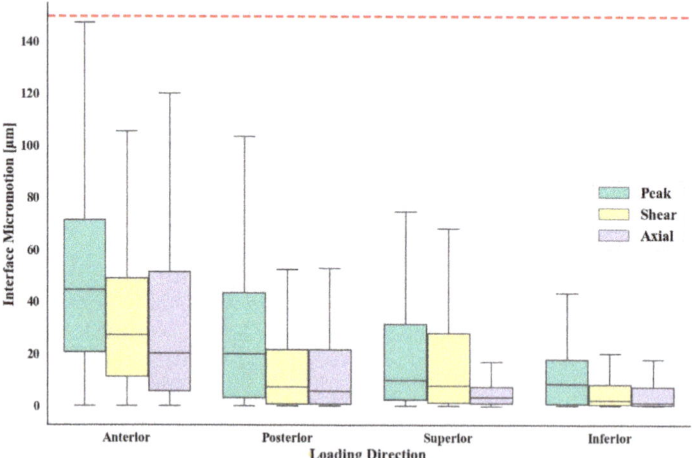

Figure 5. Interface micromotion. Shear and axial components of the total micromotion (peak) was evaluated for all the loading directions. The red dashed line represents the 150 μm threshold.

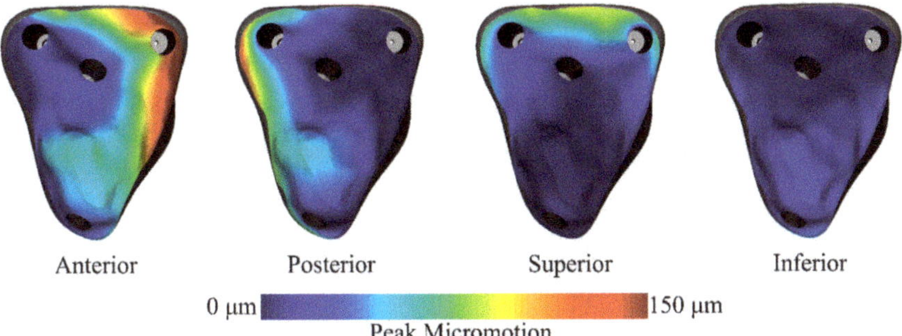

Figure 6. Back view of the implant. Peak micromotion map at the bone–implant interface for all the loading directions.

The sensitivity of the model to input parameters showed a peak micromotion for the baseline model which was significantly different ($p < 0.01$) when compared to the model with reduced and increased elastic moduli of bone substitute, for all the loading directions (Figure 7). For the anterior loading, which reported the highest micromotion values, significant differences were also found between the baseline model and the one with reduced/increased compression screws pretension.

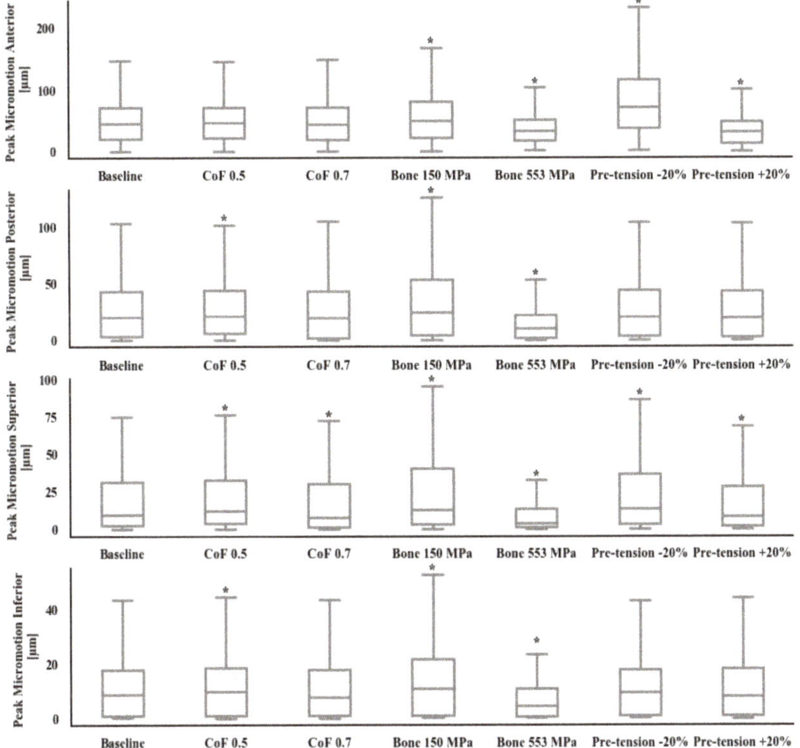

Figure 7. Change of the interface peak micromotion due to modification of different model parameters: bone Young's modulus (150 and 553 MPa), coefficient of friction (CoF = 0.5 and 0.7) and screw pretension (load ±20%). *: paired t-test, $p < 0.01$.

4. Discussion

In this study, an automated workflow to evaluate the preclinical stability of a shoulder implant through FE simulations was presented and validated. To our knowledge, this is the first work to report a full in silico modeling of ASTM F2028-14 for a custom-made prosthesis. Although previous studies [13,14,22] reported FE analysis for a similar experimental set-up, the effect of different loading directions, which in case of a custom implant cannot be neglected due to the asymmetry of the design shape, was never systematically investigated. This approach resulted in a total of eight measurements that were used to support the FE predictions.

The results of the mechanical test showed an influence of the loading direction on the implant stability. In particular, the presented design reported the lowest displacements when the shear load was applied inferiorly to the glenosphere. This is mainly due to the presence of two screws, one locking and one compression, in the superior part of the baseplate, which are almost perpendicular to the direction of the inferior load and opposite to its application point. Instead, the highest displacements were measured for the anterior loading directions, due to the absence of a good screw fixation at the anterior side. These results further corroborate the idea that each new implant should be tested in those different conditions.

All the experimental measurements showed a high variability. Although one unique design was tested with six samples, this variability is likely to reflect the variations that occurred during the production of the implants and the assembly of the different components. The 3D printed technique used for fabrication could introduce inaccuracies, especially in the porous structure, which influenced the mechanical measurements. Similarly, the bone substitute blocks were artificially carved to match the nonflat baseplate surface, possibly causing additional variation.

Direct comparison of the experimental outcomes with previous studies is not possible due to major methodological divergences. Higher mechanical loads were used to test standard implants (750 N both in axial and shear) and only the shear displacement was measured when the load was applied superiorly [12,13,15]. Under this configuration, the presented work reported slightly higher shear values (Figure 4, inferior direction), meaning that the effect of a smaller applied load was compensated by the use of a custom implant with nonstandard design (e.g., nonflat contact surface, asymmetry of the shape).

The good agreement between experimental and FE-predicted micromotions was confirmed by a Spearman's rank test, resulting in a correlation coefficient of 0.81 (high), which is lower than the one reported by Virani et al. (0.96, [13]). The lower correlation coefficient can be explained by the use of a custom design, which leads to additional complexity in the simulation. Similar to Virani et al. [13] overstiffening of the model was observed, which, in the context of this study, can be partially explained by the use of linear tetrahedral elements in the meshing process, a choice justified by the need of low computational cost.

One limit of the standard mechanical test presented here is related to the lack of micromotion measurements at the bone–implant surface. In contrast, FE modeling can provide a valuable insight on the interface behavior, although their accuracy cannot be directly evaluated against experimental outputs. As previously described, micromotion above 150 µm can jeopardize bone ingrowth and lead to an unstable fixation [9]. Design engineering should take into account this aspect when looking for possible design adaptations. For this reason, interface micromotion was estimated through the FE model. When evaluating the two separated components, higher median values were reported for the shear component. These results are in accordance with previous studies indicating that micromotion of reverse implants occurs mainly in shear [34]. For none of the loading directions peak micromotion was found to be higher than 150 µm, suggesting that the implant design does not jeopardize bone ingrowth. Additionally, the highest values were calculated at the edge of the interface, where osseointegration is less likely to happen.

The interface micromotions predicted by the FE model were sensitive to changes in some of the input parameters: the FE model was sensitive under all the loading directions to a change in bone quality (150 MPa and 553 MPa), similarly to what has been reported in the literature [14]. Moreover, this study corroborates the idea that the impact of an adequate modeling of the compression screws cannot be neglected [29]. A change in the screw pretension can lead to very different micromotion, thus suggesting that pretension should always be included in the simulation and its value estimated or derived through experimental measurements.

The generalizability of these results is subject to certain limitations which need to be addressed. Major assumptions were made during the creation of the in-silico model, looking for a trade-off between accuracy and computational cost. The bone substitutes were modeled with homogeneous isotropic material properties, a simplification commonly accepted and implemented in the literature [13,14,16,22], although not fully representative of the behavior of the bone substitute. The porous structure of the implant was not explicitly modelled to reduce the complexity of the model. As an alternative, a lower elastic modulus was used for the corresponding elements. While this assumption impacts the frictional behavior at the interface, the sensitivity showed that a change in this parameter did not substantially influence the micromotion estimations (at least in the configurations where highest values were reported).

While 150 μm is the ASTM accepted threshold to promote osseointegration [20], its application has been challenged in the literature. Other studies [15,35] referred to lower values (20 μm–50 μm) during the evaluation of interface micromotion. When lowering the threshold, the presented model would still predict bone ingrowth in the inner region of the prosthesis, however these results should be interpreted carefully and always considering the simplifications of the study.

The automated workflow was built to replicate only the static analysis described in the ASTM standard and additional efforts should be made to include the dynamic loading, which are probably not compatible with the requirement of a low computational workflow. However, it can be assumed that minimizing the initial static displacement with an optimized design, will also lead to better fatigue outcome.

Validation of the model was obtained only for a single design and under a relatively limited degree of freedom. It is believed that a more complete experimental set of tests is necessary, at least to assess the impact of additional design changes (e.g., number and type of screws) and to ensure the validity of the assumptions made. To further strengthen the predictive power of the simulation, alternative micromotion metrics would be necessary since the current mechanical set-up fails to provide a direct measure of the full-field interface micromotion [29,35].

In summary, the automated workflow presented in this study was able to replicate the mechanical condition of a standard test for a patient-specific shoulder implant. The finite element analysis can potentially support the engineers during the design phase, by virtually comparing different implants. Moreover, the minimization of the interface micromotion would lead to an improved initial stability and hence to a better clinical outcome, by allowing for secondary fixation through bone ingrowth and reducing the risk of revision surgery due to mechanical loosening. Finally, the presented tool could be used to define which configurations need to be tested when looking for worst case scenarios, thus reducing the amount of required mechanical experiments.

Author Contributions: Conceptualization, J.P.; methodology, J.P., M.W., G.H.v.L., M.A.P.; software, J.P.; writing—original draft preparation, J.P.; writing—review and editing, J.P., M.W., G.H.v.L., M.A.P.; supervision, G.H.v.L., M.A.P.; project administration, M.W., M.A.P. All authors have read and agreed to the published version of the manuscript.

Funding: This project has received funding from the European Union's Horizon 2020 Research and Innovative program under the Marie Skłodowska-Curie grant agreement No 722535.

References

1. Walker, M.; Brooks, J.; Willis, M.; Frankle, M. How Reverse Shoulder Arthroplasty Works. *Clin. Orthop. Relat. Res.* **2011**, *469*, 2440–2451. [CrossRef] [PubMed]
2. Jarrett, C.D.; Brown, B.T.; Schmidt, C.C. Reverse Shoulder Arthroplasty. *Orthop. Clin. N. Am.* **2013**, *44*, 389–408. [CrossRef] [PubMed]
3. Williams, G.R.; Iannotti, J.P. Options for glenoid bone loss: Composites of prosthetics and biologics. *J. Shoulder Elb. Surg.* **2007**, *16*, S267–S272. [CrossRef] [PubMed]
4. Hoffelner, T.; Moroder, P.; Auffarth, A.; Tauber, M.; Resch, H. Outcomes after shoulder arthroplasty revision with glenoid reconstruction and bone grafting. *Int. Orthop.* **2014**, *38*, 775–782. [CrossRef] [PubMed]
5. Rodríguez, J.A.; Entezari, V.; Iannotti, J.P.; Ricchetti, E.T. Pre-operative planning for reverse shoulder replacement: The surgical benefits and their clinical translation. *Ann Jt.* **2019**, *4*, 4. [CrossRef]
6. Chammaa, R.; Uri, O.; Lambert, S. Primary shoulder arthroplasty using a custom-made hip-inspired implant for the treatment of advanced glenohumeral arthritis in the presence of severe glenoid bone loss. *J. Shoulder Elb. Surg.* **2017**, *26*, 101–107. [CrossRef]
7. Stoffelen, D.V.C.; Eraly, K.; Debeer, P. The use of 3D printing technology in reconstruction of a severe glenoid defect: A case report with 2.5 years of follow-up. *J. Shoulder Elb. Surg.* **2015**, *24*, e218–e222. [CrossRef]
8. Dines, D.M.; Gulotta, L.; Craig, E.V.; Dines, J.S. Novel Solution for Massive Glenoid Defects in Shoulder Arthroplasty: A Patient-Specific Glenoid Vault Reconstruction System. *Am. J. Orthop.* **2017**, *46*, 104–108.
9. Jasty, M.; Bragdon, C.; Burke, D.; O'Connor, D.; Lowenstein, J.; Harris, W.H. In Vivo Skeletal Responses to Porous-Surfaced Implants Subjected to Small Induced Motions. *J. Bone Jt. Surg.* **1997**, *79*, 707–714. [CrossRef]
10. Stroud, N.; DiPaola, M.J.; Flurin, P.-H.; Roche, C.P. Reverse shoulder glenoid loosening: An evaluation of the initial fixation associated with six different reverse shoulder designs. *Bull. Hosp. Jt. Dis.* **2013**, *71* (Suppl. S2), S12–S17.
11. Formaini, N.T.; Everding, N.G.; Levy, J.C.; Santoni, B.G.; Nayak, A.N.; Wilson, C. Glenoid baseplate fixation using hybrid configurations of locked and unlocked peripheral screws. *J. Orthop. Traumatol.* **2017**, *18*, 221–228. [CrossRef] [PubMed]
12. Harman, M.; Frankle, M.; Vasey, M.; Banks, S. Initial glenoid component fixation in "reverse" total shoulder arthroplasty: A biomechanical evaluation. *J. Shoulder Elb. Surg.* **2005**, *14*, S162–S167. [CrossRef]
13. Virani, N.A.; Harman, M.; Li, K.; Levy, J.; Pupello, D.R.; Frankle, M.A. In vitro and finite element analysis of glenoid bone/baseplate interaction in the reverse shoulder design. *J. Shoulder Elb. Surg.* **2008**, *17*, 509–521. [CrossRef] [PubMed]
14. Suarez, D.R.; Valstar, E.R.; Rozing, P.M.; van Keulen, F. Fracture risk and initial fixation of a cementless glenoid implant: The effect of numbers and types of screws. *Proc. Inst. Mech. Eng. Part H* **2013**, *227*, 1058–1066. [CrossRef] [PubMed]
15. Hopkins, A.R.; Hansen, U.N.; Bull, A.M.J.; Emery, R.; Amis, A.A. Fixation of the reversed shoulder prosthesis. *J. Shoulder Elb. Surg.* **2008**, *17*, 974–980. [CrossRef]
16. Denard, P.J.; Lederman, E.; Parsons, B.O.; Romeo, A.A. Finite element analysis of glenoid-sided lateralization in reverse shoulder arthroplasty: Lateralization in RSA. *J. Orthop. Res.* **2017**, *35*, 1548–1555. [CrossRef]
17. Chae, S.-W.; Lee, H.; Kim, S.M.; Lee, J.; Han, S.-H.; Kim, S.-Y. Primary stability of inferior tilt fixation of the glenoid component in reverse total shoulder arthroplasty: A finite element study. *J. Orthop. Res.* **2016**, *34*, 1061–1068. [CrossRef]
18. Suárez, D.R.; van der Linden, J.C.; Valstar, E.R.; Broomans, P.; Poort, G.; Rozing, P.M.; van Keulen, F. Influence of the positioning of a cementless glenoid prosthesis on its interface micromotions. *Proc. Inst. Mech. Eng. Part H* **2009**, *223*, 795–804. [CrossRef]
19. Elwell, J.; Choi, J.; Willing, R. Quantifying the competing relationship between adduction range of motion and baseplate micromotion with lateralization of reverse total shoulder arthroplasty. *J. Biomech.* **2017**, *52*, 24–30. [CrossRef]
20. F04 Committee. *Standard Test Methods for Dynamic Evaluation of Glenoid Loosening or Disassociation*; ASTM International: West Conshohocken, PA, USA, 2014.
21. Lehtinen, J.T.; Tingart, M.J.; Apreleva, M.; Warner, J.J.P. Total, trabecular, and cortical bone mineral density in different regions of the glenoid. *J. Shoulder Elb. Surg.* **2004**, *13*, 344–348. [CrossRef]

22. Dharia, M.A.; Bischoff, J.E.; Schneider, D. Impact of Modeling Assumptions on Stability Predictions in Reverse Total Shoulder Arthroplasty. *Front. Physiol.* **2018**, *9*, 1116. [CrossRef] [PubMed]

23. Boampong, D.K.; Green, S.M.; Unsworth, A. N$^+$ ion Implantation of Ti6Al4V Alloy and UHMWPE for total Joint Replacement Application. *J. Appl. Biomater. Biomech.* **2003**, *1*, 164–171. [CrossRef]

24. Barui, S.; Chatterjee, S.; Mandal, S.; Kumar, A.; Basu, B. Microstructure and compression properties of 3D powder printed Ti-6Al-4V scaffolds with designed porosity: Experimental and computational analysis. *Mater. Sci. Eng. C* **2017**, *70*, 812–823. [CrossRef]

25. ASTM. *Standard Specification for Cast Cobalt-Chromium-Molybdenum Alloy for Surgical Implant Applications*; ASTM International: West Conshohocken, PA, USA, 1997; pp. 4–5.

26. F04 Committee. *Standard Practice for Finite Element Analysis (FEA) of Non-Modular Metallic Orthopaedic Hip Femoral Stems*; ASTM International: West Conshohocken, PA, USA, 2013.

27. Zhang, Y.; Ahn, P.B.; Fitzpatrick, D.C.; Heiner, A.D.; Poggie, R.A.; Brown, T.D. Interfacial Frictional Behavior: Cancellous Bone, Cortical Bone, And a Novel Porous Tantalum Biomaterial. *J. Musculoskelet. Res.* **1999**, *3*, 245–251. [CrossRef]

28. Wieding, J.; Souffrant, R.; Fritsche, A.; Mittelmeier, W.; Bader, R. Finite Element Analysis of Osteosynthesis Screw Fixation in the Bone Stock: An Appropriate Method for Automatic Screw Modelling. *PLoS ONE* **2012**, *7*, e33776. [CrossRef] [PubMed]

29. Dharia, M.; Mani, S. Impact of screw preload on primary stability in reverse shoulder arthroplasty. *Orthop. Proc.* **2019**, *101*, 48. [CrossRef]

30. Pitocchi, J.; Wirix-Speetjens, R.; van Lenthe, G.H.; Pérez, M.A. Measuring tightening torque and force of non-locking screws for reverse shoulder prosthesis. *Orthop. Proc.* **2020**, *102*, 75. [CrossRef]

31. Abaqus 6.14 Documentation. Available online: http://ivt-abaqusdoc.ivt.ntnu.no:2080/texis/search/?query=wetting&submit.x=0&submit.y=0&group=bk&CDB=v6.14 (accessed on 16 June 2020).

32. Mukaka, M. A guide to appropriate use of Correlation coefficient in medical research. *Malawi Med. J.* **2012**, *24*, 69–71.

33. Favre, P.; Henderson, A.D. Prediction of stemless humeral implant micromotion during upper limb activities. *Clin. Biomech.* **2016**, *36*, 46–51. [CrossRef]

34. Hoenig, M.P.; Loeffler, B.; Brown, S.; Peindl, R.; Fleischli, J.; Connor, P.; D'Alessandro, D. Reverse glenoid component fixation: Is a posterior screw necessary? *J. Shoulder Elb. Surg.* **2010**, *19*, 544–549. [CrossRef]

35. Favre, P.; Perala, S.; Vogel, P.; Fucentese, S.F.; Goff, J.R.; Gerber, C.; Snedeker, J.G. In vitro assessments of reverse glenoid stability using displacement gages are misleading—Recommendations for accurate measurements of interface micromotion. *Clin. Biomech.* **2011**, *26*, 917–922. [CrossRef] [PubMed]

Article

A Computational Model for Cardiomyocytes Mechano-Electric Stimulation to Enhance Cardiac Tissue Regeneration

Pau Urdeitx [1,2,3] **and Mohamed H. Doweidar** [1,2,3,*]

[1] Mechanical Engineering Department, School of Engineering and Architecture (EINA),
     University of Zaragoza, 50018 Zaragoza, Spain; purdeitx@unizar.es
[2] Aragon Institute of Engineering Research (I3A), University of Zaragoza, 50018 Zaragoza, Spain
[3] Biomedical Research Networking Center in Bioengineering, Biomaterials and Nanomedicine (CIBER-BBN),
     50018 Zaragoza, Spain
* Correspondence: mohamed@unizar.es

Received: 28 September 2020; Accepted: 22 October 2020; Published: 29 October 2020

Abstract: Electrical and mechanical stimulations play a key role in cell biological processes, being essential in processes such as cardiac cell maturation, proliferation, migration, alignment, attachment, and organization of the contractile machinery. However, the mechanisms that trigger these processes are still elusive. The coupling of mechanical and electrical stimuli makes it difficult to abstract conclusions. In this sense, computational models can establish parametric assays with a low economic and time cost to determine the optimal conditions of in-vitro experiments. Here, a computational model has been developed, using the finite element method, to study cardiac cell maturation, proliferation, migration, alignment, and organization in 3D matrices, under mechano-electric stimulation. Different types of electric fields (continuous, pulsating, and alternating) in an intensity range of 50–350 Vm^{-1}, and extracellular matrix with stiffnesses in the range of 10–40 kPa, are studied. In these experiments, the group's morphology and cell orientation are compared to define the best conditions for cell culture. The obtained results are qualitatively consistent with the bibliography. The electric field orientates the cells and stimulates the formation of elongated groups. Group lengthening is observed when applying higher electric fields in lower stiffness extracellular matrix. Groups with higher aspect ratios can be obtained by electrical stimulation, with better results for alternating electric fields.

Keywords: in-silico; 3D model; cardiac cell; cardiac muscle tissue; cardiomyocyte; electrical stimulation

1. Introduction

Electrical stimulation (ES) is an essential part of the human body physiology, which has relevant regulatory effects on cell motility, nutrient transport, and disease signaling, among others [1]. The effects due to the presence of electric fields and electric currents, generated by electric potentials, at both the cellular and tissue levels, play a key role in processes such as embryogenesis [2], tissue regeneration [3], and cancer development [4]. In the past decades, the application of cell ES has been applied to study various effects, including, among others, the development and regeneration of tissues [5,6], embryonic development [7], and tissue engineering [8]. Furthermore, ES has been shown to play a key role in maintaining a differentiated phenotype of certain cell lineages, such as neuroblasts and myocytes, keeping a close relationship between electrical activity and functionality [4,9,10]. Thus, the application of electric fields during cell maturation can influence maturation, architecture, and functionality of the developed tissues in-vitro. Different studies have shown the advantages of applying electrical stimulation in the muscle cell culture, observing benefits in cell differentiation [11,12], maturation [13,14], alignment [15,16], sarcomeric organization [17], and functional assembly [11,17]. In general, there is an improvement in the ultrastructural organization, an increase in the synchronization and amplitude of tissue contractions, in electrically stimulated tissues in comparison with non-stimulated ones.

Despite its observable effects [11], the mechanisms that regulate the cell response to electrical stimulus are still, in part, unknown, especially when it is combined with other stimuli such as mechanical cues. Cells, guided by electrotaxis, are able to detect the presence of endogenous electric fields through the extracellular matrix (ECM), polarizing and migrating in the direction of the electric field [4,6]. In the anodic zone, the cell is hyperpolarized tending to release K^+, and acquiring a locally more negative membrane potential. On the opposite side, in the cathodic side, the cell is depolarized, tending to absorb Ca^{2+} from the environment and becoming locally more positive [4]. These two effects cause migratory effects in opposite directions, being the effective direction of migration as a result of this balance. In fact, it depends, among other factors, on the cell phenotype, where both cathodic and anodic migration tendencies can be observed. For instance, Cardiac Progenitor Cells (CPC), Cardiac Fibroblasts (CF) [18] and Breast Cancer Cells [4] tend to migrate towards the cathodic direction, while Fibroblasts (FB) [18] and Keratinocytes [19] tend to migrate towards the anodic direction. However, this migratory tendency can be altered by changing ECM conditions. In fact, Frederich et al., studied the effect of the soluble Vascular Adhesion Molecules (sVCAM) on the migration of CPD and CF [18]. They observed that, in absence of sVCAM, the directionality effects produced by the electrotaxis disappeared for the CPC, while the direction was reversed for the CF. Thus, the complexity of cellular environments, which coupling several stimuli simultaneously (mechanical, electrical, and chemical), makes it difficult to study and obtain conclusions. For instance, T. A. Banks et al. have observed differences in the direction of cell migration and alignment, between Mesenchymal Stem Cells (MSCs) from different donors [20] as well as from previously published studies [21]. These differences might be due to the coupling of different stimuli [18]. Besides, H.Heidi Au et al. have compared the effect of the electric stimulation coupled with mechanical cues. They conclude that cell orientation was strongly determined by the topographical stimuli, while the electric stimuli had less relevance in cell orientation [16].

Generally, when damage is generated in cardiac tissues, the regenerative capacity is limited. This shortage is associated, among other factors, with the lack of cells proliferative capacity [18,22,23]. Stimulating and improving this capacity implies the need to know and control all the parameters that influence this process. Establishing the optimal conditions necessary to stimulate and accelerate regenerative processes implies the development of a large number of in-vitro experiments. From an experimental point of view, performing multiple tests to regulate and optimize different parameters entails a high time and economic cost, in addition to considerable technological complexity [11,23]. In this way, theoretical and computational models can offer support to study cell response in complex environments, letting us study the effect of multiple parameters on cell behavior. Thus, multiple experiments can be designed and evaluated, with reasonable economical and temporary costs, to obtain suitable conditions for a given objective, such as the stimulation of cell proliferation and tissue regeneration. Through these models, it is possible to purpose different hypotheses, breaking down and studying simple cases whose effects are known. After adequate calibration, it is possible to consider the coupling of different stimuli to analyze their effects in more complex cases. These models can be especially useful to study the formation of cellular architecture in complex tissues, where the architecture is closely related to the functionality [24,25]. Despite the advantages that these models could offer, as far as we know, up to date, there are few published models related to this topic. N. Ogawa et al. 2006 presented a computational model for cell reorientation in paramecium cells due to the galvanotaxis stimulus [26]. They considered the electrical effect on ciliary beating to be the main cue of cell reorientation. Despite their interesting point of view, cell-cell interactions as well as mechanical effects of the cellular environment were not considered in their model. In previous works, our group presented a new approach for cell galvanotaxis considering coupled effects of mechanical, electrical, thermal, and chemical stimuli in cell migration [27–29] and cell morphology [30] in 3D ECMs. However complex cell-cell interactions as cell junctions and collective cell behavior for Cardiomyocytes (CMs) were not considered. In the current work, we present a 3D mechano-electric model for the study of cell architecture based on the orientation and cellular migration of cardiac cells. This model allows to study the processes of migration, maturation, and proliferation as well as the formation of stable cell chains of Cardiomyocytes (CMs), depending on the applied electrical and mechanical stimuli.

2. Methods

In order to study the cellular response to the ES as well as the mechanical stimulus, a 3D computational model has been developed. The present model includes cell processes such as migration, maturation, and proliferation, as well as cell interaction and adhesion. The model has been developed based on the Finite Element Method (FEM) and depending on the cell internal deformations.

2.1. Cell Migration

Cell migration described in this section is based on the contractile effect of the actin-myosin (AM) machinery [31–33]. During cell migration, the cell, which is anchored to ECM through the focal adhesions, is contracted by the effect of the AM assembly. This contraction has two effects, to evaluate the mechanical environment of the cell (stiffness) and to impulse the cell. After evaluating the conditions of the ECM, the front part of the cell (preferential migration direction) generates new adhesions, while the rear part releases them [34,35]. This effect, together with the cellular contraction, drives the cell towards a new location. The direction in which the cell migrates is defined by a set of events that include actin random [36] or guided [37,38] polymerization [34,39]. Thus, the relationship between the cell internal stresses generated by the contraction of the AM and the produced deformation in the ECM (Figure 1) has been established by the following equation [40,41]:

$$
\sigma_i = \begin{cases}
K_{pas}\,\varepsilon_i & \varepsilon_i < \varepsilon_{min} \; or \; \varepsilon_i > \varepsilon_{max} \, , \\[2mm]
\frac{K_{act}\sigma_{max}(\varepsilon_{min}-\varepsilon_i)}{K_{act}\varepsilon_{min}-\sigma_{max}} + K_{pas}\,\varepsilon_i & \varepsilon_{min} \leq \varepsilon_i \leq \tilde{\varepsilon} \, , \\[2mm]
\frac{K_{act}\sigma_{max}(\varepsilon_{max}-\varepsilon_i)}{K_{act}\varepsilon_{max}-\sigma_{max}} + K_{pas}\,\varepsilon_i & \tilde{\varepsilon} \leq \varepsilon_i \leq \varepsilon_{max} \, ,
\end{cases}
\tag{1}
$$

where σ_i is the internal stresses generated by a cellular deformation ε_i at each evaluated FE point of the cell membrane. K_{pas} and K_{act}, correspond to the stiffness of the passive and active elements of the cell, respectively. σ_{max}, ε_{max}, and ε_{min}, correspond to the maximum stress generated by the contraction of the AM motor, the maximum and minimum deformation, respectively, for which the AM generates active stresses. Finally, $\tilde{\varepsilon}$, is the cellular strain for which maximum effort is generated, and is defined by $\tilde{\varepsilon} = \sigma_{max}/K_{act}$.

During the migration process, the internal stresses are transmitted to the ECM through the multiple focal adhesions as traction forces, $\mathbf{F}^i_{trac}$. The magnitude of these forces depends, in addition to the internal stresses, σ_i, on the ligands concentration, ψ, and the number of the available receptors, n_r, on the cell membrane as [27,30]:

$$
\mathbf{F}^i_{trac} = \sigma_i\, S\, k\, n_r\, \psi\, \mathbf{e}_i \, ,
\tag{2}
$$

where S is the membrane surface, k is the binding constant, and $\mathbf{e}_i$ is a unit vector that points from the evaluated membrane point towards the cell centroid. Then, the resultant traction force, $\mathbf{F}_{trac}$, is obtained through the contribution of the n forces of the cell as [42,43]:

$$
\mathbf{F}_{trac} = \sum_{i=1}^{n} \mathbf{F}^i_{trac} \, .
\tag{3}
$$

In addition to the traction forces, the model considers the effect of the forces generated by the electric field, $\mathbf{F}_{elec}$, the forces due to protrusions generation, $\mathbf{F}_{prot}$, and the drag forces, $\mathbf{F}_{drag}$, due to the viscosity of the ECM.

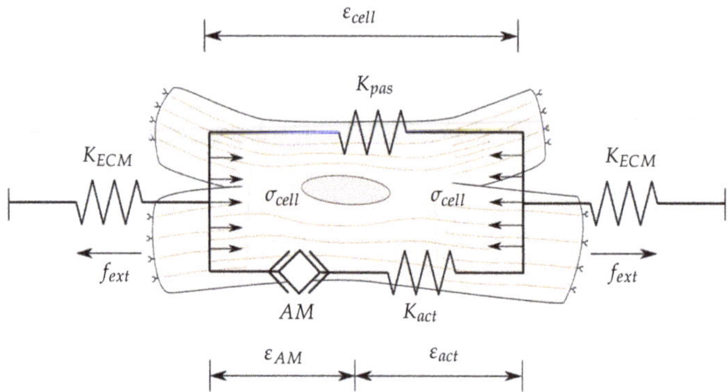

Figure 1. Mechanical model of the cell. The Actin-Myosin (AM) filaments generate cell internal contraction, ε_{AM}, which is transmitted through the active, K_{act}, and passive, K_{pas}, elements, generating cell internal deformations, ε_{cell}, and the resulting stresses, σ_{cell}. Likewise, f_{ext} corresponds to the extracellular matrix (ECM) stresses generated by the deformation of the ECM with a stiffness K_{ECM}.

Different studies have observed a linear relationship in the migratory cell behavior with respect to the electric field [17,20,44]. For instance, B. Frederich et al. studied different cardiac cells under direct current electric fields of different intensity concluding that the effect of the ES is proportional to the magnitude of the electric field [18]. Besides, C. Chen et al. showed in their review that ES stimulates cell migration and the average displacement is increased as the intensity of the ES increases, being a useful tool for regulating cell behavior [44]. This performance is attributed to the influence of Ca^{2+}, which generates a hyperpolarization of the cell in the direction of the electric field (Figure 2). Therefore, the force, $\mathbf{F}_{EF}$, with which the cell is dragged by the electric field, E, can be defined as:

$$\mathbf{F}_{EF} = -E\,\Omega\,S\,\mathbf{e}_{EF}\,, \tag{4}$$

where Ω is the cell surface charge density, S is the surface of the cell membrane, and $\mathbf{e}_{EF}$ is the direction of the electric field. The surface charge density can be obtained using the Gouy-Chapman membrane charge equations as a function of the resting potential of the membrane [45].

Different experiments conclude that, although there is a linear relationship between the intensity of the electric field and the velocity of the cell migration, cells also exhibit a threshold for which this velocity does not increase anymore. This value of EF for which saturation of the electric forces appears, E_{sat}, seems to be dependent on the analyzed cell type [18,19,21,44]. This saturation effect has been defined in the calculation of the electrical charge density of the cell as:

$$\Omega = \begin{cases} \Omega(z,\psi) & E \leq E_{sat}\,, \\ \Omega_{sat} & E > E_{sat}\,, \end{cases} \tag{5}$$

where Ω_{sat} is the saturation charge of the cell surface, and E_{sat} is the electric field for which the electric cell forces show a saturation.

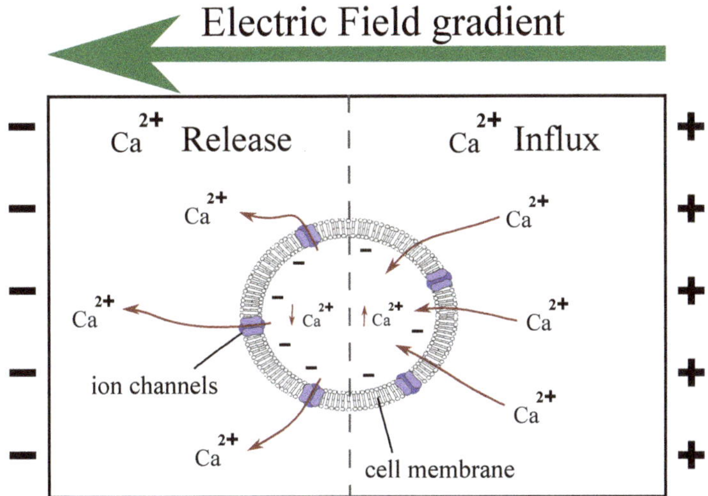

Figure 2. Cell electrotaxis. Cell membrane towards the anode is hyperpolarized allowing the influx of Ca^{2+} by passive electromechanical diffusion. In the cathodic side, the cell is depolarized, and its ion channels open releasing Ca^{2+}. The net electrotactic force depends on the balance of the two opposing attraction forces.

The present model also considers the local repulsion force produced by the individual charge of the cells. The repulsion electric force, $\mathbf{F}_{EF}^{ij}$, experienced by the cells i and j, is proportional to the electric charge of the cells, Ω_i and Ω_j, and inversely proportional to the distance between them, r_{ij}. It can be calculated by [27,28]:

$$\mathbf{F}_{EF}^{ij} = \frac{k_e}{\epsilon_r} \frac{\Omega_i S_i \Omega_j S_j}{r_{ij}^2} \mathbf{e}_{ij} \, , \tag{6}$$

where k_e is the coulomb constant, ϵ_r is the relative permittivity of the ECM, and $\mathbf{e}_{ij}$ is the direction from the jth cell towards the ith cell. With the generated forces by each jth neighbor cell and the forces due to the electric field, the total force on the ith cell, $\mathbf{F}_{elec}$, can be obtained by:

$$\mathbf{F}_{elec} = \mathbf{F}_{EF} + \sum_{j=1}^{n-1} \mathbf{F}_{EF}^{ij} \, . \tag{7}$$

Protrusion forces, due to the extension and retraction of protrusions of the cell, generate extensions of the cell which increase cell penetration. In general, it is considered as a random process. Thus, the magnitude and direction of the protrusion forces, $\mathbf{F}_{prot}$, have been calculated as [40,43]:

$$\mathbf{F}_{prot} = \kappa \parallel \mathbf{F}_{trac} \parallel \mathbf{e}_{rnd} \, , \tag{8}$$

where κ is a random value between $0 \leq \kappa < 1$, and $\mathbf{e}_{rnd}$ is a random unit vector.

Finally, the drag force effect, $\mathbf{F}_{drag}$, has been considered as a force that opposes the movement of the cell due to the medium viscosity, η. It has been defined by the Strokes law as the resistance force to a cell of radius r, which is moving at velocity $\mathbf{v}$, as:

$$\mathbf{F}_{drag} = 6\pi r \eta \mathbf{v} \, . \tag{9}$$

Proposing a balance of forces on the cell, and neglecting inertial effects due to the scale of the problem, we obtain:

$$\mathbf{F}_{trac} + \mathbf{F}_{elec} + \mathbf{F}_{prot} = \mathbf{F}_{drag} \, , \tag{10}$$

through which the direction and velocity of cell migration are obtained.

2.2. Cell Interaction

Cells show a collective response different from their individual behavior. Cell-cell interaction has a high impact on processes such as cell proliferation [46,47] and migration [48,49]. Through cell-cell interactions, cells establish intercellular connections by binding, for example, their cytoskeleton through desmosomes, or communicating electronically through gap junctions [50–52]. By counterpoint, the cells lose some of the ability to interact with the ECM along the contact surface between two cells. In muscle cells, cell-cell interaction is particularly important, where the final functionality of the tissue depends drastically on the union quality between the cells [24,50]. So, to develop in-vitro muscle tissues, a correct cell guidance to appropriate architectures is desired. Thus, the cell contact vector is defined, for any pair of cells, through the position vectors of these cells (Figure 3a), as [41,43]:

$$\mathbf{X}_{ij} = \mathbf{X}_i - \mathbf{X}_j \, , \tag{11}$$

where $\mathbf{X}_i$ and $\mathbf{X}_j$ are the position vectors of the ith and jth cells, respectively. To avoid cells overlapping, $\mathbf{X}_{ij}$ must fulfill $\mathbf{X}_{ij} \geq 2r$.

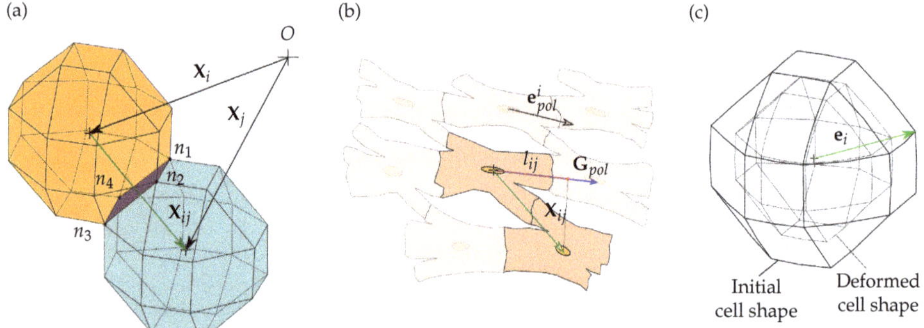

Figure 3. Cell interaction. (**a**) $\mathbf{X}_i$ and $\mathbf{X}_j$ are the coordinate vectors of the ith and jth cells, respectively, and $\mathbf{X}_{ij}$ is the contact vector, which satisfies $\| \mathbf{X}_{ij} \| \geq 2r$. The contact face, defined by the nodes $(n_1 : n_4)$ loses the capacity to interact with the ECM. (**b**) The global polarization direction $\mathbf{G}_{pol}$, is defined through the cell polarization e^i_{pol}. The projection of the cell contact, l_{ij}, is defined by the projection of the contact vector, $\mathbf{X}_{ij}$, in the $\mathbf{G}_{pol}$ direction. Being the cell junction possible when $l_{ij} \geq l_{adh}$, and $\| \mathbf{X}_{ij} \| = 2r$. (**c**) The direction of the internal cell deformation, e_i, is considered to establish the direction of the cell mechanical stimulus.

The direction of cell contact can be defined as:

$$\mathbf{e}_{ij} = \frac{\mathbf{X}_{ij}}{\| \mathbf{X}_{ij} \|} \, , \tag{12}$$

while the direction of cell polarization can be determined by the mechanical, e^i_{mech}, and electrical, e^i_{elec}, stimuli to which the cell is subjected, as:

$$\mathbf{e}^i_{pol} = \frac{\mathbf{e}^i_{mech} + \mathbf{e}^i_{elec}}{\| \mathbf{e}^i_{mech} + \mathbf{e}^i_{elec} \|} \, , \tag{13}$$

where $\mathbf{e}^i_{mech}$ and $\mathbf{e}^i_{elec}$ are the direction of the mechanical and electrical stimuli, respectively, calculated as:

$$\mathbf{e}^i_{mech} = \frac{\mathbf{F}_{trac}}{\parallel \mathbf{F}_{trac} \parallel} , \tag{14}$$

and

$$\mathbf{e}^i_{elec} = \frac{\mathbf{F}_{elec}}{\parallel \mathbf{F}_{elec} \parallel} . \tag{15}$$

Cardiac tissues are composed of highly ordered myofibrils, which increase in size, number and complexity during tissue development [3,23,53]. Thus, cells guided by different stimuli in the ECM, polarize and join other cells forming myotubule-like structures. For instance, V. Planat-Bénard et al. studied cardiomyocyte differentiation, observing how cardiac cells differentiate and form myotubule structures after 14 days of cell culture [54]. Besides, as exposed by N. Tahara et al., cardiac precursor migration showed that CM become connected to form coherent epithelia in bilateral cardiac precursor populations [55]. This behavior can be observed in different in-vitro studies [3,15,56,57]. In this context, we define the global polarization direction, $\mathbf{G}_{pol}$, is an indicator of the degree of alignment of the cells, which indicates the major direction on which the cells are structured (Figure 3b). This direction is obtained by evaluating the polarization direction of all cells as:

$$\mathbf{G}_{pol} = \frac{\mathbf{R}_{pol}}{\parallel \mathbf{R}_{pol} \parallel} , \tag{16}$$

where

$$\mathbf{R}_{pol} = \sum_{i=1}^{n} \frac{\mathbf{e}^i_{pol}}{\parallel \mathbf{e}^i_{pol} \parallel} . \tag{17}$$

This direction is compared to the cell-cell contact direction to determine the quality of the cell adhesions. Thus, to compare the direction of cell contact, $\mathbf{e}_{ij}$, with the direction of global polarization, $\mathbf{G}_{pol}$, the projection parameter, l_{ij}, has been defined as:

$$l_{ij} = \frac{Proj(\mathbf{e}_{ij}, \mathbf{G}_{pol})}{\parallel \mathbf{G}_{pol} \parallel} , \tag{18}$$

where its value is limited within the range $0 < l_{ij} \leq 1$, being 1 if the cell contact vector, $\mathbf{e}_{ij}$, and the global polarization direction, $\mathbf{G}_{pol}$, have the same orientation, and 0 if they are in perpendicular directions.

In addition, let us define the cell junction (CJ) as a parameter that represents the union of appropriately oriented two, or more cells. Thus, when two cells are in contact ($\parallel \mathbf{X}_{ij} \parallel = 2r$) and the direction of cell polarization is consistent with the polarization direction of the whole cells ($l_{ij} \geq l_{adh}$), CJ represents a strong cell contact. l_{adh} represents the minimum bound of the projection parameter to consider cell adhesion which is proposed based on the ultrastructure of cardiac tissues [24,43,50,58].

Each cell, through the previously described mechanosensing process, tends to move into a new location (Figure 4a). Nevertheless, cells attached with strong CJ, form groups that tend to remain attached during migration [22,59]. Each cell tends to drag the cells to which it is attached. This effect generated by each cell in the group causes a collective migration behavior (Figure 4b). In this case, a new equilibrium is established by [43]:

$$\mathbf{F}^{grp}_{drag} = \sum_{i=1}^{n} \mathbf{F}^i_{trac} + \mathbf{F}^i_{elec} + \mathbf{F}^i_{prot} , \tag{19}$$

where $\mathbf{F}^i_{trac}$, $\mathbf{F}^i_{elec}$, $\mathbf{F}^i_{prot}$ are corresponding to the contributions of the mechanical, electrical, and protrusion forces, respectively, of each ith cell in the group. $\mathbf{F}^{grp}_{drag}$ corresponds to the drag force of the group, which can be calculated by [43]:

$$\mathbf{F}_{drag}^{grp} = f_{sh} 6\pi r_{grp} \eta \mathbf{v}_{grp} \, , \tag{20}$$

where r_{grp} and $\mathbf{v}_{grp}$ are the equivalent radius and the velocity of the group, respectively. f_{sh} is a shape factor, due to the irregular shape of the group, calculated as [30,43,60]:

$$f_{sh} = \left[\frac{l_{max} l_{med}}{l_{min}^2} \right]^{0.09} , \tag{21}$$

where l_{max}, l_{med}, and l_{min}, correspond to the maximum, medium, and minimum dimensions of the group, respectively, which is defined in an orthogonal coordinate system.

Furthermore, cells can also be relocated to new positions, more favorable, within the same group. After evaluating the translocation of the group, if the group does not move, the individual migration of the cells of this group is evaluated, $\mathbf{v}_i$. In this case, if a cell has the capacity to migrate to a new, and available, position within the group, then, it is relocated into that new position (Figure 4c). Thus, the cells belonging to a group can migrate with the group, or relocate within it [22,43,61].

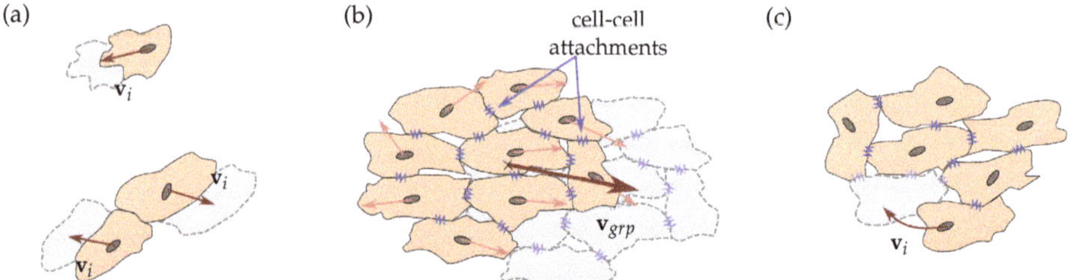

Figure 4. Cell migration. (**a**) Individual cell migration is considered when cells are separated or for cells which are not attached to another cell by CJ. $\mathbf{v}_i$ is the individual cell velocity. (**b**) Collective cell migration is considered for each group of cells attached by cell junctions. Group velocity, $\mathbf{v}_{grp}$, is defined from the migratory tendency of the cells in the group. (**c**) Cell relocation is considered when group velocity is insufficient to consider the movement. A cell can migrate to a new position, with its individual velocity, $\mathbf{v}_i$, without leaving the group.

2.3. Cell Fate

The mechanical properties of the ECM not only affect cell migration but also are important for processes such as cell maturation, proliferation, and apoptosis. In the case of cardiac cells, the mechano-electric conditions to which they are subjected during their maturation are key in the development of functional tissues [11]. For instance, under different mechanical stimuli, cells mature at different rates, showing faster maturation in stiffer ECMs [62–64]. To include the effect of the mechanical stimulus, $\gamma_c(t)$, to which a cell is subjected at each instant of time, t, based on its internal deformation, ε_i, the mechanical stimulus can be defined as (see Figure 3c) [28,65]:

$$\gamma_c(t) = \frac{1}{n} \sum_{i=1}^{n} \mathbf{e}_i : \varepsilon_i : \mathbf{e}_i^T , \tag{22}$$

where ε_i and $\mathbf{e}_i$ are the cell internal deformation and the position vector, respectively, of the ith node of the cell, and n is the number of nodes in which the cell has been discretized.

The cell maturation time, $t_{mat}(\gamma_c, t)$, which is the time necessary for a cell to reach the necessary level of maturity to proliferate, is obtained for each cell at each time step, considering the mechanical stimulus, $\gamma_c(t)$, as:

$$t_{mat}(\gamma_c, t) = t_{min} + t_p \gamma_c(t), \tag{23}$$

where t_{min} is the minimum time needed to maturate. t_p is a time proportionality factor, which depends on the mechanical stimulus.

To define the status of maturation of each cell, we have defined a Maturation Index (MI) as:

$$\text{MI} = \begin{cases} \frac{t}{t_{mat}} & t \leq t_{mat}, \\ 1 & t > t_{mat}. \end{cases} \tag{24}$$

When maturity is reached, MI $= 1$, the cell has the possibility of proliferating. However, the proliferation capacity of cardiac cells is limited [66,67], and closely related to the cell-cycle arrest and cell junctions [23,68]. In this model, an adaptive cell phenotype is considered. In this way, cardiac cells, initially considered as CM in the early stages of maturation (early CM), and upon reaching the maturity state, MI $= 1$, can advance in the maturity of their cardiac phenotype, and reach the state of adult CM (late CM). This phenotype change is associated with the ability of cells to form stable cell-cell adhesions [23,54,68]. In fact, adult CM are highly ordered in stable myofibrils, which prevents cell division [23]. Thus, CM proliferation is closely related to cell maturation and cell-cell adhesions [23]. In this way, early CM retains the capability of proliferating where an adult CM is considered post-mitotic and do not proliferate [23,67,69].

Thus, cell proliferation has been defined as a function of the number of CJ, which defines the formation of cell-cell stable adhesions, and the MI, which defines the cell-cycle status. Whereas, a cell fully incorporated into a chain undergoes cell-cycle arrest, which blocked cell proliferation. In contrast, free cells, or partially attached to a chain, maintain their proliferation capacity due to its consideration as early CM phenotype. This process is defined by the following equation:

$$\text{Cell proliferation} = \begin{cases} 1 \text{ mother} \rightarrow 2 \text{ daughters} & \text{CJ}_i < \text{CJ}_{max} \ \& \ \text{MI} = 1, \\ \text{no proliferation} & \text{otherwise}, \end{cases} \tag{25}$$

where CJ_i is the number of cell junctions of the ith cell, and CJ_{max} is the number of cell junctions that promotes cell-cycle arrest [70]. Thus, if the cell is partially surrounded, means attached to at least 4 other cells ($\text{CJ}_{max} = 4$), which corresponds to the 50% of the maximum possible CJ due to the model discretization, cardiac cell phenotype is considered to achieve an adult phenotype and cell proliferation is inhibited [43].

Once a cell proliferates, it generates two daughter cells. The locations of these new cells have been defined as:

$$\begin{aligned} \mathbf{x}_{daut}^{(1)} &= \mathbf{x}_{moth}, \\ \mathbf{x}_{daut}^{(2)} &= \mathbf{x}_{moth} + 2r\mathbf{e}_{rand}, \end{aligned} \tag{26}$$

where $\mathbf{x}_{moth}$, $\mathbf{x}_{daut}^{(1)}$, and $\mathbf{x}_{daut}^{(2)}$, are the coordinates vectors of the mother cell, the first and the second daughters, respectively. $\mathbf{e}_{rand}$ is a randomly generated unit vector.

2.4. Computational Model

The model has been implemented via a user-defined subroutine within the commercial Finite Element software Abaqus Dassault Systems (UELMAT) [71]. Within this subroutine, the cell has been defined as a discretized quasi-spherical element with 24 nodes located in the cell membrane (see Figure 3a). The ECM, with which the cell interacts, is defined through trilinear hexahedral elements, with dimensions of $800 \times 400 \times 400$ μm. The model has 128,000 elements and 136,161 nodes. Each calculation step is equivalent to 0.8 h of cell-ECM interaction, analyzing a total of 160 h in each experiment. The ECM has been considered as a linear elastic material. The model algorithm is described in Figure 5, and the employed parameters have been detailed in Table 1.

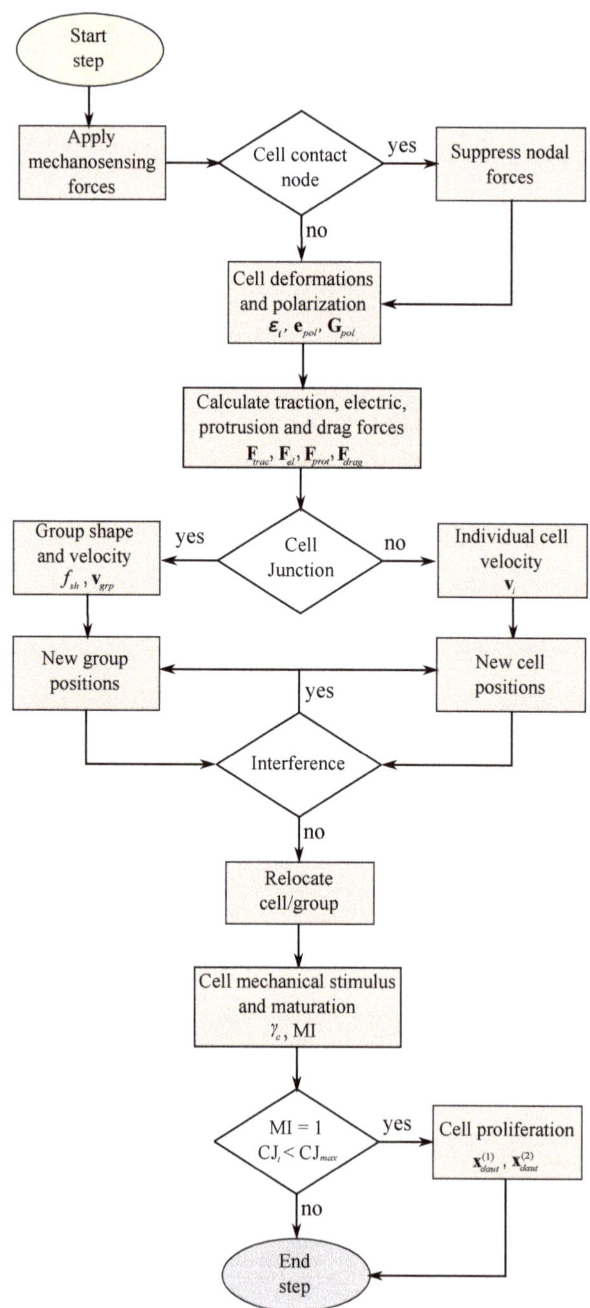

Figure 5. Algorithm of the model implemented for each time step.

Table 1. Mechanical parameters considered in the model.

Parameter	Description	Value	Refs.
K_{pas}	Stiffness of the cell passive elements	2.8 kPa	[72,73]
K_{act}	Stiffness of the actin-myosin machinery	7.0 kPa	[72,74]
ε_{max}	Maximum strain of the cell	0.09	[30,75]
ε_{min}	Minimum strain of the cell	−0.09	[30,75]
σ_{max}	Maximum contractile stress exerted by the actin-myosin machinery	0.25 kPa	[76,77]
ν	ECM Poisson ratio	0.4	[78,79]
η	ECM viscosity	1000 Pa·s	[64,80]
k	Binding constant of the cell	$10^8\ \text{mol}^{-1}$	[28,80]
n_r	Number of available receptors of the cell	1.5×10^5	[28,80]
E_{sat}	Saturation value of electric field	$1200\ \text{Vm}^{-1}$	[18]
Ω_{sat}	Saturation value of cell charge density	$5^{-2}\ \text{Cm}^{-2}$	[45,81]
ψ	Concentration of the ligands at the of the cell	$10^{-5}\ \text{mol}$	[28,80]
l_{adh}	Minimum bound of projection to consider cell adhesion	0.50	[24]
t_{min}	Minimum time needed for maturation	6 days	[54,68]
t_p	Time proportionality	200 days	[72,82]
γ_{low}	Minimum level of mechanical stimuli for cardiac cell differentiation	−0.04	[11,83]
γ_{myo}	Maximum level of mechanical stimuli for cardiac cell differentiation	−0.01	[11,83]
γ_{apop}	Maximum mechanical stimuli which trigger apoptosis	0.6	[50,82]

3. Results

A series of experiments have been developed to calibrate and compare the model results with those obtained from the bibliography. For this aim, a rigidity range equivalent to that is used in the bibliography for cardiac cell culture (10–40 kPa) [11,50,78,83] has been considered. Additionally, a determined range of electric field (50–350 Vm^{-1}) has been chosen based on the data available in the bibliography, avoiding intensities that could cause cell damage [18,84,85]. In all the cases, the electric field is applied in the longitudinal direction.

To study the variability of the results, for each case, 10 repetitions with different initial random cell distributions have been generated. For representations issue, the average value of the results of these 10 repetitions has been calculated.

3.1. Continuous Mechano-Electric Stimulation

In the first experiment, the effect induced by a constant electric field on cells is studied. Initially, 40 cells have been randomly distributed in the ECM. Their behavior has been monitored for 160 h. Different experiments with stiffnesses of 10, 20, 30, and 40 kPa combined with electric field strengths of 50, 150, 250, and 350 Vm^{-1}, with a total of 16 different configurations have been prepared. In this case, the interaction of two effects on the cell and its variation is observed by changing the stiffness and the intensity of the electric field. As observed in previous works of our group [40,43], cells tend to occupy the center of the ECM guided by mechanical stimulation, being the area where the cells undergo less internal deformations. The presence of a unidirectional ES guides the cells in the direction of the electric field, while the different stiffness generate a wide range of possible results. For high stiffness, the effects related to the mechanotaxis tend to be high, with a greater tendency of the cells to migrate towards the center of the substrate. In the same way, for high ES cases, a high drag effect of the cells is generated by the galvanotaxis. The final location of the cells depends on the intensity of both stimuli.

A differentiated effect is observed for different stiffnesses, with a higher impact of the ES in lower rigidity ECMs. As the ES increases, cells show an increase in the alignment in the longitudinal direction as well as a high migration tendency in the direction of the electric field. For the maximum ES, cells are rapidly dragged by the ES, hindering cell-cell interaction (Figure 6).

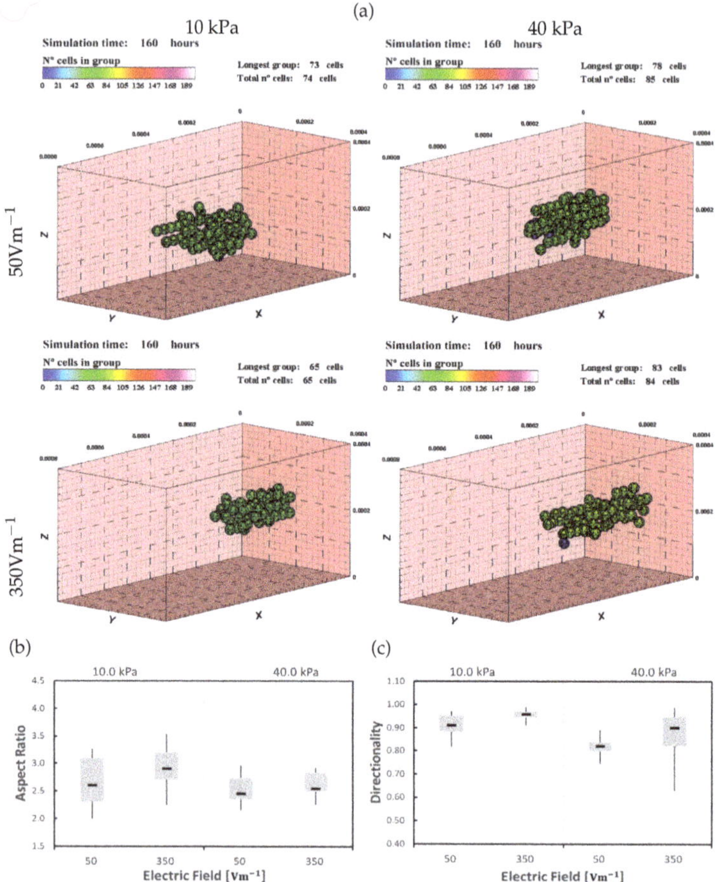

Figure 6. Cardiomyocytes (CMs) in 10 and 40 kPa ECMs, under ES of 50 and 350 Vm^{-1}, after 160 h of simulation (see also Supplementary Materials video 1). (**a**) Cells tend to form a longer group in the longitudinal direction as the electric field increases. Under low electrical stimulus (50 Vm^{-1}), cells remain at the center of the ECM (top). As the stimulus increases, cells tend to move to the outer surface of the ECM (bottom). Numerical results for Aspect Ratio (AR) (**b**) and directionality (**c**).

As the stiffness increases, the impact of the ES seems to decrease slightly, but general behavior is maintained. For the 10 kPa ECM, cells only remain in the center of the substrate for the minimum ES (50 Vm^{-1}). In 40 kPa ECM, where the mechanical stimulus is higher, the cells remain in the center of the ECM for higher electric field strengths (above 150 Vm^{-1}). Thus, as stiffness increases the ES and the mechanical stimuli seem to be better balanced, obtaining, in general, better results (Figure 6). In all the cases, the increase in the ES generates a higher cells attraction in the direction of the electric field. Thus the electric field increases the directionality of the cell migration toward the direction of the ES, which is consistent with the bibliography [18,84]. The speed with which the cells are attracted to the electric field also depends on the stiffness of the ECM, being the lower the stiffness the higher the ES attraction. Besides, directionality is dependent on the ES strength as was reported by B. Frederich et al. [18]. At the same time, the effect of mechanical stimulation on maturation increases the number of cells at the end of the simulation as the stiffness increases. The coupled of these two effects generate groups with a higher number of cells, maintaining a good degree of cell alignment.

To evaluate the alignment effects in group's morphology, we define an Aspect Ratio (AR) parameter by comparing the geometry of the groups. Thus, AR is defined, considering the main group's geometry in each simulation, as:

$$AR = \left[\frac{l_x^2}{l_y l_z} \right]^{0.5}, \tag{27}$$

where l_x, l_y, and l_z are the longitudinal group length, in X direction, and the transversal group length, in Y and Z directions, respectively. In addition, the directionality of the cells is controlled by the global polarization vector, $\mathbf{G}_{pol}$.

Analyzing the effect of the stiffness and ES on the AR, it is observed that the electrical effects are much higher than the stiffness. Likewise, the variation of AR seems to have a linear dependence with the electric field intensity (Figure 7a). The effects of the stiffness on the AR are comparatively low, with higher AR being observed for low stiffness. Similarly, in the directionality analysis, a greater effect of the ES is observed in comparison with the effect of stiffness (Figure 7b). As with the AR, the directionality is shown to be higher for cases of less rigidity and a higher electric field.

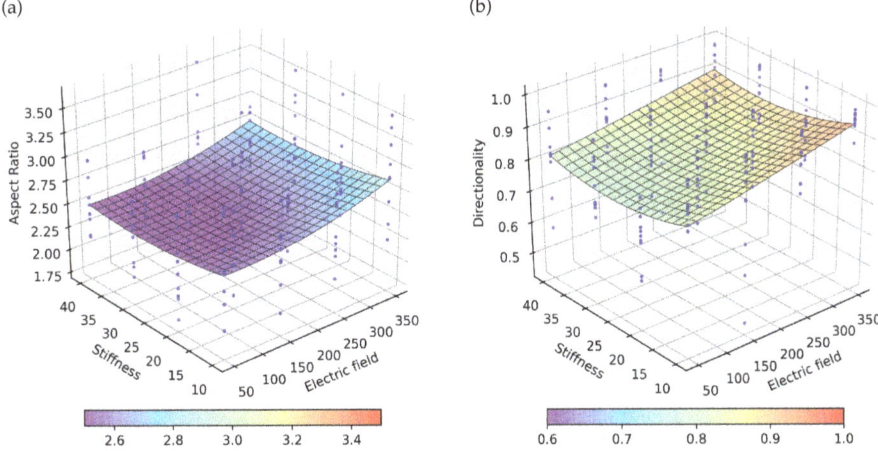

Figure 7. Representation of the obtained numerical results applying a continuous electric field with different intensities in ECM with different stiffnesses. Blue dots represent 10 experiment repetitions, on which the surface with the best fit to the results is represented. (**a**) Aspect ratio for the different combination of stiffness and electric stimulus. (**b**) Directionality for the different combination of stiffness and electric stimulus.

3.2. Pulsatory Mechano-Electric Stimulation

In in-vitro experiments, the ES can be applied to mimic the physiological electrical currents of the heart [17,84,85]. This is a primarily intended to activate and coordinate the spontaneous contraction of the AM apparatus. These currents are of a pulsating type and low frequency. For instance, S. Pietronave et al. observed an increase in the cell alignment and the expression of specific cardiac markers while stimulating CPCs with monophasic and biphasic electric fields [84]. This is also supported by M. Radisic et al. work, where cardiac myocytes were stimulated with monophasic electric fields, showing an increase of the cell alignment, with myofibers aligned in the direction of the electric field application, and ultrastructural improvement with formed GAP junctions and contractile activity within the cells after five days [17]. In this way, we also tried to study the stimulation of the cells with monophasic electric fields. So, an electric pulsatory field is applied during the simulation to study and compare the effect of different ways of stimulation on cell behavior and group's morphologies.

In this case, the electric field is applied discontinuously with the same configurations of stiffnesses (10, 20, 30, and 40 kPa), electric fields (50, 150, 250, and 350 Vm^{-1}), and initially random distributed 40 cells. The electric field is initially active and then it alternates its activation at each step (on/off).

The general behavior of the cells, observed in the previous experiment, is maintained (Figure 8). During the steps in which the electric field is active, the cells tend to migrate in the direction of the electric field, which is consistent with the bibliography [17,18,84]. In the steps in which the electric field is disabled, the cells, due to mechanical stimulation, tend to migrate toward the center of the ECM as it was seen in a previous work of our group [43]. These two effects slow down the migration to the outer surface of the ECM. Thus, compared with the previous experiment, the effects of the electric field are less pronounced.

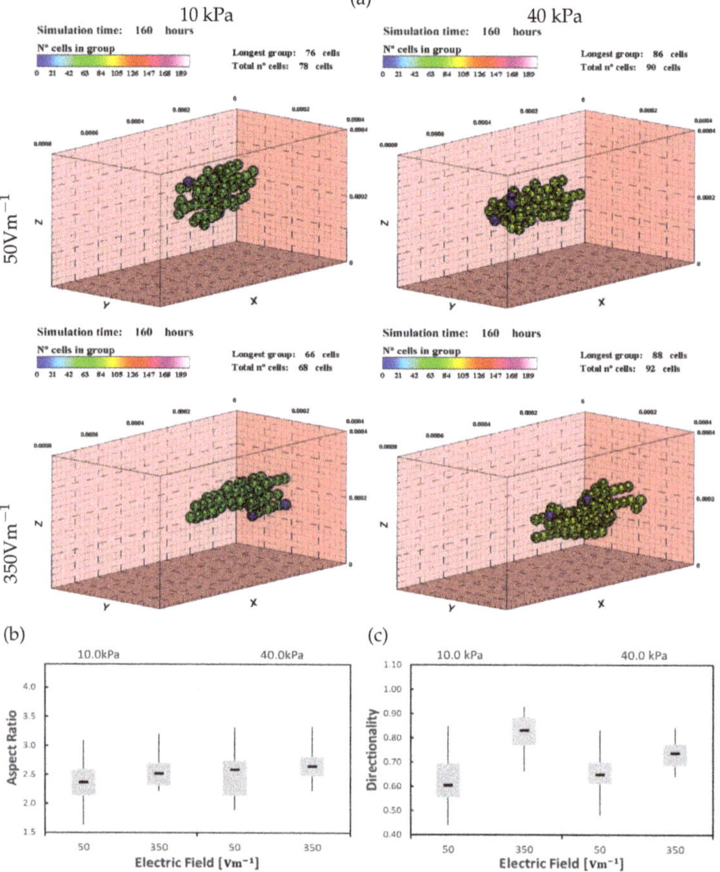

Figure 8. CMs in 10 and 40 kPa ECMs, under electrical pulsatory stimulation of 50 and 350 Vm^{-1}, after 160 h of simulation (see also Supplementary Materials video 2). (**a**) Less effect of the electric field is observed with a greater tendency of the cells to remain close to the center of the ECM. AR (**b**) and directionality (**c**) increase as the Electrical stimulation (ES) increases.

As in the previous case, the ES is more intensive in low stiffness ECM, showing higher values of the directionality (Figures 8c and 9b). On the contrary, the results for the AR seems to be balanced for all ECM stiffness (Figure 9a). In this second experiment, the dependence of the AR with the ES seems to follow a linear tendency (Figure 9a). The maximum values of the AR, obtained with the maximum ES, seem to be reduced compared with the previous experiment. However, in the visual comparison of the results, a good level of cell alignment is observed for high levels of ES, avoiding excessive drag of the cells towards the surface of the ECM (Figure 8).

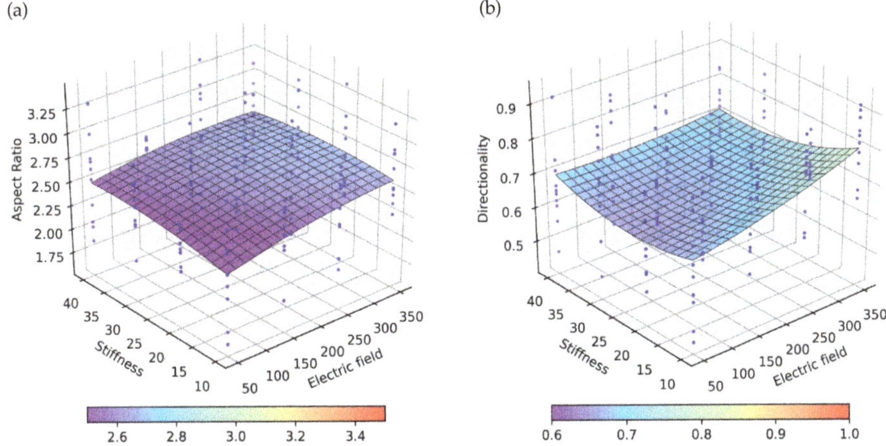

Figure 9. Representation of the obtained numerical results applying a pulsatory electric field with different intensities in ECM with different stiffnesses. Blue dots represent 10 experiment repetitions, on which the surface with the best fit to the results is represented. (**a**) Aspect ratio for the different combination of stiffness and electric stimulus. (**b**) Directionality for the different combination of stiffness and electric stimulus.

3.3. Alternating Mechano-Electric Stimulation

The cells not only can be stimulated by a discontinuous electric field, as mentioned in the previous case, but also by alternating electric field [11]. Thus, in the third experiment, the application of an alternating electric field is considered. For this purpose, the direction of the electric field has been reversed at each step. As in the previous cases, stiffnesses of 10, 20, 30, and 40 kPa, and electric fields of 50, 150, 250, and 350 Vm^{-1} are applied on 40 cells initially randomly distributed. Having in account that prolonged exposure to high-intensity electric fields can trigger cell apoptosis [18,20], cell-cell and cell-ECM interactions are studied during 160 h of simulation.

In this case, the electric field guides cells to move alternately in the longitudinal direction while the mechanical stimulation guides cells toward the center of the ECM. Although the cells are moving in the longitudinal direction, the migration direction is reversed as the electric field reverses. This is consistent with experimentally studied cases in the bibliography, where the direction of cell migration was evaluated when the electric field was reversed [18,21]. This process reduces the effective migration (total translocation) of the cells in the longitudinal direction. In low ES cases (50 Vm^{-1}), the cells migrate easily to the center of the ECM, which shows that the electric stimulus effect is lower than the mechanical stimuli. On the contrary, in high ES (350 Vm^{-1}), cells approach the center of the ECM moving slightly in the longitudinal direction. Initially, cells' effective movement is almost perpendicular to the electric field direction, toward the center of the ECM with slight movement in the direction of X axis. As the cells reach the central position of the ECM, the presence of other cells increases the mechanical stimuli, guiding the cells to migrate towards the other cells. The combination of both stimuli introduce an improvement in the length of the formed groups.

Compared with the previous cases, the groups are bigger and more elongated (Figure 10). As the cells remain in the center of the ECM, the region of higher rigidity, the cell proliferation increases. When the cells join a group, cell proliferation is considered to be inhibited. Unlike the previous cases, as the cells have a combination of two different directions stimuli, groups are formed later, which gives de cells extra time to proliferate. As in the previous experiments, the effects of ES are more pronounced in ECM of less stiffness, where the mechanical stimulus is lower. Besides, better results of AR and directionality are obtained (Figure 11). In this third case, it is observed that the AR tendency follows a behavior that can be considered linear, both for the mechanical and electrical stimulus, individually and/or combined effects (Figure 11a). On the contrary, the directionality follows a clearly non-linear tendency. For values greater than 150 Vm^{-1} and for all stiffness, saturation in the directionality of the cells is observed (Figure 11b).

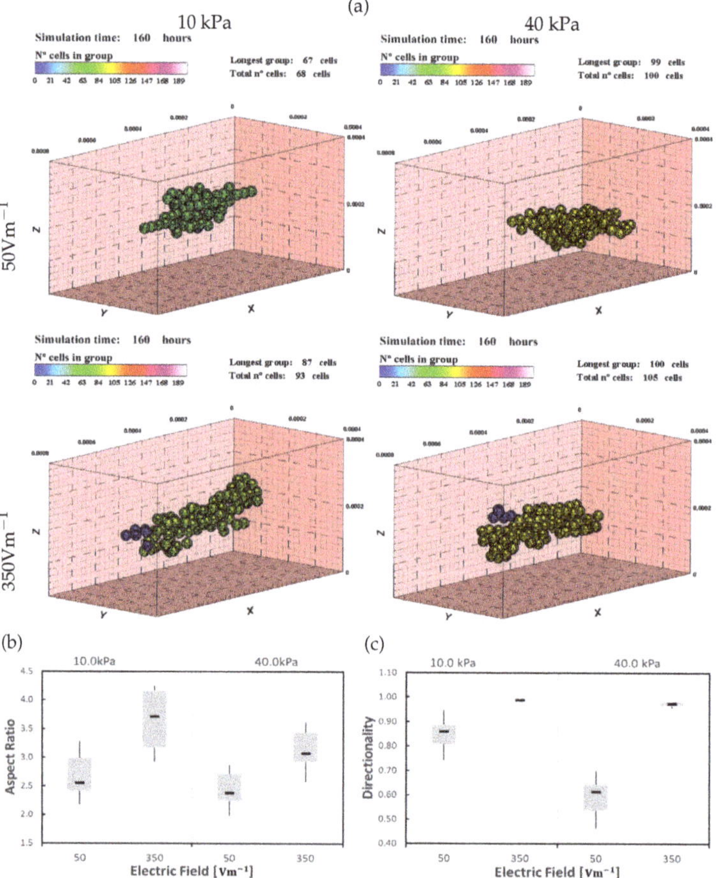

Figure 10. CMs in 10 and 40 kPa ECMs, under an alternating electric field of 50 and 350 Vm^{-1}, after 160 h of simulation (see also Supplementary Materials video 3). (**a**) Higher tendency to form a group in the longitudinal direction is observed. Cells remain close to the center of the ECM where the cells mature faster and the proliferation rate increases. Larger and better-oriented groups are formed by applying an alternating electric field, with better results for AR (**b**) and directionality (**c**).

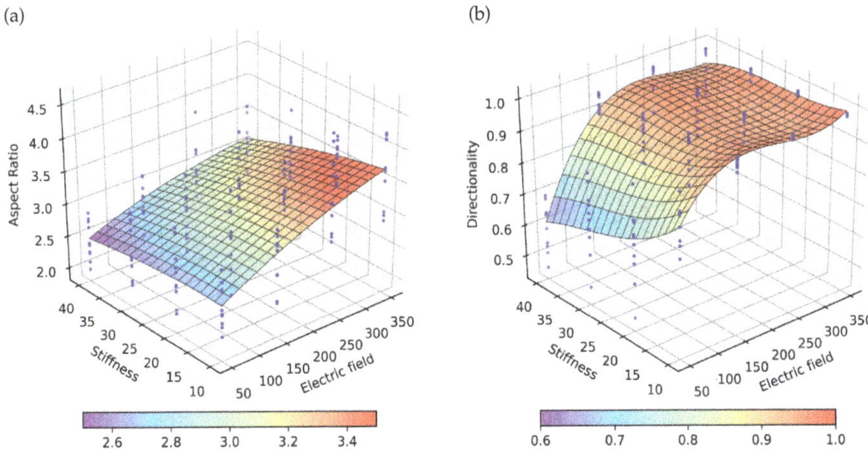

Figure 11. Representation of the obtained numerical results by applying an alternating electric field with different intensities in ECM with different stiffnesses. Blue dots represent 10 experiment repetitions, on which the surface with the best fit to the results is represented. (**a**) Aspect ratio for the different combination of stiffness and electric stimulus. (**b**) Directionality for the different combination of stiffness and electric stimulus.

4. Discussion

ES is essential for the development of engineered heart tissues, which preserving cells' mature phenotype and improving the contractile properties [11,15,17,18]. Likewise, its intensity, frequency, and duration of application can be key for the development of highly functional tissues [15]. The coupling effects of electrical and mechanical stimuli have been shown to be relevant in cardiac tissue development [11]. This strongly encourages us to study these effects in computational models where such stimuli can be studied, modeled, and balanced, saving time, costs, and pain in experimental studies. Furthermore, through the parametrization of cell behavior, it is possible to evaluate different cell and ECM conditions, including those associated with different pathologies [10,11,58].

In the proposed model, the cells initially align and migrate in the direction of the electric field which is consistent with the bibliography [11,18,21,44,85]. It has seen that different electric field strengths, as well as different stiffnesses of the ECM, generate differences in the direction and speed of cell migration as was observed in experimental models [18,85]. In general, as observed in the present model, high values of ES increase the degree of cell alignment and the groups elongated morphology. Likewise, the greater the stiffness, the faster the cell maturation, which increases cell proliferation until the cells join in groups where cell proliferation is inhibited. Their coupling effect is also extended to cell-cell interaction. If the electric stimulus is higher than the mechanical stimulus, cells can be dragged by the electric field which can delay or impede cell-cell contact and group formation. As it has been observed in the obtained results, the formation of correctly organized groups depends on the correct balance between the electrical and the mechanical stimulus.

As observed in the present model, cell response varies depending not only on the intensity but also on the type of the applied electric field. Comparing the results of the applied different modes of ES, significant differences are observed in cell response. For instance, in the cases of continuous and pulsative electric fields, cells are dragged in the direction of the electric field and tend to migrate towards the external face of the ECM (cathodic zone) [18,20,44]. On the other hand, in the case of the alternating electric field, cell migration direction is reversed as the electric field changes its direction, as can be observed in different experimental models [18,21,44].

Better results are observed when applying a continuous electric field (Figure 6). When applying ES discontinuously (Figure 8), the results are worse. This can be attributed to the decrease in the effective electric stimulus which is only active half of the time. Furthermore, by disabling the electric field, the cells only get stimulated by the mechanical cue. So, they lose some of the directionality induced by the electric field. On the other hand, when an alternating field is applied, cells maintain the directionality induced by the electric field. Due to the changing of the electric field polarity, the cell drag towards the electric field is reduced, keeping the cells in the center of the ECM. Cells alternate the direction of the migration which reduces the effective cell motility in the longitudinal direction. Thus, cells migrate towards the center of the ECM, due to the existence of strong mechanical stimulus and the presence of other cells, which stimulate the cells to form more elongated groups and improve the cell-cell interaction.

In the first and second experiments, when stiffness increases, slight differences are observed, in both AR and directionality. On the contrary, higher differences can be observed in the third experiment. A significant change in the tendency of cell directionality is observed, where a saturation point is detected when the applied electric field is greater than 150 Vm^{-1} (Figure 11b). These differences in the results obtained for the different modes of application of the electric field, which highlights the relevance of properly understanding the processes that trigger the electrical stimulation of the cells. In this way, the hypothesis of the use of computational models is reinforced, to support the experimental work, which allows advancing in the understanding of these processes.

For more rigid ECMs, larger groups can be observed. This can be attributed to, as mentioned before, the increase in stiffness, which leads to faster maturation of the cells. Besides, as the cells are considered initially at the stage of early CM, an increase in cell proliferation has been noted. As the number of cells increases, the AR ratio may be decreased due to the thickening effect of the groups. This justifies that, although visually the groups are larger in stiffer ECMs, the graphs show a slight improvement in the results. This effect of stiffness can also be seen when comparing the number of cells observed in the third experiment with the previous ones. As the cells are kept in the center of the ECM, which is the stiffest zone, the maturation of the cells is faster, and the total number of cells after 160 h of simulation is higher than in other cases.

5. Conclusions

We have developed a computational model to study cardiac cell behavior in 3D matrices, with different stiffnesses, under the effects of different external electric fields. The model has been applied to study cell migration, polarization, organization, and formation of groups through cell junctions. With this model, we studied the effect of the electric field, with different modes of application which include continuous, pulsative, and alternating electric fields. Furthermore, different combinations of ECM stiffness and electric field intensity were simulated to study the effect of the coupling of the mechanical and electric stimuli. The obtained results are qualitatively consistent with the bibliography [2,17,18,44,84,85]. Thus, cells tend to migrate, in the direction of the electric field, proportionally to the intensity of the electric stimulus [18,44]. Besides, cells polarize towards the electric field direction and tend to form aligned groups in the direction of the electric field which depends on its intensity [17,18,44,85].

The stimulation of the cells through the application of an external electric field improves the directionality of the cells in the longitudinal direction, which corresponds to the electric field direction. As the electric field increases, cells tend to form elongated groups with higher AR. These effects are more pronounced in less ECM's stiffness, revealing the effects of the coupling of electrical and mechanical stimuli. While the ES induce the cells to migrate in the longitudinal direction, the mechanical stimulation guides cells to the center of the ECM. As a result, an improvement of the AR and directionality is observed by decreasing stiffness and thereby mechanical stimulation. Similarly, by increasing stiffness, the effects of ES decrease. In case of lower electric field (50 Vm^{-1}), cells tend to remain at the center of the ECM, indicating that the mechanical stimulus dominates the electric stimulus. Thus, the best directionality and AR results are obtained by applying the maximum electric field combined with the minimum stiffness. On the other hand, cell maturation is faster in more rigid ECMs, which increases the maturation speed as well as increasing the proliferation rate of the early CM. The effect of stiffness on the proliferation can also be observed in the third experiment. In this case, cells keep at the center of the ECM, which corresponds to the stiffest zone of the ECM, increasing cell maturation and showing higher proliferation.

The mode of application of the electric field changed the response of the cells. Thus, a considerable increase in directionality and AR has been observed when applying an alternating electric field. In this case, a saturation point of the directionality is observed for ES above 150 Vm^{-1}, where this is not observed in the other modes of the electric field. Significant differences can be observed due to the coupling of electrical and mechanical stimuli, with substantial variations in the results. Cells are guided by stimuli with different effects and the cellular response depends on the incidence of each stimulus. Thus, when a high mechanical stimulus is observed, the effect produced by the electrical stimulus tends to decrease. In the same way, for high electrical stimuli, the effect of the mechanical stimulus loses relevance.

In conclusion, groups with higher AR can be obtained by applying higher ES. The increase in AR seems to increase linearly by increasing the intensity of the electric field. Different electric field applications (continuous, pulsating, or alternating) show a different influence in AR and directionality, with better results when the alternating electric field is applied. Likewise, an increase in stiffness is favorable to promote cell proliferation.

The presented model has been elaborated establishing a series of simplifications that must be considered with the aim of simplifying and providing stability to the calculations. Among them, cell morphology, which is considered spherical along with the simulation. Likewise, cellular interaction with certain growth factors, which can modify or inhibit certain cellular behaviors, has been simplified by considering neutral cell culture. Despite the model limitations, it is capable of evaluating the coupling of different stimuli, electrical and mechanical, which is relevant in the case of cardiac cells. Furthermore, the results indicate that the mode of application of the electric field can significantly change the cell behavior. This strengthens the idea of using computational models to study the appropriate conditions for cell culture, giving support to the in-vitro and in-vivo assays. Likewise, computational models can reproduce a large number of culture conditions with reduced time and economic cost, thus being able to establish preliminary studies that reduce dramatically the number of experiments. In this case, it has been seen that the coupling of the electrical and mechanical stimulation notably increases and accelerates the cell proliferation process. Consequently, it can be considered a highly recommended combination for the in-vitro and in-vivo experiments.

Supplementary Materials: Supplementary Materials (video 1, video 2, video 3) can be downloaded at http://www.mdpi.com/2227-7390/8/11/1875/s1.

Author Contributions: Conceptualization, P.U., M.H.D.; methodology, P.U., M.H.D.; software, P.U., M.H.D.; validation, P.U., M.H.D.; formal analysis, P.U., M.H.D.; investigation, P.U., M.H.D.; resources, M.H.D.; data curation, P.U., M.H.D.; writing—original draft preparation, P.U.; writing—review and editing, M.H.D.; visualization, P.U.; supervision, M.H.D.; project administration, M.H.D.; funding acquisition, M.H.D. All authors have read and agreed to the published version of the manuscript.

Funding: This research was funded by the Spanish Ministry of Science and Innovation (PID2019-106099RB-C44/AEI/10.13039/501100011033), the Government of Aragon (DGA-T24_20R) and the Biomedical Research Networking Center in Bioengineering, Biomaterials and Nanomedicine (CIBER-BBN).

Acknowledgments: The authors gratefully acknowledge the financial support from the Spanish Ministry of Science and Innovation (PID2019-106099RB-C44/AEI/10.13039/501100011033), the Government of Aragon (DGA-T24_20R) and the Biomedical Research Networking Center in Bioengineering, Biomaterials and Nanomedicine (CIBER-BBN). CIBER-BBN is financed by the Instituto de Salud Carlos III with assistance from the European Regional Development Fund.

Conflicts of Interest: The authors declare no conflict of interest.

References

1. Hart, F.X.; Palisano, J.R. The Application of Electric Fields in Biology and Medicine. In *Electric Field*; InTech: Vienna, Austria, 2018. [CrossRef]
2. Sauer, H.; Rahimi, G.; Hescheler, J.; Wartenberg, M. Effects of electrical fields on cardiomyocyte differentiation of embryonic stem cells. *J. Cell. Biochem.* **1999**, *75*, 710–723. [CrossRef]
3. Ghafar-Zadeh, E.; Waldeisen, J.R.; Lee, L.P. Engineered approaches to the stem cell microenvironment for cardiac tissue regeneration. *Lab Chip* **2011**, *11*, 3031. [CrossRef] [PubMed]
4. Mycielska, M.E.; Djamgoz, M.B.A. Cellular mechanisms of direct-current electric field effects: Galvanotaxis and metastatic disease. *J. Cell Sci.* **2004**, *117*, 1631–1639. [CrossRef] [PubMed]
5. Pullar, C.E. *The Physiology of Bioelectricity in Development, Tissue Regeneration and Cancer*; CRC Press: Boca Raton, FL, USA, 2016; p. 308. [CrossRef]
6. Robinson, K.R.; Messerli, M.A. Left/right, up/down: The role of endogenous electrical fields as directional signals in development, repair and invasion. *BioEssays* **2003**, *25*, 759–766. [CrossRef] [PubMed]
7. Levin, M. Molecular bioelectricity: How endogenous voltage potentials control cell behavior and instruct pattern regulation in vivo. *Mol. Biol. Cell* **2014**, *25*, 3835–3850. [CrossRef]
8. Meng, S.; Rouabhia, M.; Zhang, Z. Electrical stimulation modulates osteoblast proliferation and bone protein production through heparin-bioactivated conductive scaffolds. *Bioelectromagnetics* **2013**, *34*, 189–199. [CrossRef]
9. Walker, C.A.; Spinale, F.G. The structure and function of the cardiac myocyte: A review of fundamental concepts. *J. Thorac. Cardiovasc. Surg.* **1999**, *118*, 375–382. [CrossRef]
10. Ferrari, R. Healthy versus sick myocytes: Metabolism, structure and function. *Eur. Heart J. Suppl.* **2002**, *4*, G1–G12. [CrossRef]
11. Stoppel, W.L.; Kaplan, D.L.; Black, L.D. Electrical and mechanical stimulation of cardiac cells and tissue constructs. *Adv. Drug Deliv. Rev.* **2016**, *96*, 135–155. [CrossRef]
12. Llucià-Valldeperas, A.; Sanchez, B.; Soler-Botija, C.; Gálvez-Montón, C.; Prat-Vidal, C.; Roura, S.; Rosell-Ferrer, J.; Bragos, R.; Bayes-Genis, A. Electrical stimulation of cardiac adipose tissue-derived progenitor cells modulates cell phenotype and genetic machinery. *J. Tissue Eng. Regen. Med.* **2015**, *9*, E76–E83. [CrossRef]
13. Sun, Y.; Wang, H.; Liu, M.; Lin, F.; Hua, J. Resveratrol abrogates the effects of hypoxia on cell proliferation, invasion and EMT in osteosarcoma cellsthrough downregulation of the HIF-1α protein. *Mol. Med. Rep.* **2015**, *11*, 1975–1981. [CrossRef]
14. Tandon, N.; Cannizzaro, C.; Chao, P.P.H.G.H.G.; Maidhof, R.; Marsano, A.; Au, H.T.H.; Radisic, M.; Vunjak-Novakovic, G. Electrical stimulation systems for cardiac tissue engineering. *Nat. Protoc.* **2009**, *4*, 155–173. [CrossRef] [PubMed]
15. Hirt, M.N.; Boeddinghaus, J.; Mitchell, A.; Schaaf, S.; Börnchen, C.; Müller, C.; Schulz, H.; Hubner, N.; Stenzig, J.; Stoehr, A.; et al. Functional improvement and maturation of rat and human engineered heart tissue by chronic electrical stimulation. *J. Mol. Cell. Cardiol.* **2014**, *74*, 151–161. [CrossRef] [PubMed]
16. Heidi Au, H.T.; Cui, B.; Chu, Z.E.; Veres, T.; Radisic, M. Cell culture chips for simultaneous application of topographical and electrical cues enhance phenotype of cardiomyocytes. *Lab Chip* **2009**, *9*, 564–575. [CrossRef] [PubMed]
17. Radisic, M.; Park, H.; Shing, H.; Consi, T.; Schoen, F.J.; Langer, R.; Freed, L.E.; Vunjak-Novakovic, G. Functional assembly of engineered myocardium by electrical stimulation of cardiac myocytes cultured on scaffolds. *Proc. Natl. Acad. Sci. USA* **2004**, *101*, 18129–18134. [CrossRef] [PubMed]
18. Frederich, B.J.; Timofeyev, V.; Thai, P.N.; Haddad, M.J.; Poe, A.J.; Lau, V.C.; Moshref, M.; Knowlton, A.A.; Sirish, P.; Chiamvimonvat, N. Electrotaxis of cardiac progenitor cells, cardiac fibroblasts, and induced pluripotent stem cell-derived cardiac progenitor cells requires serum and is directed via PI3K pathways. *Heart Rhythm* **2017**, *14*, 1685–1692. [CrossRef] [PubMed]
19. Nishimura, K.Y.; Isseroff, R.R.; Nucciteili, R. Human keratinocytes migrate to the negative pole in direct current electric fields comparable to those measured in mammalian wounds. *J. Cell Sci.* **1996**, *109*, 199–207.
20. Banks, T.A.; Luckman, P.S.B.; Frith, J.E.; Cooper-White, J.J. Effects of electric fields on human mesenchymal stem cell behaviour and morphology using a novel multichannel device. *Integr. Biol.* **2015**, *7*, 693–712. [CrossRef]
21. Zhao, Z.; Watt, C.; Karystinou, A.; Roelofs, A.; McCaig, C.; Gibson, I.; De Bari, C. Directed migration of human bone marrow mesenchymal stem cells in a physiological direct current electric field. *Eur. Cells Mater.* **2011**, *22*, 344–358. [CrossRef]

22. Sassoli, C.; Pini, A.; Mazzanti, B.; Quercioli, F.; Nistri, S.; Saccardi, R.; Orlandini, S.Z.; Bani, D.; Formigli, L. Mesenchymal stromal cells affect cardiomyocyte growth through juxtacrine Notch-1/Jagged-1 signaling and paracrine mechanisms: Clues for cardiac regeneration. *J. Mol. Cell. Cardiol.* **2011**, *51*, 399–408. [CrossRef] [PubMed]

23. Ahuja, P.; Sdek, P.; MacLellan, W.R. Cardiac Myocyte Cell Cycle Control in Development, Disease, and Regeneration. *Physiol. Rev.* **2007**, *87*, 521–544. [CrossRef] [PubMed]

24. Costa, K.D.; Lee, E.J.; Holmes, J.W. Creating Alignment and Anisotropy in Engineered Heart Tissue: Role of Boundary Conditions in a Model Three-Dimensional Culture System. *Tissue Eng.* **2003**, *9*, 567–577. [CrossRef] [PubMed]

25. Gumbiner, B.M. Cell Adhesion: The Molecular Basis of Tissue Architecture and Morphogenesis. *Cell* **1996**, *84*, 345–357. [CrossRef]

26. Ogawa, N.; Oku, H.; Hashimoto, K.; Ishikawa, M. A physical model for galvanotaxis of Paramecium cell. *J. Theor. Biol.* **2006**, *242*, 314–328. [CrossRef] [PubMed]

27. Mousavi, S.J.; Doblaré, M.; Doweidar, M.H. Computational modelling of multi-cell migration in a multi-signalling substrate. *Phys. Biol.* **2014**, *11*, 026002. [CrossRef]

28. Mousavi, S.J.; Doweidar, M.H. Encapsulated piezoelectric nanoparticle–hydrogel smart material to remotely regulate cell differentiation and proliferation: A finite element model. *Comput. Mech.* **2019**, *63*, 471–489. [CrossRef]

29. Urdeitx, P.; Farzaneh, S.; Mousavi, S.J.; Doweidar, M.H. Role of oxygen concentration in the osteoblasts behavior: A finite element model. *J. Mech. Med. Biol.* **2020**, *20*, 1950064. [CrossRef]

30. Mousavi, S.J.; Doweidar, M.H. Three-Dimensional Numerical Model of Cell Morphology during Migration in Multi-Signaling Substrates. *PLoS ONE* **2015**, *10*, e0122094. [CrossRef]

31. Bernheim-Groswasser, A.; Prost, J.; Sykes, C. Mechanism of Actin-Based Motility: A Dynamic State Diagram. *Biophys. J.* **2005**, *89*, 1411–1419. [CrossRef]

32. Mogilner, A. Mathematics of cell motility: Have we got its number? *J. Math. Biol.* **2009**, *58*, 105–134. [CrossRef]

33. Selmeczi, D.; Mosler, S.; Hagedorn, P.H.; Larsen, N.B.; Flyvbjerg, H. Cell Motility as Persistent Random Motion: Theories from Experiments. *Biophys. J.* **2005**, *89*, 912–931. [CrossRef] [PubMed]

34. Reig, G.; Pulgar, E.; Concha, M.L. Cell migration: From tissue culture to embryos. *Development* **2014**, *141*, 1999–2013. [CrossRef] [PubMed]

35. Ridley, A.J.; Schwartz, M.A.; Burridge, K.; Firtel, R.A.; Ginsberg, M.H.; Borisy, G.; Parsons, J.T.; Horwitz, A.R. Cell Migration: Integrating Signals from Front to Back. *Science* **2003**, *302*, 1704–1709. [CrossRef] [PubMed]

36. Mogilner, A.; Rubinstein, B. The Physics of Filopodial Protrusion. *Biophys. J.* **2005**, *89*, 782–795. [CrossRef] [PubMed]

37. Neilson, M.P.; Veltman, D.M.; van Haastert, P.J.; Webb, S.D.; Mackenzie, J.A.; Insall, R.H. Chemotaxis: A feedback-based computational model robustly predicts multiple aspects of real cell behaviour. *PLoS Biol.* **2011**, *9*, e1000618. [CrossRef]

38. Bosgraaf, L.; van Haastert, P.J. Navigation of chemotactic cells by parallel signaling to pseudopod persistence and orientation. *PLoS ONE* **2009**, *4*, e6842. [CrossRef]

39. Lange, J.R.; Fabry, B. Cell and tissue mechanics in cell migration. *Exp. Cell Res.* **2013**, *319*, 2418–2423. [CrossRef]

40. Mousavi, S.J.; Doweidar, M.H.; Doblaré, M. 3D computational modelling of cell migration: A mechano-chemo-thermo-electrotaxis approach. *J. Theor. Biol.* **2013**, *329*, 64–73. [CrossRef]

41. Mousavi, S.J.; Doweidar, M.H.; Doblaré, M. Computational modelling and analysis of mechanical conditions on cell locomotion and cell–cell interaction. *Comput. Methods Biomech. Biomed. Eng.* **2014**, *17*, 678–693. [CrossRef]

42. Mousavi, S.J.; Doweidar, M.H. Role of mechanical cues in cell differentiation and proliferation: A 3D numerical model. *PLoS ONE* **2015**, *10*, e0124529. [CrossRef]

43. Urdeitx, P.; Doweidar, M.H. Mechanical stimulation of cell microenvironment for cardiac muscle tissue regeneration: A 3D in-silico model. *Comput. Mech.* **2020**, *66*, 1003–1023. [CrossRef]

44. Chen, C.; Bai, X.; Ding, Y.; Lee, I.S. Electrical stimulation as a novel tool for regulating cell behavior in tissue engineering. *Biomater. Res.* **2019**, *23*, 25. [CrossRef]

45. Huang, Y.X.; Zheng, X.J.; Kang, L.L.; Chen, X.Y.; Liu, W.J.; Huang, B.T.; Wu, Z.J. Quantum dots as a sensor for quantitative visualization of surface charges on single living cells with nano-scale resolution. *Biosens. Bioelectron.* **2011**, *26*, 2114–2118. [CrossRef] [PubMed]

46. Aragona, M.; Panciera, T.; Manfrin, A.; Giulitti, S.; Michielin, F.; Elvassore, N.; Dupont, S.; Piccolo, S. A mechanical checkpoint controls multicellular growth through YAP/TAZ regulation by actin-processing factors. *Cell* **2013**, *154*, 1047–1059. [CrossRef] [PubMed]

47. Low, B.C.; Pan, C.Q.; Shivashankar, G.V.; Bershadsky, A.; Sudol, M.; Sheetz, M. YAP/TAZ as mechanosensors and mechanotransducers in regulating organ size and tumor growth. *FEBS Lett.* **2014**, *588*, 2663–2670. [CrossRef]

48. Zhao, M. Electrical fields in wound healing—An overriding signal that directs cell migration. *Semin. Cell Dev. Biol.* **2009**, *20*, 674–682. [CrossRef] [PubMed]

49. Abercrombie, M. Contact inhibition and malignancy. *Nature* **1979**, *281*, 259–262. [CrossRef]

50. Chen, Q.Z.; Harding, S.E.; Ali, N.N.; Lyon, A.R.; Boccaccini, A.R. Biomaterials in cardiac tissue engineering: Ten years of research survey. *Mater. Sci. Eng. R Rep.* **2008**, *59*, 1–37. [CrossRef]

51. Zimmermann, W.H.; Schneiderbanger, K.; Schubert, P.; Didié, M.; Münzel, F.; Heubach, J.F.; Kostin, S.; Neuhuber, W.L.; Eschenhagen, T. Tissue engineering of a differentiated cardiac muscle construct. *Circ. Res.* **2002**, *90*, 223–230. [CrossRef] [PubMed]

52. Kresh, J.Y.; Chopra, A. Intercellular and extracellular mechanotransduction in cardiac myocytes. *Pflüg. Arch. Eur. J. Physiol.* **2011**, *462*, 75–87. [CrossRef]

53. Lee, E.J.; Holmes, J.W.; Costa, K.D. Remodeling of engineered tissue anisotropy in response to altered loading conditions. *Ann. Biomed. Eng.* **2008**, *36*, 1322–1334. [CrossRef]

54. Planat-Bénard, V.; Menard, C.; André, M.; Puceat, M.; Perez, A.; Garcia-Verdugo, J.M.; Pénicaud, L.; Casteilla, L. Spontaneous Cardiomyocyte Differentiation from Adipose Tissue Stroma Cells. *Circ. Res.* **2004**, *94*, 223–229. [CrossRef]

55. Tahara, N.; Brush, M.; Kawakami, Y. Cell migration during heart regeneration in zebrafish. *Dev. Dyn.* **2016**, *245*, 774–787. [CrossRef]

56. Nunes, S.S.; Miklas, J.W.; Liu, J.; Aschar-Sobbi, R.; Xiao, Y.; Zhang, B.; Jiang, J.; Massé, S.; Gagliardi, M.; Hsieh, A.; et al. Biowire: A platform for maturation of human pluripotent stem cell–derived cardiomyocytes. *Nat. Methods* **2013**, *10*, 781–787. [CrossRef] [PubMed]

57. Schmelter, M.; Ateghang, B.; Helmig, S.; Wartenberg, M.; Sauer, H. Embryonic stem cells utilize reactive oxygen species as transducers of mechanical strain-induced cardiovascular differentiation. *FASEB J.* **2006**, *20*, 1182–1184. [CrossRef] [PubMed]

58. Camelliti, P.; McCulloch, A.D.; Kohl, P. Microstructured Cocultures of Cardiac Myocytes and Fibroblasts: A Two-Dimensional In Vitro Model of Cardiac Tissue. *Microsc. Microanal.* **2005**, *11*, 249–259. [CrossRef]

59. Holtzman, N.G.; Schoenebeck, J.J.; Tsai, H.J.; Yelon, D. Endocardium is necessary for cardiomyocyte movement during heart tube assembly. *Development* **2007**, *134*, 2379–2386. [CrossRef] [PubMed]

60. Mousavi, S.J.; Doweidar, M.H. A novel mechanotactic 3D modeling of cell morphology. *Phys. Biol.* **2014**, *11*, 046005. [CrossRef] [PubMed]

61. Ye, J.; Boyle, A.J.; Shih, H.; Sievers, R.E.; Wang, Z.E.; Gormley, M.; Yeghiazarians, Y. CD45-positive cells are not an essential component in cardiosphere formation. *Cell Tissue Res.* **2013**, *351*, 201–205. [CrossRef] [PubMed]

62. Wu, Q.Q.; Chen, Q. Mechanoregulation of chondrocyte proliferation, maturation, and hypertrophy: Ion-channel dependent transduction of matrix deformation signals. *Exp. Cell Res.* **2000**, *256*, 383–391. [CrossRef]

63. Delaine-Smith, R.M.; Reilly, G.C. Mesenchymal stem cell responses to mechanical stimuli. *Muscles Ligaments Tendons J.* **2012**, *2*, 169–180. [PubMed]

64. Ulrich, T.A.; De Juan Pardo, E.M.; Kumar, S. The mechanical rigidity of the extracellular matrix regulates the structure, motility, and proliferation of glioma cells. *Cancer Res.* **2009**, *69*, 4167–4174. [CrossRef]

65. Mousavi, S.J.; Doweidar, M.H. Numerical modeling of cell differentiation and proliferation in force-induced substrates via encapsulated magnetic nanoparticles. *Comput. Methods Programs Biomed.* **2016**, *130*, 106–117. [CrossRef] [PubMed]

66. Foglia, M.J.; Poss, K.D. Building and re-building the heart by cardiomyocyte proliferation. *Development* **2016**, *143*, 729–740. [CrossRef]

67. Yutzey, K.E. Cardiomyocyte Proliferation. *Circ. Res.* **2017**, *120*, 627–629. [CrossRef]

68. Asumda, F.Z.; Asumda, F.Z. Towards the development of a reliable protocol for mesenchymal stem cell cardiomyogenesis. *Stem Cell Discov.* **2013**, *3*, 13–21. [CrossRef]

69. Zhang, R.; Han, P.; Yang, H.; Ouyang, K.; Lee, D.; Lin, Y.F.; Ocorr, K.; Kang, G.; Chen, J.; Stainier, D.Y.R.; et al. In vivo cardiac reprogramming contributes to zebrafish heart regeneration. *Nature* **2013**, *498*, 497–501. [CrossRef] [PubMed]

70. McClatchey, A.I.; Yap, A.S. Contact inhibition (of proliferation) redux. *Curr. Opin. Cell Biol.* **2012**, *24*, 685–694. [CrossRef]

71. Dassault Systemes. Abaqus 6.14. 2014. Available online: https://www.3ds.com/products-services/simulia/services-support/support/documentation/ (accessed on 15 September 2020).

72. Schäfer, A.; Radmacher, M. Influence of myosin II activity on stiffness of fibroblast cells. *Acta Biomater.* **2005**, *1*, 273–280. [CrossRef]

73. Darling, E.M.; Topel, M.; Zauscher, S.; Vail, T.P.; Guilak, F. Viscoelastic properties of human mesenchymally-derived stem cells and primary osteoblasts, chondrocytes, and adipocytes. *J. Biomech.* **2008**, *41*, 454–464. [CrossRef] [PubMed]

74. Discher, D.E. Tissue Cells Feel and Respon to the Stiffness of Their Substrate. *Science* **2005**, *310*, 1139–1143. [CrossRef] [PubMed]

75. Ramtani, S. Mechanical modelling of cell/ECM and cell/cell interactions during the contraction of a fibroblast-populated collagen microsphere: Theory and model simulation. *J. Biomech.* **2004**, *37*, 1709–1718. [CrossRef] [PubMed]

76. Rodriguez, M.L.; Graham, B.T.; Pabon, L.M.; Han, S.J.; Murry, C.E.; Sniadecki, N.J.; Pabon, L.M.; Murry, C.E.; Graham, B.T.; Han, S.J.; et al. Measuring the Contractile Forces of Human Induced Pluripotent Stem Cell-Derived Cardiomyocytes With Arrays of Microposts. *J. Biomech. Eng.* **2014**, *136*, 051005. [CrossRef] [PubMed]

77. Gardel, M.L.; Sabass, B.; Ji, L.; Danuser, G.; Schwarz, U.S.; Waterman, C.M. Traction stress in focal adhesions correlates biphasically with actin retrograde flow speed. *J. Cell Biol.* **2008**, *183*, 999–1005. [CrossRef]

78. Bhana, B.; Iyer, R.K.; Chen, W.L.K.; Zhao, R.; Sider, K.L.; Likhitpanichkul, M.; Simmons, C.A.; Radisic, M. Influence of substrate stiffness on the phenotype of heart cells. *Biotechnol. Bioeng.* **2010**, *105*, 1148–1160. [CrossRef]

79. Mathur, A.B.; Collinsworth, A.M.; Reichert, W.M.; Kraus, W.E.; Truskey, G.A. Endothelial, cardiac muscle and skeletal muscle exhibit different viscous and elastic properties as determined by atomic force microscopy. *J. Biomech.* **2001**, *34*, 1545–1553. [CrossRef]

80. Zaman, M.H.; Kamm, R.D.; Matsudaira, P.; Lauffenburger, D.A. Computational Model for Cell Migration in Three-Dimensional Matrices. *Biophys. J.* **2005**, *89*, 1389–1397. [CrossRef]

81. Fearnley, C.J.; Roderick, H.L.; Bootman, M.D. Calcium Signaling in Cardiac Myocytes. *Cold Spring Harb. Perspect. Biol.* **2011**, *3*, a004242. [CrossRef] [PubMed]

82. Kang, K.T.; Park, J.H.; Kim, H.J.; Lee, H.Y.H.M.; Lee, K.I.; Jung, H.H.; Lee, H.Y.H.M.; Jang, J.W. Study of Tissue Differentiation of Mesenchymal Stem Cells by Mechanical Stimuli and an Algorithm for Bone Fracture Healing. *Tissue Eng. Regen. Med.* **2011**, *8*, 359–370.

83. Li, Z.; Guo, X.; Palmer, A.F.; Das, H.; Guan, J. High-efficiency matrix modulus-induced cardiac differentiation of human mesenchymal stem cells inside a thermosensitive hydrogel. *Acta Biomater.* **2012**, *8*, 3586–3595. [CrossRef]

84. Pietronave, S.; Zamperone, A.; Oltolina, F.; Colangelo, D.; Follenzi, A.; Novelli, E.; Diena, M.; Pavesi, A.; Consolo, F.; Fiore, G.B.; et al. Monophasic and biphasic electrical stimulation induces a precardiac differentiation in progenitor cells isolated from human heart. *Stem Cells Dev.* **2014**, *23*, 888–898. [CrossRef] [PubMed]

85. Baumgartner, S.; Halbach, M.; Krausgrill, B.; Maass, M.; Srinivasan, S.P.; Sahito, R.G.A.; Peinkofer, G.; Nguemo, F.; Müller-Ehmsen, J.; Hescheler, J. Electrophysiological and morphological maturation of murine fetal cardiomyocytes during electrical stimulation in vitro. *J. Cardiovasc. Pharmacol. Ther.* **2015**, *20*, 104–112. [CrossRef]

Publisher's Note: MDPI stays neutral with regard to jurisdictional claims in published maps and institutional affiliations.

Article

Analysis of the Parametric Correlation in Mathematical Modeling of In Vitro Glioblastoma Evolution Using Copulas

Jacobo Ayensa-Jiménez [1,2], Marina Pérez-Aliacar [1,2], Teodora Randelovic [2,3], José Antonio Sanz-Herrera [4], Mohamed H. Doweidar [1,2,5] and Manuel Doblaré [1,2,3,5,*]

1   Mechanical Engineering Department, School of Engineering and Architecture (EINA), University of Zaragoza, 50018 Zaragoza, Spain; jacoboaj@unizar.es (J.A.-J.); 722195@unizar.es (M.P.-A.); mohamed@unizar.es (M.H.D.)
2   Aragon Institute of Engineering Research (I3A), University of Zaragoza, 50018 Zaragoza, Spain; 753388@unizar.es
3   Aragón Institute of Health Research (IIS Aragón), 50009 Zaragoza, Spain
4   Department of Mechanics of Continuous Media and Theory of Structures, School of Engineering, University of Seville, 41092 Sevilla, Spain; jsanz@us.es
5   Centro de Investigación Biomédica en Red en Bioingeniería, Biomateriales y Nanomedicina (CIBER-BBN), 50018 Zaragoza, Spain
*   Correspondence: mdoblare@unizar.es

Abstract: Modeling and simulation are essential tools for better understanding complex biological processes, such as cancer evolution. However, the resulting mathematical models are often highly non-linear and include many parameters, which, in many cases, are difficult to estimate and present strong correlations. Therefore, a proper parametric analysis is mandatory. Following a previous work in which we modeled the in vitro evolution of Glioblastoma Multiforme (GBM) under hypoxic conditions, we analyze and solve here the problem found of parametric correlation. With this aim, we develop a methodology based on *copulas* to approximate the multidimensional probability density function of the correlated parameters. Once the model is defined, we analyze the experimental setting to optimize the utility of each configuration in terms of gathered information. We prove that experimental configurations with oxygen gradient and high cell concentration have the highest utility when we want to separate correlated effects in our experimental design. We demonstrate that copulas are an adequate tool to analyze highly-correlated multiparametric mathematical models such as those appearing in Biology, with the added value of providing key information for the optimal design of experiments, reducing time and cost in in vivo and in vitro experimental campaigns, like those required in microfluidic models of GBM evolution.

Keywords: copulas; design of experiments; glioblastoma multiforme; mathematical modelling

MSC: 62H20; 62K05; 62P10

Citation: Ayensa-Jiménez, J.; Pérez-Aliacar, M.; Randelovic, T.; Sanz-Herrera, J.A.; Doweidar, M.H.; Doblaré, M. Analysis of the Parametric Correlation in Mathematical Modeling of In Vitro Glioblastoma Evolution Using Copulas. *Mathematics* **2021**, *9*, 27. https://dx.doi.org/10.3390/math9010027

Received: 24 November 2020
Accepted: 21 December 2020
Published: 24 December 2020

1. Introduction

Biological processes usually involve several cell populations interacting in a complex, dynamic, and multiple interactive micro-environment [1]. Understanding these interactions between cells and microenvironment is crucial in many physiological and pathological processes [2]. However, progressing in this understanding with only in vivo experiments is difficult. Despite them being more realistic, isolating effects or achieving particular conditions is complex in such experiments due to technical and/or ethical reasons.

In vitro experiments permit better control of the variables, while reducing costly and ethically-questioned animal assays. Nonetheless, the predictive power of currently available in vitro models is still poor due to the strong difficulties that we face in reproducing the structure and distribution of the different cell populations as well as the particular environmental conditions in which cells live, adapt and react (e.g., three-dimensionality) [3]. Microfluidics is a new in vitro technique that allows more precise reproductions of the

61

microenvironment and cell distribution [4,5], including three-dimensionality, thus making in vitro tests much closer to the actual in vivo conditions. This permits, for example, a more reliable and efficient drug testing [6,7].

Finally, mathematical models allow to separate and quantify the effects of each mechanism or parameter, as well as to predict the outcome in "what if" situations, which are sometimes impossible to achieve in in vivo or in vitro experiments [8,9]. Nevertheless, these models are mostly non-linear, involve highly-coupled multiphysic interactions, and include many parameters. In many occasions, those parameters are difficult to measure and have strong hidden correlations. Moreover, it is usual to have a lack of data both for quantification and validation of the parameters and results [10]. Therefore, they are fitted only for the results available, which usually correspond to very specific conditions. This may lead to trivial conclusions that could have been directly derived from the model assumptions, making the results only useful for those particular experiments, with the obtained conclusions impossible to generalize.

In a previous paper [11], we addressed this parametric analysis in a particular problem—the mathematical modeling of the in vitro (using microfluidic devices) evolution of glioblastoma multiforme (GBM), the most aggressive and lethal among primary glioma tumors [12]. In Ref. [11], we presented a general framework in which the main cell processes involved (proliferation, chemotaxis, random migration, apoptosis, and necrosis), in response to changes in the oxygen concentration, were mathematically formulated. We then analyzed three different experimental configurations, reproducing the main GBM migratory structures (pseudopalisade and necrotic core formation). An extensive analysis of all model parameters was performed, both from literature and by fitting the associated in silico results with those derived from the experiments. As main results of that work, we identified a unique set of parameters able to accurately reproduce the quantitative results for the three case-studies. However, we also found two model limitations: (i) the sensitivity analysis showed that the model is strongly affected by small variations in the oxygen cell consumption and diffusion and (ii) a strong correlation was found between the parameters associated with those two mechanisms.

The objective of the present work is to present the possibilities in this context offered by a methodology that is able to separate the correlated effects found in that study, and to get a more accurate and reliable representation of the experimental results in the parametric space. With that purpose, we approximate the multidimensional probability density function of the parameters by means of appropriate copulas. Copulas allow considering separately the marginal distributions and the dependence between variables in multivariate statistical problems, including those with high correlation. This permits using general models for the marginal distributions, while the variable dependence model can be different [13]. Copulas are today used in a wide range of areas in Economic sciences and Engineering. The most recent models have been successfully applied in portfolio management and optimization [14], actuarial analysis [15], quantitative finance and risk theory [16,17]. A particularly hot topic is the study of climate-agent time series [18,19], hydrology [20,21] and weather and climate research [22,23]. Some efforts have been made in transportation research [24] and traffic policy [25]. Recently, copulas have been successfully applied in reliability analysis in civil [26], mechanical and structural [27], offshore [28] and software [29] engineering. In Biology, copulas have been used in the field of genetics [30] to model gene dependencies.

Up to the authors' knowledge, there is no work using copulas for the parametric analysis of evolution processes in Biology, where, as commented, many of the parameters involved are unknown and uncontrolled, and high correlations between parameters are common. We prove here that copulas are an adequate tool to improve the analysis of highly-correlated multiparametric mathematical models such as those appearing in Biology, with the added value of providing key information for the optimal design of new experiments with the highest information possible, thus reducing time and cost not only in in vitro experiments but also in scarce and costly in vivo cases.

2. Rationale of the Approach

2.1. Deterministic and Stochastic Models

Let us suppose that our problem may be represented by the following mathematical relationship:

$$u = F(\lambda, \theta), \tag{1}$$

with

- u (an m-dimensional vector) the output variable, that is, the outcome of the experiments, that we measure.
- λ the variables which we can control when performing the experiments (such as environmental variables, geometric parameters, or boundary conditions).
- θ the model parameters, that we cannot control and whose values must be determined ($\theta \in \Omega$, with Ω the parametric space of dimension n).
- F the mathematical model, that relates the experimental configuration λ with the output variables u in terms of the set of parameters θ.

In relation to the accuracy and precision of the model, it is possible to define three levels of analysis: (1) the model is perfect and the experimental measures are noise-free; (2) the model is perfect and the experimental measures are noisy; and (3) the model is not perfect and the measurements are noisy. Only the third case is, in general, realistic in complex problems as the one here analyzed.

In addition, it is difficult to define universal values for the parameters in biological problems, since they are highly-dependent on the particular experimental context.

As a consequence of all the previous observations, it is more appropriate to consider a stochastic approach, and reformulate Equation (1) as:

$$U = F(\lambda, \Theta), \tag{2}$$

where U and Θ are now random vectors of dimensions m and n respectively.

The proposed approach is therefore suitable when the following conditions are satisfied:

- Many coupled phenomena are present, being difficult to design experiments able to isolate each of them (complexity).
- The measurement space is large and it is possible to perform a sufficiently big number of experiments N (data availability).

From a mathematical point of view, these two statements may be reformulated as:

- The model F includes many parameters ($n \gg 1$) and/or is non-separable.
  The separability of a model is evaluated by the possibility of approximating F as:

$$F(\lambda, \theta) \simeq F^M(\lambda, \theta) = \sum_{i=1}^{M} \prod_{j=1}^{n} F_{i,j}(\lambda, \theta_j). \tag{3}$$

The lower M, the easier to define a set of different experimental configurations $\mathcal{S} = \{\lambda^j\}_{j=1,\dots,k}$ to isolate each of the parameters θ_j by solving separately each equation $u^j = F^M(\lambda, \theta)$. Although this separability definition is not very rigorous, it is enlightening enough for our purposes.

- The dimension of the measurement space is high ($m \gg 1$) and/or the sample size is large enough ($N \gg 1$). Without loss of generality, we consider that m is, actually, the reduced dimensionality of the space or in other words that all variables of the ambient space are independent.

2.2. Case Study: In Vitro GBM Evolution

There have been many attempts to develop mathematical models to describe how tumors grow and respond to therapies [10,31]. In particular, in previous works, we demonstrated the possibility of developing GBM pseudopalisades [32] and necrotic cores [33] in vitro. Figure 1 illustrates one of such experiments in which a high density cell culture is exposed to oxygen flow by two lateral channels but, due to self-induced hypoxia, the formation of a necrotic core in the central part of the chamber is observed.

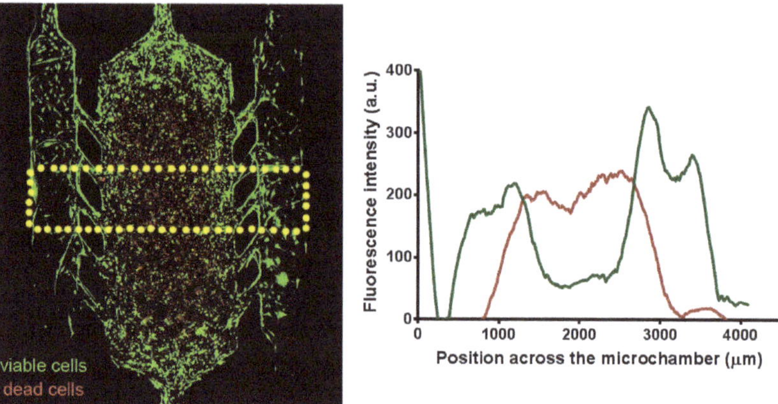

Figure 1. Formation of a necrotic core in the microfluidic device.

One of the main problems in these models is the lack of reliable values for the many parameters involved that forces many times to rely on values fitted from different situations, leading sometimes to unreliable conclusions. We recently proposed a mathematical model for GBM in vitro evolution [11], together with an extensive parameter discussion. This model enables the simulation of different stages of GBM evolution under several experimental conditions, showing robustness, while keeping a small uncertainty range in the results. It is established in terms of three advection-reaction-diffusion equations and the associated parameters that are expressed as:

$$\frac{\partial C_a}{\partial t} = \frac{\partial}{\partial x}\left(D_a \frac{\partial C_a}{\partial x} - K_a \chi_a^{O_2}(O_2) \chi_a^{C_a}(C_a) C_a \frac{\partial O_2}{\partial x} \right)$$
$$+ \frac{1}{\tau_a}\beta_a(O_2)G_a(C_a, C_d)C_a - \frac{1}{\tau_{ad}}S_{ad}(O_2)C_a \tag{4}$$

$$\frac{\partial C_d}{\partial t} = \frac{1}{\tau_{ad}}S_{ad}(O_2)C_a \tag{5}$$

$$\frac{\partial O_2}{\partial t} = D_{O_2}\frac{\partial^2 O_2}{\partial x^2} - \alpha_a H_a(O_2)C_a. \tag{6}$$

Equation (4) quantifies the evolution of the cell normoxic phenotype concentration, C_a, with three terms: random diffusion, growth-death source, and chemotaxis. Equation (5) models the evolution of the necrotic phenotype concentration, C_d, which contains only the dead cells derived from the normoxic phenotype. Finally, Equation (6) defines the O_2 concentration evolution in the hydrogel in which cells are embedded, considering both oxygen diffusion and cell consumption. Functions β_a, G_a, $\chi_a^{O_2}$, $\chi_a^{C_a}$, S_{ad} and H_a are nonlinear corrections accounting for cell metabolic behavior:

$\chi_a^{O_2}$ defines a chemotaxis correction accounting for the oxygen concentration. It has been shown that GBM cells present what is called the *go or grow* behavior [34]: cells spend resources in proliferating when they are enough oxygenated and activate migration mechanisms under hypoxia conditions, that is, when the oxygen concentration is under a certain hypoxia threshold O_2^H. Therefore, we state:

$$\chi_a^{O_2}(O_2) = \begin{cases} 1 - O_2/O_2^H & \text{if} \quad 0 \le O_2 \le O_2^H \\ 0 & \text{if} \quad O_2 > O_2^H. \end{cases} \tag{7}$$

$\chi_a^{C_a}$ defines a chemotaxis correction accounting for the cell concentration. We assume that cellular motility is only possible when the cell concentration is below the saturation capacity of the hydrogel C^M:

$$\chi_a^{C_a}(C_a) = \begin{cases} 1 - C_a/C^M & \text{if} \quad 0 \le C_a \le C^M \\ 0 & \text{if} \quad C_a > C^M. \end{cases} \tag{8}$$

β_a accounts for the dependence of the proliferation activity on the oxygen concentration, in agreement with the *go or grow* paradigm [34]. Cell proliferation decreases when the oxygen concentration is under the hypoxia threshold, O_2^H, and is totally inactivated under total lack of oxygen:

$$\beta_a(O_2) = \begin{cases} O_2/O_2^H & \text{if} \quad 0 \le O_2 \le O_2^H \\ 1 & \text{if} \quad O_2 > O_2^H. \end{cases} \tag{9}$$

G_a is a logistic growth correction accounting for space and nutrients availability [35]. Cell proliferation decreases when the cell concentration approaches the hydrogel saturation capacity, C^M:

$$G_a(C_a, C_d) = \left(1 - \frac{C_a + C_d}{C^M}\right). \tag{10}$$

S_{ad} is a death activation function accounting for the oxygen concentration. Cell death is a complex phenomenon that can be due to two different cell mechanisms, necrosis, and apoptosis [36,37]. Cell necrosis is highly dependent on the oxygen concentration, while cell apoptosis is not. Therefore, we have chosen a soft transition function for S_{ad} depending on two parameters—a location parameter, O_2^A, identifying the anoxia oxygen concentration and a spread parameter, ΔO_2^A, associated with the death stochastic nature:

$$S_{ad}(O_2) = \frac{1}{2}\left(1 - \tanh\left(\frac{O_2 - O_2^A}{\Delta O_2^A}\right)\right). \tag{11}$$

Finally, H_a is the Michaelis-Menten correction factor in oxygen consumption, related to the oxidative phosphorylation kinetics [38]. The consumption rate is constant for high oxygen concentrations, but decreases to zero with a homographic shape. The value of the oxygen concentration for which the consumption rate is halved is the so-called Michaelis-Menten constant, O_2^M. The function H_a is then stated as:

$$H_a(O_2) = \frac{O_2}{O_2^M + O_2}. \tag{12}$$

Equations (4)–(6) are complemented with the boundary and initial conditions. For the experiments carried out in our microfluidic devices, we assume total impermeability (Neumann boundary conditions) for the cell populations and a fixed value for the oxygen concentration at both sides of the channel (Dirichlet boundary conditions). Therefore, if L is the chamber length, we may write:

$$\begin{aligned} \frac{\partial C_a}{\partial x} &= 0, \quad x = 0, L \\ \frac{\partial C_d}{\partial x} &= 0, \quad x = 0, L \\ O_2 &= O_2^l, \quad x = 0 \\ O_2 &= O_2^r, \quad x = L, \end{aligned} \tag{13}$$

with O_2^l and O_2^r the oxygen levels at the left and right channels of the chip.

The initial oxygen concentration is assumed to be homogeneous over the whole chamber and equal to the maximum of both lateral oxygen concentrations, that is $O_2(x, t = 0) = O_2^0 = \max(O_2^l, O_2^r)$.

The resulting experimental parametric space consists, therefore, of three parameters, corresponding to the concentration at the boundaries of the chip, (O_2^l, O_2^r), and the initial cell concentration, (C_0), assumed constant throughout the chip. That is:

$$\lambda = [O_2^l, O_2^r, C_0]. \tag{14}$$

O_2^H, O_2^A, ΔO_2^A and O_2^M have a clear meaning in terms of cell metabolism and are assumed to be known and constant for all cell cultures used in our experiments, at least from an illustrative point of view. Besides, although C^M is very dependent on the experimental conditions (hydrogel mechanical properties, nutrients, ...), we shall assume it is constant, for the sake of simplicity. The values for these parameters were taken from a previous work [11].

Previous research in computational biology has mainly focused on the value of the parameters or, in the best case, in their (individual) uncertainty. However, in many cases, the fitting process is very complex and the parameters are highly correlated due to, at least, two facts:

- **Samples variability**: Different physical phenomena may have an inherent correlation supported by physical considerations, being this correlation independent of the experiments performed or the model used. For example, when working with GBM cellular models, cell motility is induced by the random motion inherent to any cell and several taxis effects driven by external physical or chemical stimuli. Mathematical parameters related to these phenomena (e.g., diffusion and chemotaxis coefficients) appearing in the model equations will present, therefore, a strong correlation in the different experimental samples.
- **Model complexity**: The non-separability of the model and/or the experiments does not allow to isolate the particular mechanisms. For example, when working with GBM cellular models, without further measurements of cell oxygen consumption or oxygen flux, it is impossible to establish if a lack of oxygen in a certain region is due to high cell consumption or due to low oxygen diffusion. The mathematical parameters related to these phenomena (e.g., oxygen diffusion and cell oxygen consumption coefficients) should present a strong correlation, although this correlation does not have a physical meaning, being inherent to the model or to the experimental set-up.

Thanks to the flexibility, portability, automation, integration, and miniaturization of the microfluidic experiments, a huge amount of data may be generated. Accordingly, this type of experiments is a perfect domain of application for the framework presented herein.

3. Methods

3.1. Data Generation and Numerical Solution

As the methodology is based on the availability of sufficient data, the data set used for illustrating the methodology was generated synthetically using numerical simulation. For this purpose, the assumed values for the parameters were extracted from Ref. [11] and a data set of "experimental" measurements was generated by simulation, using randomly generated boundary and initial conditions.

The summary of the model parameters is shown in Table 1, together with the value used for data generation.

Table 1. Model parameters and values used for data generation.

Parameter	Symbol	Value Used for Data Generation [11]
Normoxic cell diffusion coefficient	D_a	$5 \times 10^{-10}\,\mathrm{cm}^2/\mathrm{s}$
Normoxic cell chemotaxis coefficient	K_a	$7.5 \times 10^{-9}\,\mathrm{cm}^2/\mathrm{mmHg{\cdot}s}$
Oxygen diffusion coefficient	D_{O_2}	$1 \times 10^{-5}\,\mathrm{cm}^2/\mathrm{s}$
Oxygen consumption coefficient	α_a	$1 \times 10^{-9}\,\mathrm{mmHg{\cdot}cm}^3/\mathrm{cell{\cdot}s}$
Growth characteristic time	τ_a	$200\,\mathrm{h}$
Death characteristic time	τ_{ad}	$48\,\mathrm{h}$
Hypoxia activation threshold	O_2^H	$7\,\mathrm{mmHg}$
Growth saturation capacity	C^M	$5 \times 10^7\,\mathrm{cell/mL}$
Anoxia activation location parameter	O_2^A	$1.8\,\mathrm{mmHg}$
Anoxia activation spread parameter	ΔO_2^A	$0.1\,\mathrm{mmHg}$
Michaelis-Menten constant	O_2^M	$2.5\,\mathrm{mmHg}$

With respect to the simulated virtual experiment, we set a chip length of $L = 0.1\,\mathrm{cm}$, a mesh size of $\Delta x = 0.0025\,\mathrm{cm}$ and a time step of $\Delta t = 1000\,\mathrm{s}$. $N = 400$ different experiments, $\{\lambda^i\}_{i=1,\dots,400}$, were simulated varying the boundary conditions: the left and right channel oxygen concentrations were set randomly between 0 and $7\,\mathrm{mmHg}$ using two independent uniform distributions while the initial oxygen concentration was set to the maximum of both values, as mentioned. The initial cell profile is supposed to be uniform and randomly sampled from a reciprocal distribution (to take into account both the exponential and saturated growth regimes) between 4×10^6 and $5 \times 10^7\,\mathrm{cell/mL}$. The numerical solutions are obtained for $t_m = 8\,\mathrm{d}$ and the output variable associated to the experiment i, $u^i = u_s(x, t_m; \lambda^i)$, is the numerical solution of the model equations (the mathematical approach and numerical procedures and algorithms are detailed in Ref. [11]), with boundary and initial conditions defined by λ, at time t_m and at points given by the defined mesh x. Here, $x_j = j\Delta x$, $j = 1, \dots, 41$. The computed data were all perturbed with a uniform noise $\epsilon_j = 0.2 \times u_j \times V$ with V a random uniform distribution $V \sim \mathcal{U}[-1, 1]$. Consequently, $u_j^i = u_s(x_j, t_m, \lambda^i) + \epsilon_j^i$, $j = 1, \dots, 41$ and $i = 1, \dots, 400$.

Within the framework presented in Section 2.1, $u = F(\lambda, \theta)$ are the numerical solutions obtained, with λ the control parameters, θ the unknown parameters and F the mathematical model presented.

3.2. Copula-Based Parametric Model Analysis

3.2.1. Concept of Copulas

In Probability and Statistics, a copula is an n-multivariate probability distribution function U whose marginals, U_i, are uniform distributions on $[0, 1]$ [39]. They were introduced by Sklar in 1959 [40]. As the marginal distributions are known, a copula describing the structural dependence between variables is enough to perfectly define the model.

Mathematical definition.

As mentioned, a copula is a function $C : I^n \to I$, where $I = [0; 1]$ such that:

- For $u_1, \dots, u_n \in I$, and if $u_i = 0$ for some $1 \le i \le n$:

$$C(u_1, \dots, u_n) = 0. \tag{15}$$

- For $u_j \in I, 1 \le j \le n$:

$$C(1, \dots, 1, u_j, 1, \dots, 1) = u_j. \tag{16}$$

- C is n-non decreasing, that is, for each $B = \prod_{i=1}^{n}[x_i; y_i] \subset I^n$, the C-volume of B is non-negative:

$$\int_B dC(u) = \sum_{z \in \times_{i=1}^{n}\{x_i; y_i\}} (-1)^{\#\{k:z_k=x_k\}} C(z) \geq 0. \tag{17}$$

We can distinguish between parametric and non-parametric copulas. In this work, we use a hybrid approach, as we fit the marginal distributions by means of kernel estimators [41] of the probability density functions and use a parametric copula. With this approach, the required data-set grows as $\mathcal{O}(n)$ where n is the space dimension.

3.2.2. Fitting and Model Validation

Let us suppose we have a data-set of values for different experiments, λ^i, characterized in terms of a resultant mean value μ^i and a covariance matrix Σ^i, $i = 1, \ldots, N$, obtained from different measurements associated to the configuration i. As the assumed model F is known, it is possible, for each piece of data u_i, to obtain the set of parameters θ^i which best fits it.

In order to avoid pathological numerical convergence, we only take into account those sets of parameters θ^i which lie inside the bibliography ranges considered in Ref. [11], amplified by 50% to avoid considering the parameters bounds as deterministic values, that, as shown in Table 2, are very large ranges. Therefore, the resulting intervals are $[(1 - \kappa)x_{\inf}, (1 + \kappa)x_{\sup}]$, being $x_{\inf}$ and $x_{\sup}$ the lower and upper bounds detailed in Ref. [11] and $\kappa = 0.5$, as summarized in Table 2. As a result of this process, we obtain a dataset with $n = 6$ (number of parameters), $N = 111$ (dataset size) and $m = 41$ (measurement space dimension), so we are under the scope of the presented framework: $N \times m \gg n > 1$.

Table 2. Parameter ranges considered in the analysis.

Parameter	Lower Bound	Upper Bound	Units
D_a	3.3×10^{-12}	7.5×10^{-5}	cm^2/s
K_a	1×10^{-10}	1.1×10^{-3}	$cm^2/mmHg{\cdot}s$
D_{O_2}	5×10^{-6}	3×10^{-5}	cm^2/s
α_a	5×10^{-10}	1.1×10^{-6}	$mmHg{\cdot}cm^3/cell{\cdot}s$
τ_a	8	3000	h
τ_{ad}	24	917	h

Once θ^i, $i = 1, \ldots, N$ are obtained, the next step is the adjustment of the marginal distributions. The values θ^i_j, $j = 1, \ldots, n$, are used for fitting the marginal random variable Θ_j whose cumulative distribution is assumed to be G_j. Here, we can follow either a parametric (that is, $G_j(x) = G_j(x; \alpha_j)$) or a non-parametric approach (which is the one followed in this work). The values θ^i_j are therefore transformed into uniformly distributed ones via the standard transformation $y^i_j = G_j(\theta^i_j)$. As y^i are considered uniformly distributed with a joint dependence, it is possible to fit this structural dependence using parametric copulas.

To summarize, the steps of the training process are:

1. Problem minimization to obtain θ^i. We have to minimize the residual function R^i:

$$R^i(\theta) = \left(F(\lambda^i, \theta) - \mu^i\right)^T (\Sigma^i)^{-1} \left(F(\lambda^i, \theta) - \mu^i\right), \tag{18}$$

where the Mahalanobis distance has been used to take into account the sample variability. Assuming that $\Sigma^i = \sigma^{i^2} I$, Equation (18) can be rewritten as:

$$R^i(\theta) = \frac{1}{\sigma^{i^2}} \left\| F(\lambda^i, \theta) - \mu^i \right\|^2. \tag{19}$$

2. Kernel density estimation of the marginal distributions from the data θ_j^i.
3. Transformation into uniformly distributed values y_j^i.
4. Copula fitting of the y data to capture the joint dependence.

The presented sequence of steps allows moving from a dataset $\mathcal{S} = \{\boldsymbol{\theta}^i\}_{i=1,\ldots,N}$ to a probabilistic model for the random vector $\boldsymbol{\Theta}$ (the marginal kernel densities and the copula parameters encoding the structural dependence), as it is the aim of statistical procedures.

To avoid overfitting, we follow a typical train-test approach: we divide the datasets $\lambda^i - \boldsymbol{u}^i$ (where $\boldsymbol{u}^i$ includes $\boldsymbol{\mu}^i$ and $\boldsymbol{\Sigma}^i$) in two separate subsets, one used for training and the other used for testing.

If we consider now the test data-set, the procedure is:

1. Problem minimization to obtain $\boldsymbol{\theta}^i$.
2. Testing the statistical fitting:
   - Marginal fitting: q-q plots, histograms, empirical cumulative distribution functions (ecdf), boxplots, parametric or non-parametric statistical tests [42].
   - Joint 2 vs. 2 correlations: correlations, scatterplots, parametric statistical tests for correlations [42].
   - Whole joint structural dependence: multivariate parametric and non-parametric statistical tests [43].

3.2.3. Model Analysis and Parameter Estimation

Once the distribution of the random vector $\boldsymbol{\Theta}$ is learned, the model is known from a probabilistic point of view. The first straightforward application is parameter estimation It is important to emphasize that with "parameter estimation" we refer to the parameters of the mathematical model, not to the parameters of the distributions used in the statistical characterization (actually, the statistical characterization may be non-parametric), that may be estimated via common statistical inference techniques. A point estimate of the model parameters is given by:

$$\hat{\theta} = \mathbb{P}[\boldsymbol{\Theta}], \tag{20}$$

where $\mathbb{P}$ is a central tendency operator, for example, the expectation operator $\mathbb{E}$, minimizing the L^2 squared norm dispersion (its minimum is the variance), or the geometric median operator $\mathbb{M}$, minimizing the L^2 norm dispersion (its minimum is the mean absolute deviation).

However, it is more interesting to perform a confidence region estimation. As suggested in Ref. [44], in this work, we use the so-called Highest Density Regions (HDR) because of their easy interpretation, straightforward generalization to multi-dimensional spaces and direct computation. Recall that, under some distributional assumptions (e.g., normality assumption), HDR computation is reduced to other standard confidence region computation techniques (e.g., χ^2 quantile tolerance ellipsoids). HDR computation enables reliable parameter estimation since, given a significant level threshold α, it is possible to define an HDR region in which the parameters are located with a $p = 1 - \alpha$ probability. This may be performed for single parameters, or, in general, k-tuples of parameters.

This methodology is also applicable to conditional distributions. Let us suppose that we know the value of a certain subset of parameters θ^* and let us define $\theta = (\theta', \theta^*)$. Knowing the distribution $\boldsymbol{\Theta}$, that is obtained after the fitting-validation procedure, it is possible to define the conditioned distribution of $\boldsymbol{\Theta}$ given $\boldsymbol{\Theta}^* = \theta^*$ by its density f' defined in terms of the density f of θ:

$$f'(\theta'|\theta^*) = \frac{f(\theta', \theta^*)}{\int f(\eta, \theta^*)\, d\eta}, \tag{21}$$

so all HDR computations are now applied to the distribution of $\boldsymbol{\Theta}$ given $\boldsymbol{\Theta}^* = \theta^*$ by replacing f by f'.

3.2.4. Design of Experiments

Design of experiments techniques aim to maximize the information obtained from each performed experiment, in order to reduce the number of them required [45]. In particular, in this work, we use the techniques within the Bayesian Experimental Design (BED), based on the Bayesian interpretation of probability.

BED aims to maximize the expected utility of the experiment outcome [46]. The utility function expresses how useful is the information provided by an experiment. Of course, the optimal experiment design depends on the chosen utility criterion. In this work, the definition of the utility function is based on the Shannon entropy or Information entropy [47].

Under these assumptions, the utility of an experiment λ is defined as the prior-posterior gain in Shannon information. That is, the additional information that the experimental configuration λ provides about our model parameters. The utility $U(\lambda)$ then writes:

$$U(\lambda) = \int \int f(\boldsymbol{\theta}, \boldsymbol{u}|\lambda) \log f(\boldsymbol{u}|\boldsymbol{\theta}, \lambda) \, d\boldsymbol{\theta} d\boldsymbol{u} - \int f(\boldsymbol{u}|\lambda) \log f(\boldsymbol{u}|\lambda) \, d\boldsymbol{u}, \qquad (22)$$

where $\boldsymbol{u}$ is the experimental observation and $\boldsymbol{\theta}$ is a vector of parameters to be determined. $f(\boldsymbol{u}|\boldsymbol{\theta}, \lambda)$ is the probability density of obtaining an experimental outcome $\boldsymbol{u}$ given the experimental configuration λ and the model parameters $\boldsymbol{\theta}$ and $f(\boldsymbol{\theta}, \boldsymbol{u}|\lambda)$ is obtained as follows, being $f(\boldsymbol{\theta})$ the prior PDF over the parameters $\boldsymbol{\theta}$:

$$f(\boldsymbol{\theta}, \boldsymbol{u}|\lambda) = f(\boldsymbol{\theta}) f(\boldsymbol{u}|\boldsymbol{\theta}, \lambda). \qquad (23)$$

If we assume that $\boldsymbol{u}$ has a multivariate normal distribution (what is indeed not necessary but has been here considered for illustration purposes) with covariance matrix $\Sigma = \sigma^2 I$, and knowing that the entropy of a multivariate normal distribution of dimension n is only dependent on the standard deviation σ [48], we have the following expression for the utility:

$$U(\lambda) = -\frac{n}{2} \log \left(2\pi e \sigma^2 \right) - \int f(\boldsymbol{u}|\lambda) \log f(\boldsymbol{u}|\lambda) \, d\boldsymbol{u}. \qquad (24)$$

We assume that we measure the alive cell concentration at 5 given points: $u_k = C_a(x = x_k)$, $k = 1, \ldots, 5$, where $x_1 = 0.015\,\mathrm{cm}$, $x_2 = 0.035\,\mathrm{cm}$, $x_3 = 0.050\,\mathrm{cm}$, $x_4 = 0.065\,\mathrm{cm}$, $x_5 = 0.085\,\mathrm{cm}$. We work under the homoscedasticity and independence assumption so that each concentration measurement is assumed to be normally distributed with $\mu_i = u_i$ and $\sigma_i = \sigma$, $i = 1, \ldots, 5$. The uncertainty associated with the measurement of the cell concentration is assumed to be $\sigma = 1 \times 10^6\,\mathrm{cell/mL}$.

As we work under the assumptions detailed above, Equation (24), representing the utility of an experimental configuration λ, may be computed via numerical integration. A convergence analysis was performed, justifying the use of a given value of N_θ (number of sampling points for the model parameter) and N_u (number of sampling points for the experimental outcome) for each computation in the numerical integration process.

The simulations were performed for ten different oxygen levels at each side of the chip , $O_2^l = O_2(x = 0)$ and $O_2^r = O_2(x = L)$ (from 0 to 9 mmHg) and four different initial cell concentrations ($1 \times 10^6\,\mathrm{cell/mL}$, $5 \times 10^6\,\mathrm{cell/mL}$, $1 \times 10^7\,\mathrm{cell/mL}$ and $5 \times 10^7\,\mathrm{cell/mL}$).

In order to avoid numerical problems, in all simulations the uniform distributions of the parameters were sampled from $\epsilon = 0.01$ to $1 - \epsilon = 0.99$.

4. Results

4.1. Copula Fitting

4.1.1. Marginal Distributions

First of all, we obtain the fitting of the univariate marginal distributions. Figure 2 shows the kernel estimation of the marginal distribution of the different parameters. We have chosen a Gaussian kernel for all the estimations with variable bandwidths ($w_1 = 7.46 \times 10^{-11}\,\mathrm{cm^2/s}$, $w_2 = 9.52 \times 10^{-10}\,\mathrm{cm^2/(mmHg \cdot s)}$, $w_3 = 1.66 \times 10^{-6}\,\mathrm{cm^2/s}$,

$w_4 = 2.17 \times 10^{-10}\,\text{mmHg·cm}^3/(\text{cell·s})$, $w_5 = 9.57 \times 10^4\,\text{s}$ and $w_6 = 2.74 \times 10^4\,\text{s}$). The values are generally concentrated around the one used for the data generation, although the distributions present a variable uncertainty, related to the model complexity and its influence on the minimization procedure. For example, it is interesting to observe that all distributions present a multimodal feature, surely related to the existence of several local minima in the minimization procedure.

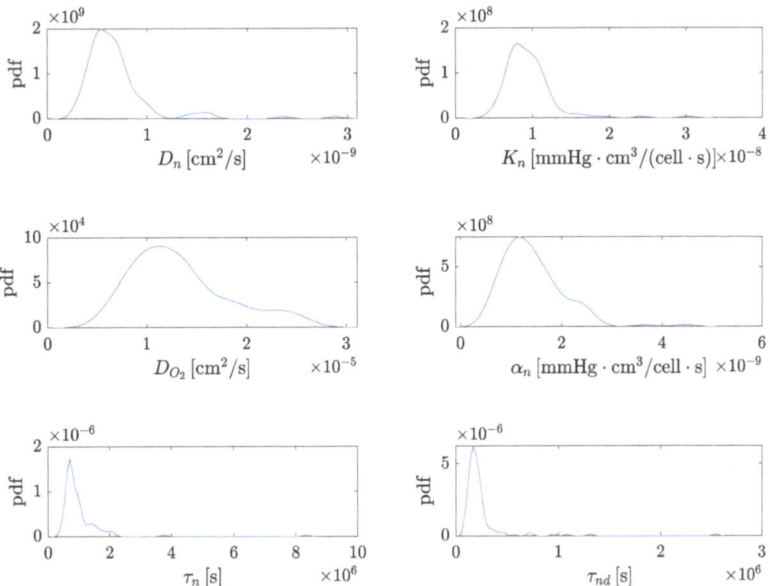

Figure 2. Kernel density estimation of the marginal distributions.

4.1.2. Parametric Copula Structure

Then, the data are transformed into uniformly distributed values using the cumulative distribution function (CDF) associated to this kernel estimation and a t-Student copula fitted by means of maximum likelihood (ML) estimation. The use of a t-Student copula is justified as it allows a different structural dependence for each of the variable pairs considered [16] and, besides, it outperforms Gaussian copula when estimating the co-occurrence of extreme events [49]. We obtain a copula with $\nu = 1.8$ degrees of freedom and a Pearson correlation matrix of:

$$P = \begin{bmatrix} 1.00 & 0.93 & 0.71 & 0.77 & 0.70 & 0.40 \\ 0.93 & 1.00 & 0.74 & 0.74 & 0.77 & 0.38 \\ 0.71 & 0.51 & 1.00 & 0.91 & 0.61 & 0.20 \\ 0.77 & 0.74 & 0.91 & 1.00 & 0.54 & 0.26 \\ 0.70 & 0.77 & 0.61 & 0.54 & 1.00 & 0.24 \\ 0.40 & 0.38 & 0.20 & 0.26 & 0.24 & 1.00 \end{bmatrix} \tag{25}$$

Note that the value obtained for ν is far from the Gaussian limit ($\nu \to \infty$), justifying the use of the t-Student model.

4.1.3. Complete Probabilistic Model and Bayesian *a Posteriori* Corrections

In order to briefly analyze the aspect of the whole model, we represent in Figure 3a the bivariate joint distribution of (D_{O_2}, α_a). Knowing the whole joint distribution function allows us to make *a posteriori* corrections using Bayesian theory and conditional probability as explained in Section 3.2.3. If we are interested in the joint distribution of two parameters (e.g., D_{O_2} and α_a), assuming that we know the rest ($D_a, K_a, \tau_a, \tau_{ad}$), the uncertainty of the parameter estimation obviously decreases, as can be seen in Figure 3b. In order to compare the impact of setting *a posteriori* the rest of the parameters, contour plots of both distributions, absolute and conditional (normalized between 0 and 1 to compare them more easily) are depicted in Figure 3c.

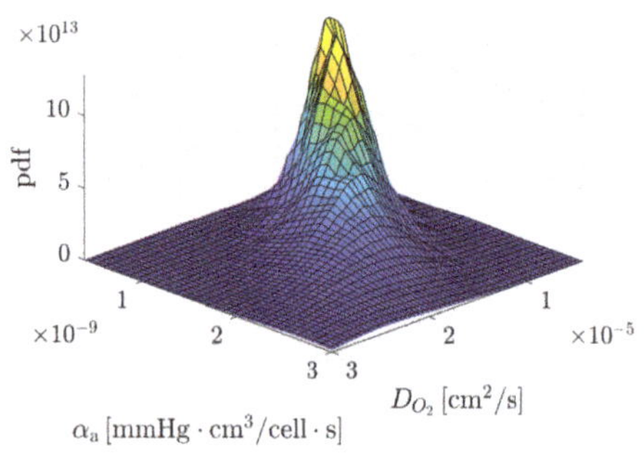

(**a**) Bivariate joint distribution of (D_{O_2}, α_a).

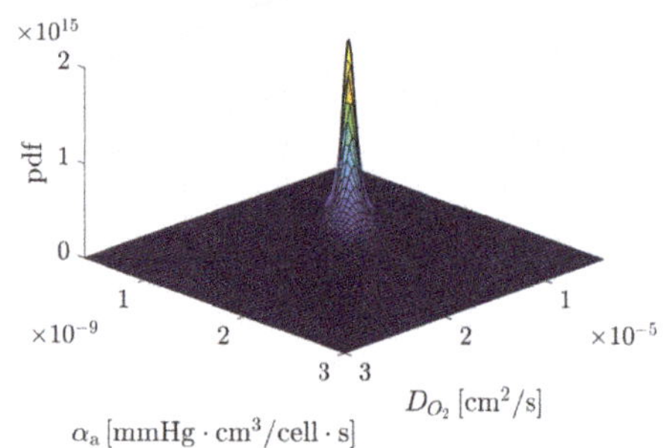

(**b**) Bivariate joint distribution of (D_{O_2}, α_a) assuming we know the rest of parameters.

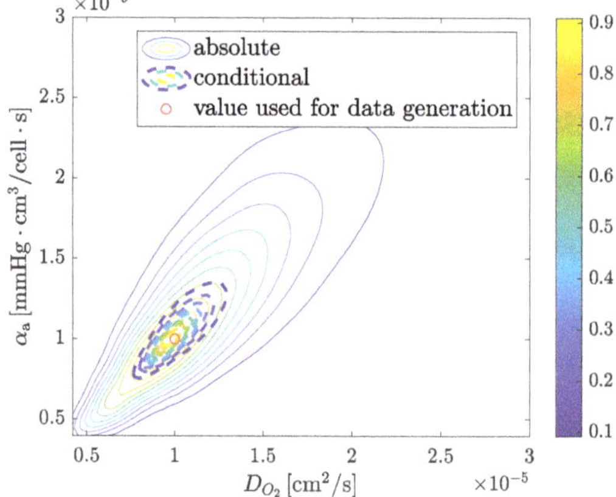

(**c**) Comparison between the distribution shape of (**a**) and (**b**).

Figure 3. Bivariate joint distribution functions of D_{O_2} and α_a.

4.2. Validation of the Results Using Test Data

Over-fitting is one of the main problems in any statistical or numerical parametric fitting. In our methodology, this is avoided by using a sub-set of the data as test data for validating the models.

4.2.1. Marginal Distributions

Marginal distributions are validated as pointed out in Section 3.2. To do so, new "experimental" data are compared to the data generated from the multivariate model. It is important to note that the original data are not used, but, on the contrary, a new data-set is strictly generated from the parametric copula and marginal densities, using the same procedure described for the generation of the original data. The histogram of data, the ecdf of the test data (with 95% confident interval) compared to the model data, the boxplot of both test and model data and the Q-Q plot of the test data, when compared to the model, are shown in Figure 4 for D_a as an illustrative example. The validation of the whole set of variables has been performed and good agreement was found between the model and test data except, if at all, for the extreme values, at the tail values of the distributions. In Figure 5, the ecdf of the test data for each model parameter is shown.

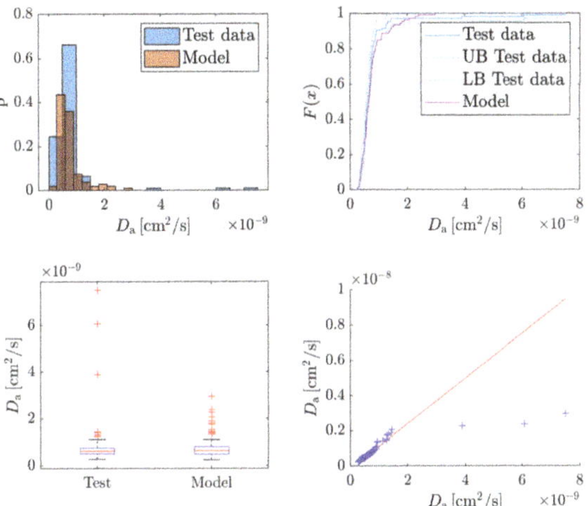

Figure 4. Validation of the marginal distributions for the parameter D_a.

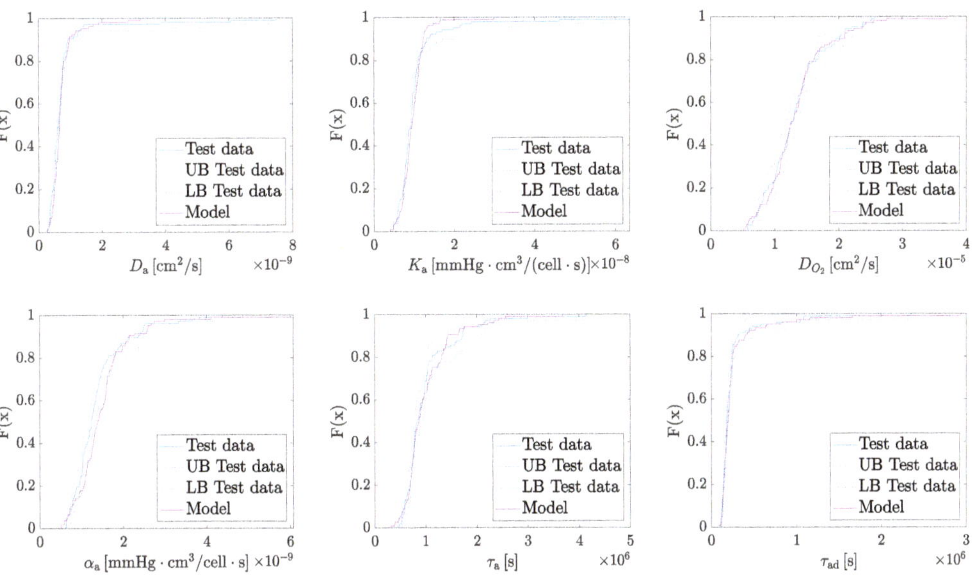

Figure 5. Empirical cumulative distribution functions (ecdf) of the test data for each parameter.

4.2.2. Joint Dependencies

Testing the structural dependence between parameters is not trivial. In Section 3.2, a multivariate statistical test was referenced. However, here we evaluate merely the differences in the correlation coefficients between the model-based and the test data. In Figure 6b, we represent the Kendall τ correlation index between the variables for the model and test data. We observe again a good agreement between the model values of the correlation coefficients (Figure 6a) and those obtained from the sample of the test data

(Figure 6b), even though the test sample is finite, which can cause differences between the model and the statistical values.

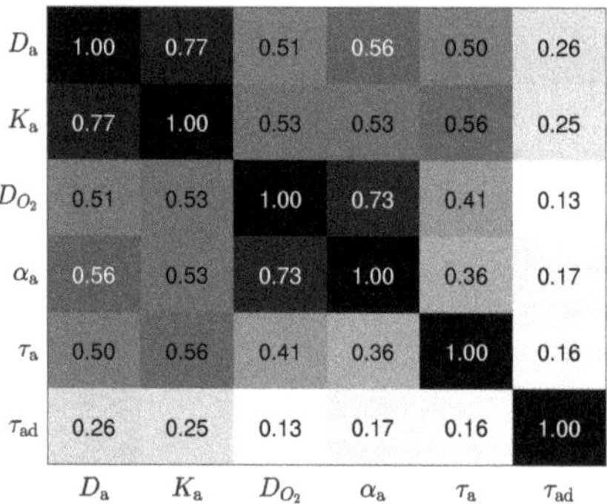

(a) Kendall τ for the training data.

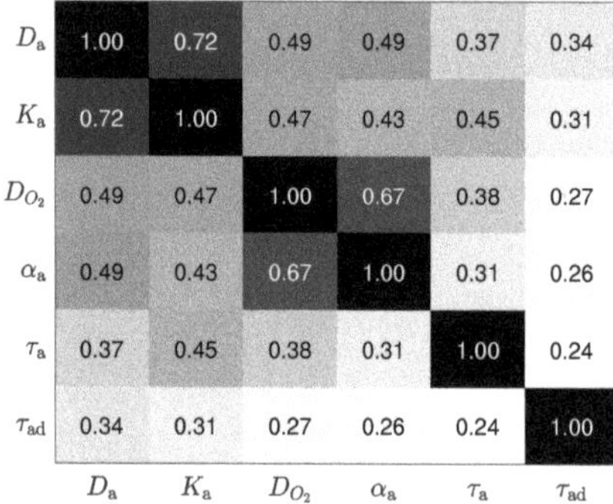

(b) Kendall τ for the test data.

Figure 6. Kendall τ correlation coefficient for each pair of variables for the training and test data.

4.3. Parameter Estimation

In Figure 7, we show p-confident HDR regions for $p = 0.90$, $p = 0.95$ and $p = 0.99$ for the pair of variables $D_{O_2} - \alpha_\mathrm{a}$. We present the results for the absolute distribution and the conditional distribution when the rest of parameters are known. The results are compared with the classical ellipsoid estimation, which is based on the normality assumption. The differences, both in the shape and the size of the regions, are clear and are explained by the complex dependence structure between variables.

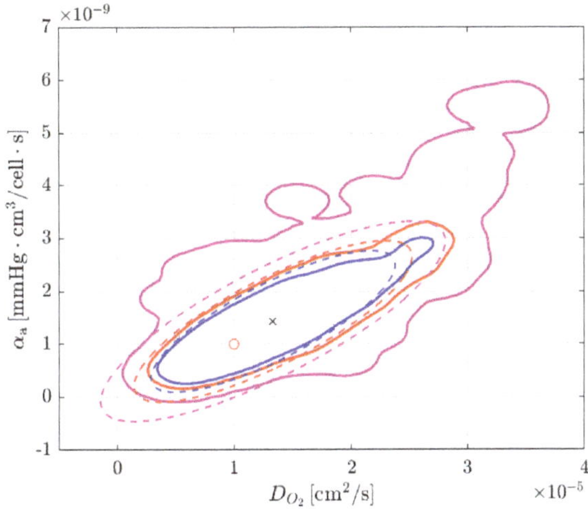

(**a**) Absolute distribution.

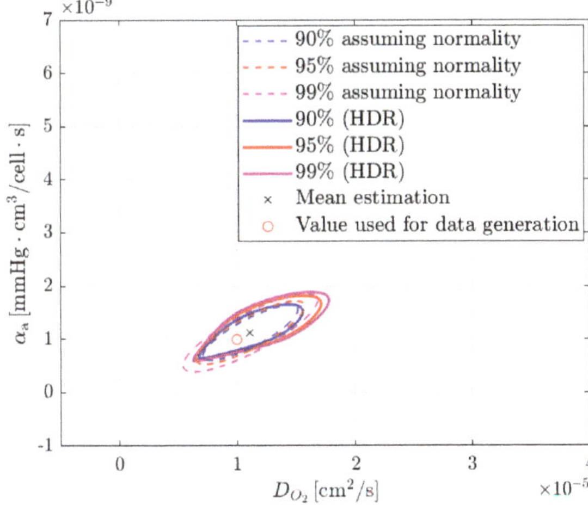

(**b**) Conditional distribution.

Figure 7. $D_{O_2} - \alpha_a$ point (mean) and region (HDR) estimations.

4.4. Estimation of the Output Variables

Once the multivariate distribution of the random vector Θ is characterized, we know the distribution of the random vectors $U = F(\lambda, \Theta)$. In Figure 8, we show the distribution of the vector U for three experiments, which illustrates completely different behaviors corresponding to the main histopathological features of GBM. For the first one, the oxygen flow is set to 2 mmHg in the left channel and 0 in the right channel and the initial concentration of cells is $C_0 = 4 \times 10^6$ cell/mL (pseudopalisade experiment in Ref. [11]). For the second one, the oxygen flow is set to 7 mmHg in both channels and the initial concentration of cells is $C_0 = 40 \times 10^6$ cell/mL (necrotic core experiment in Ref. [11]). Finally, for the third one, the oxygen flow is set to 7 mmHg in both channels and the initial concentration of cells is $C_0 = 4 \times 10^6$ cell/mL (double pseudopalisade experiment in oxygenated conditions in Ref. [11]).

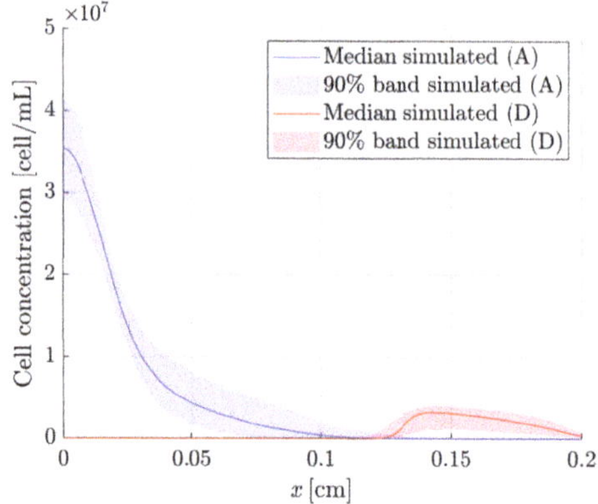

(**a**) Pseudopalisade experiment.

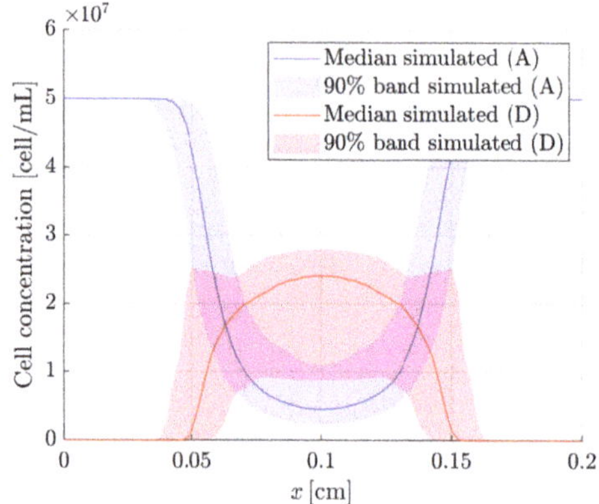

(**b**) Necrotic core experiment.

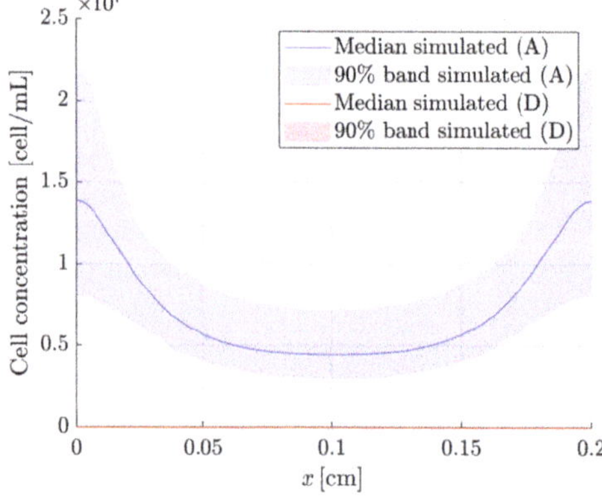

(**c**) Double pseudopalisade experiment.

Figure 8. Distribution of the measured variable for in silico experiments.

4.5. Design of Experiments

In this section, the aim is to determine the experimental configuration with the highest utility, that is, to choose both right and left oxygen flow levels and the initial cell concentration to get the maximum possible information from the new experiment. We focus here on the effect of coupling between parameters and how it affects the utility interpretation and model parameter estimation.

Two different families of simulations were carried out. In the first one, only one parameter dependence is analyzed at a time, leaving the rest fixed at the value set in Section 3.1. These figures show configurations where, if the rest of the parameters are assumed to be known, the unknown parameter will be estimated accurately. This is the case in Figure 9a,b. In the second family, two parameter dependencies are analyzed. They are considered as bivariate distributions in order to observe the effect that the parameter correlation has in characterizing these parameters, that is, how it modifies the utility values. Figure 9c, which belongs to this family, illustrates experimental configurations where the two-dimensional vector will be estimated accurately.

In Figure 9 we compare the iso-utility curves when analyzing one or two parameter dependencies for the pair of parameters related to oxygen, changes in the cell population and cell motility respectively. We assume for all figures $C_0 = 5 \times 10^7$ cell/mL. In these figures we can see the most useful experiments (those configurations corresponding to the highest utility values) and those that lead to a poor adjustment of the model parameters.

This analysis may be performed for different parameter combinations, and for different degrees of knowledge. For instance, Table 3 summarizes all possibilities when exploring the relationship between D_{O_2} and α_a, as we are interested in the estimation of these two parameters, both individually or jointly. The cases analyzed in this paper are reported in the third column.

Table 3. Different possibilities when exploring the relationship between D_{O_2} and α_a in the utility computation.

Parameters to Be Estimated	Known Parameters	Figure
D_{O_2}	None	-
D_{O_2}	$D_a, K_a, \tau_a, \tau_{ad}$	-
D_{O_2}	$D_a, K_a, \tau_a, \tau_{ad}, \alpha_a$	Figure 9a
α_a	None	-
α_a	$D_a, K_a, \tau_a, \tau_{ad}$	-
α_a	$D_a, K_a, \tau_a, \tau_{ad}, D_{O_2}$	Figure 9b
D_{O_2}, α_a	None	-
D_{O_2}, α_a	$D_a, K_a, \tau_a, \tau_{ad}$	Figure 9c

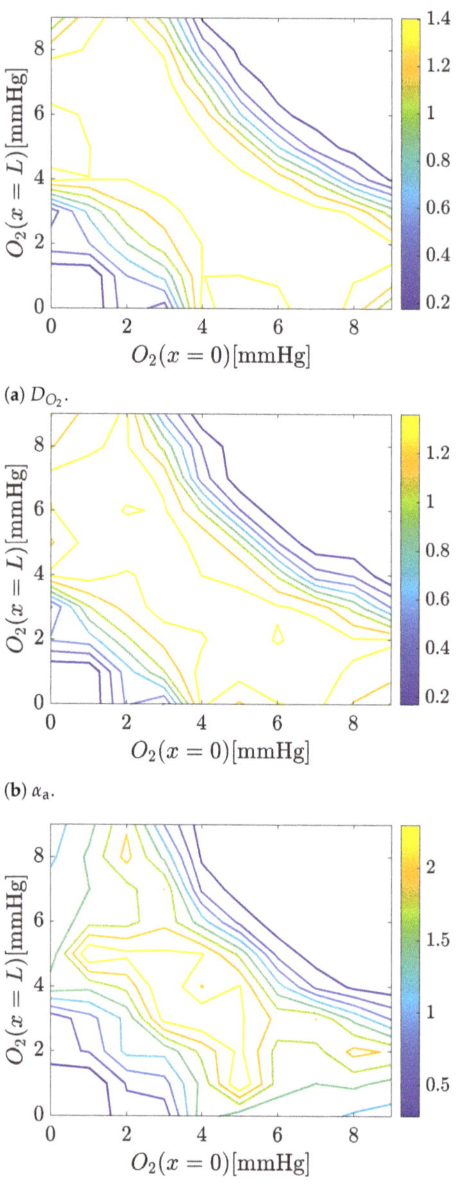

(a) D_{O_2}.

(b) α_a.

(c) (D_{O_2}, α_a).

Figure 9. Iso-utility curves for parameters related to oxygen for an initial concentration of $C_0 = 5 \times 10^7$ cell/mL.

5. Discussion

The train-test methodology based on copulas followed in the fitting process has shown that it is possible to establish a gradation in the strength of the parameter dependencies. Figure 6a illustrates the strength of this relationship, showing that there are pairs of phenomena difficult to isolate from the experimental and/or computational points of view. For example, cell random motility and chemotaxis migration ($\tau = 0.77$). Both phenomena have similar effects but in the opposite direction. Thus, it is difficult to isolate their individual effect on cell behavior if we have limited measurements available on the cell profiles. It is then only possible to evaluate, on the outcome, their combined resultant effect, that is, the average cell motility. This analysis may be done for each parameter couple, justifying the approach adopted in this work.

It is important to note that the high complexity of biological systems, resulting in coupling between pairs of variables, is moderated by the values of the rest, since the bivariate random distributions (shown for example in Figure 3a) are only a projection of the whole 6-dimensional joint distribution. Comparing Figure 3a,b, for example, we can observe the conditioning effect in location, spread, and directionality of the dependency.

Once the probabilistic model is fitted, predicting the actual value of the model parameters is easily carried out. As it is observed in Figure 7, the normality assumption for the confidence region estimation is not always a good starting hypothesis. First, it does not take into account the complexity of the relationship between the model parameters (i.e., physical phenomena) and may lead to non reliable values (meaningless physical magnitudes, such as negative oxygen diffusion). Secondly, it may mislead with respect to the uncertainty that we actually have for different significant levels. In any case, the confident region estimation using HDR and a proper probabilistic analysis are very informative about the degree of reliability of the mathematical model used for a biological explanation. These two observations become even more evident when the uncertainty of the model is reduced, as it can be seen when comparing Figure 7a,b: the chosen significant level has a major impact on the confidence region size and shape. In all the cases analyzed, this uncertainty reduction makes the confidence region to concentrate around the parameter values used in the data generation process.

Knowledge of the model parameter variation (from a probabilistic point of view) allows to predict the outcome of a given experiment. This can be used not only for model calibration and validation, but for experimental planning (deciding the appropriate material, equipment or accuracy of the measuring devices and techniques to be used). For example, in Figure 8, it may be seen that the necrotic core experiment requires less accuracy in the measurement of the cell profile in the central part of the chamber for parameter estimation, while the pseudopalisade experiment requires a measurement technique able to detect extremely low alive cell concentrations. It can also be observed that the appearance of significant alive cells at the right side of the chamber in the pseudopalisade experiment would not be explained by the model parameter variability, but rather by a model limitation.

The probabilistic knowledge of the model can be further exploited in experimental planning and design by using BED theory. In the analysis performed in this work, there are several aspects important to remark. All graphics showing the utility function are symmetric with respect to the line $O_l = O_r$. This is coherent with the symmetrical configuration of the experimental set-up (geometry and properties). The utility value should therefore not be modified by flipping the boundary conditions. Besides, it can be seen that the level curves belonging to D_{O_2} and α_a have similar shapes. This is due to the correlation between parameters, as it can be observed from the Kendall correlation coefficient τ for each pair of variables (Figure 6b). The coefficient corresponding to D_{O_2} and α_a is high and, consequently, they are strongly correlated, so the experiments needed to characterize the value of one of them are similar to the ones needed to characterize the other.

Iso-utility curves give us a picture that may be interpreted biologically and is coherent with the different phenomena occurring in the microfluidic device. However, the coupling between them makes this interpretation difficult. In this work, the utility has been computed for four different initial cell concentrations, ranging from a low concentration $C_0 = 1 \times 10^6$ cell/mL to the chip saturation concentration $C_0 = C^M = 5 \times 10^7$ cell/mL. The maximum utility is always reached for the highest initial concentration (5×10^7 cell/mL).

A summary of the analysis is presented in Table 4, where the best experimental configuration is presented for each of the parameters' calibration, together with the maximum utility value.

Table 4. Most useful experimental configuration for each of the parameters' evaluation.

Parameters to Be Estimated	Upper O_2 Concentration [mmHg]	Lower O_2 Concentration [mmHg]	Maximum Utility Value
D_{O_2}	7	0	1.58
α_a	5	2	1.53
(D_{O_2}, α_a)	5	1	2.58
τ_a	7	0	0.07
τ_{ad}	7	0	1.29
(τ_a, τ_{ad})	7	0	1.63
D_a	8	0	0.51
K_a	7	1	0.35
(D_a, K_a)	8	0	0.49

For the analyzed family of experiments, the most useful experiments are always the ones performed for high concentrated cell cultures. As most phenomena are related to cell concentrations, the higher the concentration, the more quantifiable the different biological mechanisms. Besides, it results clear that configurations with oxygen gradient are more useful for accurately characterizing the parameters related to oxygen (D_{O_2}, α_a) and cell migration (D_a, K_a), when the other parameters are assumed to be known. However, this gradient has to be moderate to avoid regions of total normoxia or total anoxia. When the aim is to perfectly discriminate between their effects, softer gradients are generally preferred (Figure 9c, Table 4). Finally, for high initial cell concentrations, growth and death parameters are also well characterized under gradient conditions: we need to induce localized hypoxic conditions in order to evaluate growth under saturation capacity and death.

6. Conclusions

Mathematical modeling of complex cell processes is very challenging due to its intrinsic non-linearity, highly-coupled multiphysic interactions, and the many correlated parameters which are difficult to measure or simply unknown. These parameters are most times obtained for a particular problem under specific conditions, leading in many cases to conclusions, directly derived from the modeling assumptions and therefore providing little new information. Also, they are difficult to generalize.

As a result, a proper and extensive parametric analysis is mandatory. This should include an extensive and detailed study of the values reported in the bibliography, a careful sensitivity analysis and a sufficient number of different experiments, not only for calibration but also for validation, avoiding parameter overfitting.

This analysis, although it allows the identification of the optimal set of parameters, is most times difficult to extend to other problems with reasonable accuracy and therefore with a certain validation of its actual physical character and its value range. It is also difficult to discriminate between correlated parameters associated to mechanisms that cannot be isolated in the experiments. Hence, we need additional information both to get a better discrimination between them, and to identify the optimal conditions for additional experiments to provide the maximum information possible in order to get such discrimination.

We have proved here that copulas are a simple and powerful tool to detect and improve highly-correlated multiparametric mathematical models such as those appearing in Biology, with the added value of providing key information for the optimal design of new experiments with the highest information possible for the problem in hands, thus reducing time and cost not only in our in vitro experiments but also in scarce and costly in vivo cases.

Author Contributions: Conceptualization M.D., M.H.D. and J.A.-J.; Methodology J.A.-J.; Software and validation J.A.-J., M.P.-A.; Writting original draft M.D., J.A.-J. and M.P.-A.; Figures and visualization J.A.-J., M.P.-A and T.R.; Supervision M.D. and M.H.D.; Project administration and funding M.D., M.H.D. and J.A.S.-H. All authors reviewed the manuscript. All authors have read and agreed to the published version of the manuscript.

Funding: The authors gratefully acknowledge the financial support from the Spanish Ministry of Economy and Competitiveness (MINECO) and FEDER, UE through the project PGC2018-097257-B-C31, the Spanish Ministry of Science and Innovation through the project PID2019-106099RB-C44/AEI/10.13039/501100011033, the Government of Aragon (DGA) and the Centro de Investigacion Biomedica en Red en Bioingenieria, Biomateriales y Nanomedicina (CIBER-BBN). CIBER-BBN is financed by the Instituto de Salud Carlos III with assistance from the European Regional Development Fund.

Institutional Review Board Statement: Not applicable.

Informed Consent Statement: Not applicable.

Data Availability Statement: Data and codes available under request to the authors.

Conflicts of Interest: The authors declare no conflict of interest. The funders had no role in the design of the study; in the collection, analyses, or interpretation of data; in the writing of the manuscript, or in the decision to publish the results.

References

1. Quail, D.F.; Joyce, J.A. Microenvironmental regulation of tumor progression and metastasis. *Nat. Med.* **2013**, *19*, 1423. [CrossRef] [PubMed]
2. Hanahan, D.; Weinberg, R.A. Hallmarks of cancer: The next generation. *Cell* **2011**, *144*, 646–674. [CrossRef] [PubMed]
3. Scannell, J.W.; Blanckley, A.; Boldon, H.; Warrington, B. Diagnosing the decline in pharmaceutical R & D efficiency. *Nat. Rev. Drug Discov.* **2012**, *11*, 191. [PubMed]
4. Sackmann, E.K.; Fulton, A.L.; Beebe, D.J. The present and future role of microfluidics in biomedical research. *Nature* **2014**, *507*, 181. [CrossRef] [PubMed]
5. Bhatia, S.N.; Ingber, D.E. Microfluidic organs-on-chips. *Nat. Biotechnol.* **2014**, *32*, 760. [CrossRef] [PubMed]
6. Boussommier-Calleja, A.; Li, R.; Chen, M.B.; Wong, S.C.; Kamm, R.D. Microfluidics: A new tool for modeling cancer–immune interactions. *Trends Cancer* **2016**, *2*, 6–19. [CrossRef] [PubMed]
7. Zervantonakis, I.K.; Hughes-Alford, S.K.; Charest, J.L.; Condeelis, J.S.; Gertler, F.B.; Kamm, R.D. Three-dimensional microfluidic model for tumor cell intravasation and endothelial barrier function. *Proc. Natl. Acad. Sci. USA* **2012**, *109*, 13515–13520. [CrossRef] [PubMed]
8. Byrne, H.; Alarcon, T.; Owen, M.; Webb, S.; Maini, P. Modelling aspects of cancer dynamics: A review. *Philos. Trans. R. Soc. Lond. A Math. Phys. Eng. Sci.* **2006**, *364*, 1563–1578. [CrossRef]
9. Kitano, H. Computational systems biology. *Nature* **2002**, *420*, 206. [CrossRef]
10. Bearer, E.L.; Lowengrub, J.S.; Frieboes, H.B.; Chuang, Y.L.; Jin, F.; Wise, S.M.; Ferrari, M.; Agus, D.B.; Cristini, V. Multiparameter computational modeling of tumor invasion. *Cancer Res.* **2009**, *69*, 4493–4501. [CrossRef]
11. Ayensa-Jiménez, J.; Pérez-Aliacar, M.; Randelovic, T.; Oliván, S.; Fernández, L.; Sanz-Herrera, J.A.; Ochoa, I.; Doweidar, M.H.; Doblaré, M. Mathematical formulation and parametric analysis of in vitro cell models in microfluidic devices: Application to different stages of glioblastoma evolution. *Sci. Rep.* **2020**, *10*, 1–21. [CrossRef] [PubMed]
12. Brat, D.J. Glioblastoma: Biology, genetics, and behavior. In *American Society of Clinical Oncology Educational Book*; American Society of Clinical Oncology: Alexandria, VA, USA, 2012; pp. 102–107._am.2012.32.102. [CrossRef]
13. Ang, A.; Chen, J. Asymmetric correlations of equity portfolios. *J. Financ. Econ.* **2002**, *63*, 443–494. [CrossRef]
14. Boubaker, H.; Sghaier, N. Portfolio optimization in the presence of dependent financial returns with long memory: A copula based approach. *J. Bank. Financ.* **2013**, *37*, 361–377. [CrossRef]
15. McNeil, A.; Frey, R.; Embrechts, P. *Quantitative Risk Management: Concepts, Techniques, and Tools*; Princeton University Press: Princeton, NJ, USA, 2017.
16. Kole, E.; Koedijk, K.; Verbeek, M. Selecting copulas for risk management. *J. Bank. Financ.* **2007**, *31*, 2405–2423. [CrossRef]
17. Meucci, A. A new breed of copulas for risk and portfolio management. *Risk* **2011**, *24*, 122–126.
18. Solari, S.; Losada, M. Non-stationary wave height climate modeling and simulation. *J. Geophys. Res. Ocean.* **2011**, *116*. [CrossRef]
19. Munkhammar, J.; Widén, J. An autocorrelation-based copula model for generating realistic clear-sky index time-series. *Sol. Energy* **2017**, *158*, 9–19. [CrossRef]
20. Arya, F.K.; Zhang, L. Copula-based Markov process for forecasting and analyzing risk of water quality time series. *J. Hydrol. Eng.* **2017**, *22*, 04017005. [CrossRef]

21. Laux, P.; Wagner, S.; Wagner, A.; Jacobeit, J.; Bardossy, A.; Kunstmann, H. Modelling daily precipitation features in the Volta Basin of West Africa. *Int. J. Climatol. A J. R. Meteorol. Soc.* **2009**, *29*, 937–954. [CrossRef]
22. Schoelzel, C.; Friederichs, P. Multivariate non-normally distributed random variables in climate research–introduction to the copula approach. *Nonlinear Process. Geophys.* **2008**, *15*, 761–772. [CrossRef]
23. Laux, P.; Vogl, S.; Qiu, W.; Knoche, H.R.; Kunstmann, H. Copula-based statistical refinement of precipitation in RCM simulations over complex terrain. *Hydrol. Earth Syst. Sci.* **2011**, *15*, 2401–2419. [CrossRef]
24. Zou, Y.; Zhang, Y. A copula-based approach to accommodate the dependence among microscopic traffic variables. *Transp. Res. Part C Emerg. Technol.* **2016**, *70*, 53–68. [CrossRef]
25. Spissu, E.; Pinjari, A.R.; Pendyala, R.M.; Bhat, C.R. A copula-based joint multinomial discrete–continuous model of vehicle type choice and miles of travel. *Transportation* **2009**, *36*, 403–422. [CrossRef]
26. Kilgore, R.T.; Thompson, D.B. Estimating joint flow probabilities at stream confluences by using copulas. *Transp. Res. Rec.* **2011**, *2262*, 200–206. [CrossRef]
27. Bartoli, G.; Mannini, C.; Massai, T. Quasi-static combination of wind loads: A copula-based approach. *J. Wind Eng. Ind. Aerodyn.* **2011**, *99*, 672–681. [CrossRef]
28. Dong, S.; Zhou, C.; Tao, S.S.; Xue, D.S. Bivariate Gumbel distribution based on Clayton Copula and its application in offshore platform design. *Period. Ocean Univ. China* **2011**, *41*, 117–120.
29. Pham, H. Recent studies in software reliability engineering. In *Handbook of Reliability Engineering*; Springer: London, UK, 2003; pp. 285–302.
30. Kim, J.M.; Jung, Y.S.; Sungur, E.A.; Han, K.H.; Park, C.; Sohn, I. A copula method for modeling directional dependence of genes. *BMC Bioinform.* **2008**, *9*, 225. [CrossRef]
31. Kim, Y.; Jeon, H.; Othmer, H. The role of the tumor microenvironment in glioblastoma: A mathematical model. *IEEE Trans. Biomed. Eng.* **2017**, *64*, 519–527. [CrossRef]
32. Ayuso, J.M.; Monge, R.; Martínez-González, A.; Virumbrales-Muñoz, M.; Llamazares, G.A.; Berganzo, J.; Hernández-Laín, A.; Santolaria, J.; Doblaré, M.; Hubert, C.; et al. Glioblastoma on a microfluidic chip: Generating pseudopalisades and enhancing aggressiveness through blood vessel obstruction events. *Neuro-Oncology* **2017**, *19*, 503–513. [CrossRef]
33. Ayuso, J.M.; Virumbrales-Muñoz, M.; Lacueva, A.; Lanuza, P.M.; Checa-Chavarria, E.; Botella, P.; Fernández, E.; Doblare, M.; Allison, S.J.; Phillips, R.M.; et al. Development and characterization of a microfluidic model of the tumour microenvironment. *Sci. Rep.* **2016**, *6*, 36086. [CrossRef]
34. Hatzikirou, H.; Basanta, D.; Simon, M.; Schaller, K.; Deutsch, A. 'Go or grow': The key to the emergence of invasion in tumour progression? *Math. Med. Biol. A J. IMA* **2012**, *29*, 49–65. [CrossRef] [PubMed]
35. Stramer, B.; Mayor, R. Mechanisms and in vivo functions of contact inhibition of locomotion. *Nat. Rev. Mol. Cell Biol.* **2017**, *18*, 43. [CrossRef] [PubMed]
36. Galluzzi, L.; Vitale, I.; Aaronson, S.A.; Abrams, J.M.; Adam, D.; Agostinis, P.; Alnemri, E.S.; Altucci, L.; Amelio, I.; Andrews, D.W.; et al. Molecular mechanisms of cell death: Recommendations of the Nomenclature Committee on Cell Death 2018. *Cell Death Differ.* **2018**, *25*, 486. [CrossRef] [PubMed]
37. Sendoel, A.; Hengartner, M.O. Apoptotic cell death under hypoxia. *Physiology* **2014**, *29*, 168–176. [CrossRef] [PubMed]
38. Chance, B.; Williams, G.R. The respiratory chain and oxidative phosphorylation. *Adv. Enzymol. Relat. Areas Mol. Biol.* **1956**, *17*, 65–134.
39. Jaworski, P.; Durante, F.; Härdle, W.K.; Rychlik, T. *Copula Theory and Its Applications: Proceedings of the Workshop Held in Warsaw, Poland, 25–26 September 2009*; Springer: Berlin, Germany, 2010; Volume 198.
40. Sklar, M. Fonctions de repartition an dimensions et leurs marges. *Publ. Inst. Statist. Univ. Paris* **1959**, *8*, 229–231.
41. Wand, M.P.; Jones, M.C. *Kernel Smoothing*; CRC Press: Boca Raton, FL, USA, 1994.
42. Kottegoda, N.T.; Rosso, R. *Applied Statistics for Civil and Environmental Engineers*; Blackwell Malden: Malden, MA, USA, 2008.
43. Fan, Y. Goodness-of-fit tests for a multivariate distribution by the empirical characteristic function. *J. Multivar. Anal.* **1997**, *62*, 36–63. [CrossRef]
44. Hyndman, R.J. Computing and graphing highest density regions. *Am. Stat.* **1996**, *50*, 120–126.
45. Fisher, R.A. *The Design of Experiments*; Oliver and Boyd: Edinburgh/London, UK, 1937.
46. Chaloner, K.; Verdinelli, I. Bayesian Experimental Design: A Review. *Stat. Sci.* **1995**, *10*, 273–304. [CrossRef]
47. Shannon, C.E. A mathematical theory of communication. *Bell Syst. Tech. J.* **1948**, *27*, 379–423. [CrossRef]
48. Ahmed, N.A.; Gokhale, D. Entropy expressions and their estimators for multivariate distributions. *IEEE Trans. Inf. Theory* **1989**, *35*, 688–692. [CrossRef]
49. Demarta, S.; McNeil, A.J. The t copula and related copulas. *Int. Stat. Rev.* **2005**, *73*, 111–129. [CrossRef]

Article

Empowering Advanced Driver-Assistance Systems from Topological Data Analysis

Tarek Frahi [1,*], Francisco Chinesta [1], Antonio Falcó [2], Alberto Badias [3], Elias Cueto [3], Hyung Yun Choi [4], Manyong Han [5] and Jean-Louis Duval [6]

1   PIMM Lab, Arts et Metiers Institute of Technology, 151 boulevard de l'Hopital, 75013 Paris, France; Francisco.Chinesta@ensam.eu

2   Departamento de Matemáticas, Física y Ciencias Tecnológicas, Universidad Cardenal Herrera-CEU, CEU Universities, San Bartolome 55, 46115 Alfara del Patriarca, Valencia, Spain; afalco@uch.ceu.es

3   I3A, Aragon Institute of Engineering Research, Universidad de Zaragoza, 50018 Zaragoza, Aragon, Spain; abadias@unizar.es (A.B.); ecueto@unizar.es (E.C.)

4   Department of Mechanical and System Design Engineering, Hongik University, 94 Wausanro, Mapogu, Seoul 04066, Korea; hychoi@hongik.ac.kr

5   Digital Human Lab, Hongik University, 94 Wausanro, Mapogu, Seoul 04066, Korea; myhan0521@gmail.com

6   ESI Group, 3bis rue Saarinen, CEDEX 1, 94528 Rungis, France; Jean-Louis.Duval@esi-group.com

*   Correspondence: tarek.frahi@ensam.eu

Abstract: We are interested in evaluating the state of drivers to determine whether they are attentive to the road or not by using motion sensor data collected from car driving experiments. That is, our goal is to design a predictive model that can estimate the state of drivers given the data collected from motion sensors. For that purpose, we leverage recent developments in topological data analysis (TDA) to analyze and transform the data coming from sensor time series and build a machine learning model based on the topological features extracted with the TDA. We provide some experiments showing that our model proves to be accurate in the identification of the state of the user, predicting whether they are relaxed or tense.

Keywords: Morse theory; topological data analysis; machine learning; time series; smart driving

check for **updates**

Citation: Frahi, T.; Chinesta, F.; Falcó, A.; Badias, A.; Cueto, E.; Choi, H.Y.; Han, M.; Duval, J.-L. Empowering Advanced Driver-Assistance Systems from Topological Data Analysis. *Mathematics* **2021**, *9*, 634. https://doi.org/10.3390/math9060634

Academic Editors: Duarte Valério and Mauro Malvè

Received: 31 January 2021
Accepted: 11 March 2021
Published: 16 March 2021

Publisher's Note: MDPI stays neutral with regard to jurisdictional claims in published maps and institutional affiliations.

1. Introduction

While there have recently been considerable advances in self-driving car technology, driving still relies mainly on human factors. Even in self-driving mode, human drivers must often make decision in a fraction of a second to avoid accidents. Therefore, it is still of utmost importance to develop systems capable of discerning if the human driver is attentive or not to the road conditions. In general, the so-called advanced driver assistance systems (ADAS) [1,2] are systems that are able to improve the driver's performance, among which, adaptive speed limiters, pedestrian detectors [3], and cruise controllers are some of the most popular systems. Fatigue alerting systems are among the most useful among ADAS systems, and the aim of this work is to contribute to the development of such a system based on a systematic analysis of drivers in actual driving conditions.

The estimation of the driver's condition (degree of attention to the road, fatigue, etc.) is a very important factor to ensure safety in driving [4,5]. A recent review on the topic can be found in [6]. The goal of this work is to extract behavior patterns from car user data to be able to accurately estimate their state. We used data obtained by the laboratory of prof. Hyung Yun Choi at Hongik University in Seoul. His experiment involved the application of mechanical stimulation to people seated in an automobile.

Our main goal is to extract patterns of behavior from experimental data so as to allow us to learn the most relevant factors affecting driver's attention to the situation of the road.

In the present work, we combine some tools from Morse theory [7] and topological data analysis (TDA) with all of the associated concepts and methods (e.g., Betti numbers,

83

homology persistence, barcodes, persistence images, etc.) [8], most of them introduced and employed later in order to analyze and classify the experimental data. This allows us to introduce concepts as barcodes, that is, persistent and life-time diagrams in a similar way to how they are used in persistent homology. Our main goal is to predict car user behavior following a supervised approach [9]. Instead of considering an original sensor signal as the quantity of interest, we focus on its topological features. In this sense, the framework proposed in this paper allows us to unveil the true dimensionality of data or, in other words, the actual number of factors affecting driver's performance. Thus, we model a sensor signal as a dynamical system, and, therefore, our approach seems to be better at describing its properties, or rather its variations, such as extrema, patterns, and self-similarity, than other approaches. We note that our approach is, in some senses, similar to that followed by Milnor and Thurston [10] in the study of the combinatorial properties of dynamical systems by combining tools from automata theory.

The structure of the paper is as follows: In Section 2, we describe the material and methods employed in this work. Particular attention is paid to the process of data acquisition and the description of time series and data curation. In Section 3, we present the main results of this work, and we discuss the main consequences in Section 4. As a complement, in Appendix A, we thoroughly illustrate the process of computing persistence images for the data of interest.

2. Material and Methods

In this section, we describe the collection and preprocessing of the experimental data. In Section 2.1, we describe the data acquisition, and in Section 2.2, we provide a description of the time series. Section 2.3 is devoted to data preprocessing. The mathematical tools used to describe the times series at a topological level are explained in Section 2.4. Finally, the image classification methodology is given in Section 2.5.

2.1. Data Acquisition

Our proposed predictor directly uses the data collected from the experiments. The data acquisition process involves measuring the response of human behavior when an excitation is applied to the seat. Figure 1 shows the location of the sensors in the experiments.

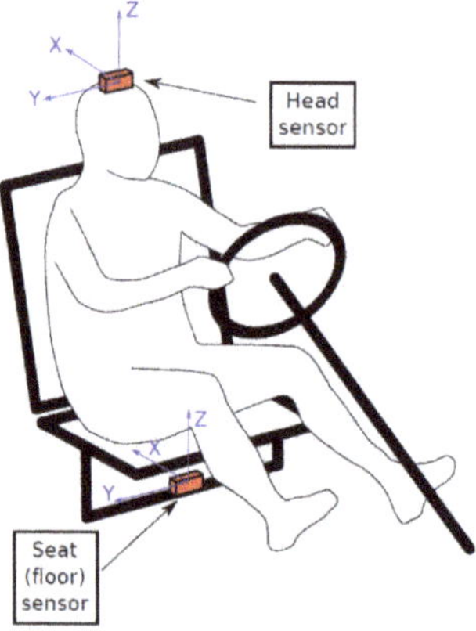

Figure 1. Scheme of the data acquisition process showing the location of the sensors.

The excitation signal is an angular acceleration imposed on the seat of the user. This acceleration is an oscillating chirp function with a frequency range of 1 to 7.5 Hz on the X axis in rotation. The linear acceleration $\mathbf{a} = (a_x, a_y, a_z)$ and angular velocity $\omega = (\omega_x, \omega_y, \omega_z)$ were measured in both the head and the seat by two IMU (Shimmer inertia measurement unit (IMU) sensors) at 256 Hz. By observing the floor excitation signals, we noted that the excitation is purely rotational around the X-axis—see Figure 2.

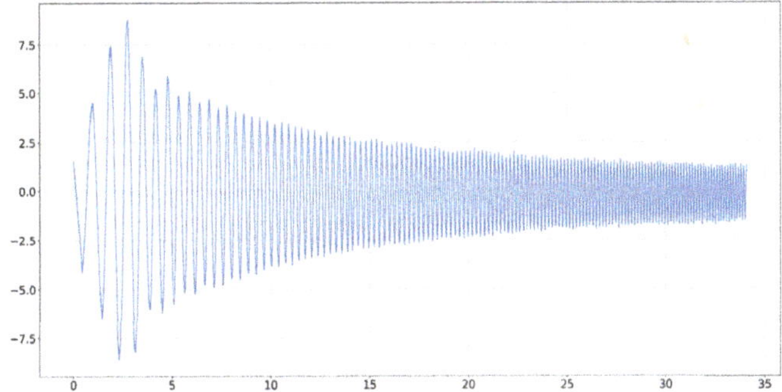

Figure 2. Floor excitation: X-axis angular velocity time series.

Several experiences were conducted by nine people by taking into account a set of six fixed states: driver, passenger, tense person, relaxed person, rigid seat, and SAV (sport activity vehicle seat). In particular, for each individual, eight experiments for the six available states were performed:

$$
\left\{
\begin{array}{ll}
\textbf{Class} & \textbf{Label} \\
1 & SAVRelaxedPassager \\
2 & SAVTensePassager \\
3 & SAVRelaxedDriver \\
4 & SAVTenseDriver \\
5 & RigidRelaxedPassager \\
6 & RigidTensePassager \\
7 & RigidRelaxedDriver \\
8 & RigidTenseDriver
\end{array}
\right.
$$

As a consequence, we worked with a sample of 72 experiences, each of them encoded in a time series (as we explain later). Our goal is to classify the behavior of a generic driver, assigning one of the two states (tense or relaxed) by using the sensor data.

2.2. Time Series Description

The data acquired from sensors (see Figures 3 and 4) were stored into six-dimensional time series, for both linear acceleration and angular velocity of the head movement. The sampling frequency of the data was 256 Hz, and the duration of the experiment was 34 s; hence, the resulting data dimensionality is $256 \times 34 = 8704$. For each times series, where $1 \leq t \leq 8704$, we constructed three new times series called the sliding window, embedding a length of 5800. The first one is given by the times values from $t = 1$ to $t = 5800$, the second is given by the times values from $t = 1450$ to $t = 7250$, and, to conclude, the third time window is defined as from $t = 2904$ to $t = 8704$. Each element in the sample $(1 \leq i \leq 72)$ was encoded by means of three six-dimensional time series representing each of the three sliding windows that we represent in matrix form as follows:

$$
TS_{3(i-1)+1} = \begin{bmatrix}
a_x^\ell(1) & a_x^\ell(2) & \cdots & a_x^\ell(5800) \\
a_y^\ell(1) & a_y^\ell(2) & \cdots & a_y^\ell(5800) \\
a_z^\ell(1) & a_z^\ell(2) & \cdots & a_z^\ell(5800) \\
\omega_x^\ell(1) & \omega_x^\ell(2) & \cdots & \omega_x^\ell(5800) \\
\omega_y^\ell(1) & \omega_y^\ell(2) & \cdots & \omega_y^\ell(5800) \\
\omega_z^\ell(1) & \omega_z^\ell(2) & \cdots & \omega_z^\ell(5800)
\end{bmatrix}, \;
TS_{3(i-1)+2} = \begin{bmatrix}
a_x^\ell(1450) & a_x^\ell(1451) & \cdots & a_x^\ell(7251) \\
a_y^\ell(1450) & a_y^\ell(1451) & \cdots & a_y^\ell(7251) \\
a_z^\ell(1450) & a_z^\ell(1451) & \cdots & a_z^\ell(7251) \\
\omega_x^\ell(1450) & \omega_x^\ell(1451) & \cdots & \omega_x^\ell(7251) \\
\omega_y^\ell(1450) & \omega_y^\ell(1451) & \cdots & \omega_y^\ell(7251) \\
\omega_z^\ell(1450) & \omega_z^\ell(1451) & \cdots & \omega_z^\ell(7251)
\end{bmatrix}
$$

and

$$
TS_{3i} = \begin{bmatrix}
a_x^\ell(2903) & a_x^\ell(2905) & \cdots & a_x^\ell(8704) \\
a_y^\ell(2903) & a_y^\ell(2905) & \cdots & a_y^\ell(8704) \\
a_z^\ell(2903) & a_z^\ell(2905) & \cdots & a_z^\ell(8704) \\
\omega_x^\ell(2903) & \omega_x^\ell(2905) & \cdots & \omega_x^\ell(8704) \\
\omega_y^\ell(2903) & \omega_y^\ell(2905) & \cdots & \omega_y^\ell(8704) \\
\omega_z^\ell(2903) & \omega_z^\ell(2905) & \cdots & \omega_z^\ell(8704)
\end{bmatrix}.
$$

Here, the matrices have a size of 6×5800 and $1 \leq i \leq 72$. This allows us to represent the information by using a third-order tensor, namely, $\mathcal{Z} \in \mathbb{R}^{216 \times 6 \times 5800}$ defined by

$$
\mathcal{Z}_{i,j,k} := (TS_i)_{j,k}
$$

for $1 \leq i \leq 216$, $1 \leq j \leq 6$ and $1 \leq k \leq 5800$. We can identify $Z_i = TS_i$ for $1 \leq i \leq 216$.

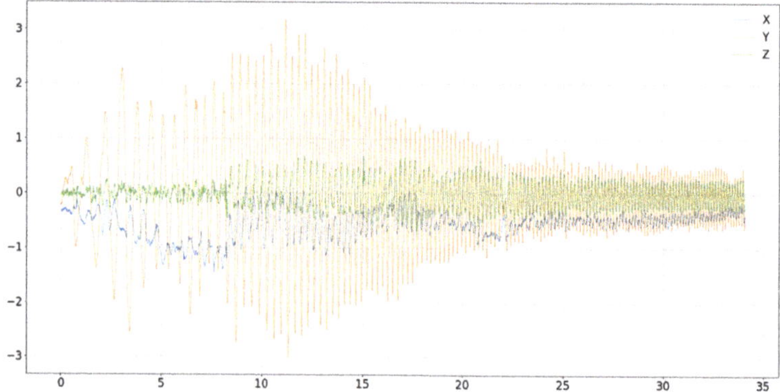

Figure 3. Sensor data: linear acceleration time series.

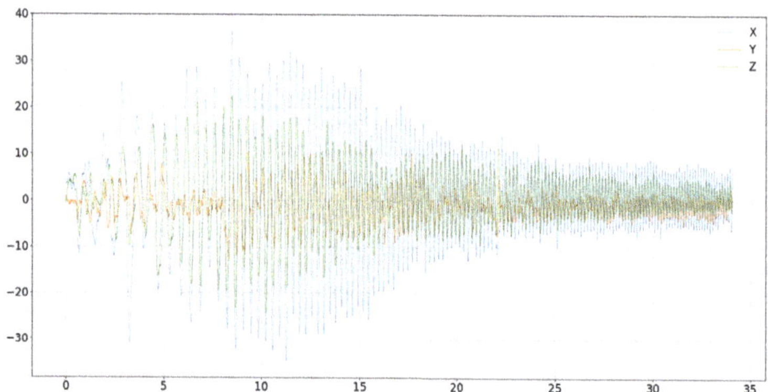

Figure 4. Sensor data: angular velocity time series.

2.3. Data Preprocessing

In order to obtain a single series for each observation, we concatenated all of the 6 time series (linear accelerations and angular velocities) for each observation horizontally and then created a data frame by stacking the 216 in sample observations.

The concatenation operation on the multidimensional time series collapsed the last two dimensions into one dimensional arrays with a length of $5800 \times 6 = 34{,}800$. The result is the two-dimensional table of concatenated time series

$$
D = \begin{bmatrix} vec(\mathcal{Z}_{1,:,:}) \\ \dots \\ vec(\mathcal{Z}_{216,:,:}) \end{bmatrix} \in \mathbb{R}^{216 \times 34800}.
$$

We chose not to filter the signals because the topological sub-level set method should filter the high-frequency features naturally. We also chose to keep working on acceleration signals in order to avoid signal deviations after two integrations in time so as to obtain positions, the sensors not always keeping a zero mean height. Thus, the approach is completely (topologically) data-based.

The six time series $\mathcal{Z}_i$ of each observation were collapsed into a single concatenated time series with a size of 34,800—see Figure 5. The concatenated time series for the 216 observations were then stacked to create the dataset D with a size of $216 \times 34{,}800$. We also used binary labels in the chained time series $\mathcal{Z}_i$ on the two target classes that we were interested in. In particular, we wrote $\mathcal{Z}_i^{(\alpha)}$ where α is "0" for a relaxed driver and "1" for a tense one.

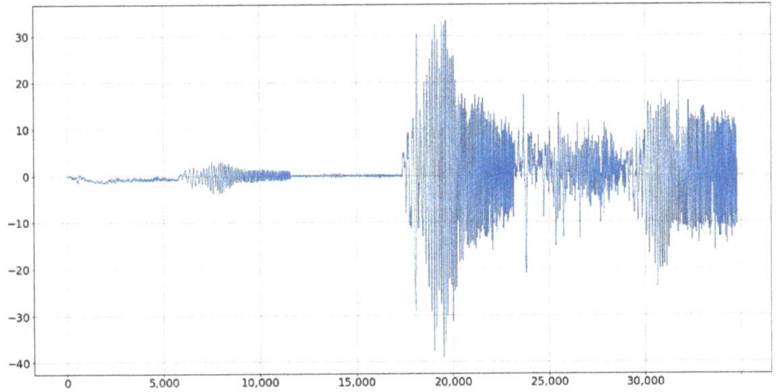

Figure 5. Tensor reduction of a sensor time series.

2.4. Extracting Topological Features from a Time Series

The idea to extract the topological information regarding the times series is to consider each sample observation as a piecewise linear continuous map from a closed interval to the real line. Therefore, we used a construction closely related to the Reeb graph [11] used in Morse theory to describe the times series at the topological level.

To this end, we consider the time series x_t for $0 \leq t \leq N - 1$ $(N \geq 3)$ given by a vector

$$
\mathbf{X} = (x_0, x_1, \dots, x_{N-1}) \in \mathbb{R}^N.
$$

we can view $\mathbf{X}$ as a function also denoted by $\mathbf{X} : \{0, 1, \dots, N - 1\} \longrightarrow \mathbb{R}$ defined by $\mathbf{X}(i) = x_i$ for $0 \leq i \leq N - 1$. Here, to study the topological features of $\mathbf{X}$ we use the sub-level set of a piecewise-linear function $f_{\mathbf{X}} : \mathbb{R} \longrightarrow \mathbb{R}$ associated with $\mathbf{X}$ satisfying that $f_{\mathbf{X}}(i) = \mathbf{X}(i) = x_i$ for $0 \leq i \leq N - 1$.

To construct this function, we consider the basis functions $\{\varphi_0, \dots, \varphi_{N-1}\}$ of continuous functions $\varphi_i : \mathbb{R} \longrightarrow \mathbb{R}$ defined by

$$
\varphi_i(s) := \begin{cases} s - i + 1 & \text{if} & i - 1 \leq s \leq i \\ i + 1 - s & \text{if} & i \leq s \leq i + 1 \\ 0 & \text{if} & s \notin \,]i - 1, i + 1[\end{cases}
$$

where $i = 1, \ldots, N - 2$ and

$$\varphi_0(s) := \begin{cases} 1 - s & \text{if} \quad 0 \leq s \leq 1 \\ 0 & \text{if} \quad s \in [0, 1] \end{cases}$$

$$\varphi_{N-1}(s) := \begin{cases} s - N + 2 & \text{if} \quad N - 2 \leq s \leq N - 1 \\ 0 & \text{if} \quad s \notin [N - 2, N - 1[\end{cases}$$

This allows us to construct a piecewise continuous map $f_{\mathbf{X}} : \mathbb{R} \longrightarrow \mathbb{R}$ by

$$f_{\mathbf{X}}(s) = \sum_{j=0}^{N-1} x_j \varphi_j(s),$$

and also to endow $\mathbb{R}^N$ with a norm given by

$$\|\mathbf{X}\| := \|f_{\mathbf{X}}\|_{L^2(\mathbb{R})} = \left(\int_{-\infty}^{\infty} |f_{\mathbf{X}}(s)|^2 ds \right)^{1/2}.$$

In particular, we prove the following result, which helps us to identify the time series given by the vector $\mathbf{X}$ in $\mathbb{R}^N$ with the function $f_{\mathbf{X}}$ in $L^2(\mathbb{R})$.

Proposition 1. *The linear map* $\Phi : (\mathbb{R}^N, \|\cdot\|) \longrightarrow (L^2(\mathbb{R}), \|\cdot\|_{L^2(\mathbb{R})})$ *given by* $\Phi(\mathbf{X}) = f_{\mathbf{X}}$ *is an injective isometry between Hilbert spaces. Furthermore,* $\Phi(\mathbb{R}^N)$ *is a closed subspace in* $L^2(\mathbb{R}^N)$.

Proof. The map is clearly isometric and injective because $\{\varphi_0, \ldots, \varphi_{N-1}\}$ is a set of linear independent functions in $L^2(\mathbb{R})$. $\square$

Here, we describe the maps $f_{\mathbf{X}} \in \Phi(\mathbb{R}^N)$ at the combinatorial level using the connected components (intervals) associated with its λ sub-level sets

$$\mathcal{LS}_\lambda(f_{\mathbf{X}}) := \{ x \in [0, N - 1] : f_{\mathbf{X}}(x) \leq \lambda \}$$

for $\lambda \in \mathbb{R}$. For this purpose, we introduce the following distinguished objects related to the $\mathrm{supp}(f_{\mathbf{X}}) = [0, N - 1] \subset \mathbb{R}$ of $f_{\mathbf{X}}$:

- The nodes or vertices denoted by

$$\mathcal{V} := \{ [0], [1], \ldots, [N - 1] \}$$

that represent the components of the vector $\mathbf{X}_{,;}$
- The faces denoted by

$$\mathcal{F} := \{ [0, 1][1, 2], \ldots, [N - 2, N - 1] \}$$

that represent the intervals used to construct the connected components of the sub-level sets of the map $f_{\mathbf{X}}$. Recall that we consider

$$[i, i + 1] := \{ z \in \mathbb{R} : z = \mu\, x_{i+1} + (1 - \mu) x_i,\, 0 \leq \mu \leq 1 \} \subset \mathbb{R}.$$

Let

$$\lambda_{\max} = \max_{s \in [0, N-1]} f_{\mathbf{X}}(s) = \max_{0 \leq i \leq N-1} \mathbf{X}(i),$$

and

$$\lambda_{\min} = \min_{s \in [0, N-1]} f_{\mathbf{X}}(s) = \min_{0 \leq i \leq N-1} \mathbf{X}(i).$$

For each $\lambda_{\min} \leq \lambda \leq \lambda_{\max}$, we introduce the following symbolic λ sub-level set for the map $f_{\mathbf{X}}$:

$$LS_\lambda(f_{\mathbf{X}}) := \{ \sigma \in \mathcal{F} : f(\sigma) \leq \lambda \}$$

For $\lambda_{\min} \le \lambda \le \lambda' \le \lambda_{\max}$, it holds

$$LS_\lambda(f_{\mathbf{X}}) \subset LS_{\lambda'}(f_{\mathbf{X}}).$$

Our next goal was to quantify the evolution of the above symbolic λ sub-level with. To this end, we introduce the notion of feature associated with the λ sub-level set $LS_\lambda(f_{\mathbf{X}})$. We define the set of features for functions in $\Phi(\mathbb{R}^N)$ as

$$\mathfrak{F}(\Phi(\mathbb{R}^N)) := \{[i,j] \subset \mathbb{R} : 0 \le i < j \le N-1\}.$$

We note that $LS_\lambda(f_{\mathbf{X}}) \subset \mathcal{F} \subset \mathfrak{F}(\Phi(\mathbb{R}^N))$. Then next definition introduces the notion of features for a symbolic λ sub-level set as the interval of $\mathfrak{F}(\Phi(\mathbb{R}^N))$ constructed by a maximal union of faces of $LS_\lambda(f_{\mathbf{X}})$.

Definition 1. *We suggest that $\mathbb{I} \in \mathfrak{F}(\Phi(\mathbb{R}^N))$ is a feature for the symbolic λ sub-level set $LS_\lambda(f_{\mathbf{X}})$ if there exists $\mathbb{I}_1, \ldots, \mathbb{I}_k \in LS_\lambda(f_{\mathbf{X}})$ such that $\mathbb{I} = \bigcup_{j=1}^{k} \mathbb{I}_k$ and for every $\mathbb{J} \in LS_\lambda(f_{\mathbf{X}})$ such that $\mathbb{J} \ne \mathbb{I}_i$ for $1 \le i \le k$ it holds that $\mathbb{I} \cap \mathbb{J} = \varnothing$. We denote by $\mathfrak{F}(LS_\lambda(f_{\mathbf{X}}))$ the set of features for the λ sub-level set $LS_\lambda(f_{\mathbf{X}})$.*

A feature for a λ sub-level set $LS_\lambda(f_{\mathbf{X}})$ is the maximal interval of $\mathfrak{F}(\Phi(\mathbb{R}^N))$ that we can construct by unions of intervals in $LS_\lambda(f_{\mathbf{X}})$. To illustrate this definition, we give the following example:

Example 1. *Let us consider the time series*

$$\mathbf{X} = (11, 14, 9, 7, 9, 7, 8, 10, 9).$$

This allows us to construct the map $f_{\mathbf{X}}$ as shown in Figure 6. Then, $\lambda_{min} = 7$ and $\lambda_{\max} = 14$, and we have the following symbolic λ sub-level sets.

$$
\begin{aligned}
LS_{\lambda=7}(f_{\mathbf{X}}) &= \varnothing \\
LS_{\lambda=8}(f_{\mathbf{X}}) &= LS_{\lambda=7}(f_{\mathbf{X}}) \cup \{[5,6]\} \\
LS_{\lambda=9}(f_{\mathbf{X}}) &= LS_{\lambda=8}(f_{\mathbf{X}}) \cup \{[2,3],[3,4],[4,5]\} \\
LS_{\lambda=10}(f_{\mathbf{X}}) &= LS_{\lambda=9}(f_{\mathbf{X}}) \cup \{[6,7],[7,8]\} \\
LS_{\lambda=11}(f_{\mathbf{X}}) &= LS_{\lambda=10}(f_{\mathbf{X}}) \\
LS_{\lambda=12}(f_{\mathbf{X}}) &= LS_{\lambda=11}(f_{\mathbf{X}}) \\
LS_{\lambda=13}(f_{\mathbf{X}}) &= LS_{\lambda=11}(f_{\mathbf{X}}) \\
LS_{\lambda=14}(f_{\mathbf{X}}) &= LS_{\lambda=11}(f_{\mathbf{X}}) \cup \{[0,1]\}.
\end{aligned}
$$

This allows us to compute the available features for each λ-value:

	$\lambda=7$	$\lambda=8$	$\lambda=9$	$\lambda=10$	$\lambda=11$	$\lambda=12$	$\lambda=13$	$\lambda=14$
$\mathfrak{F}(LS_\lambda(f_{\mathbf{X}}))$	$\varnothing$	$[5,6]$	$[2,6]$	$[2,8]$	$[2,8]$	$[2,8]$	$[2,8]$	$[0,8]$

Let $\mathfrak{F}(f_{\mathbf{X}})$ be the whole set of features for $f_{\mathbf{X}}$, that is,

$$\mathfrak{F}(f_{\mathbf{X}}) = \{\mathbb{I} : \mathbb{I} \in \mathfrak{F}(LS_\lambda(f_{\mathbf{X}})) \text{ for some } \lambda_{\min} \le \lambda \le \lambda_{\max}\}.$$

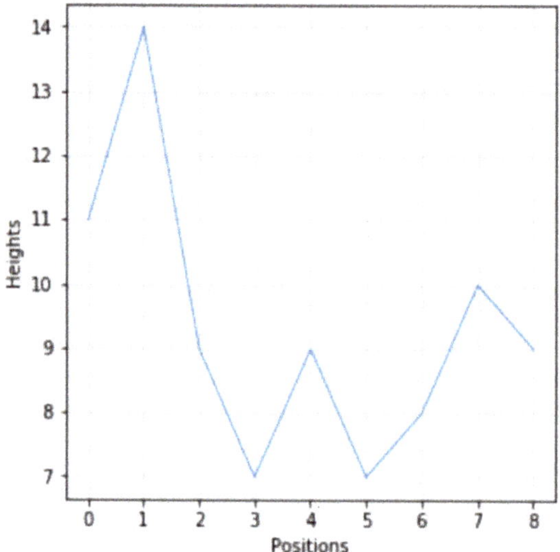

Figure 6. The map $f_\mathbf{X}$ for $\mathbf{X} = (11, 14, 9, 7, 9, 7, 8, 10, 9)$.

Example 2. *From Example 1, we obtain*

$$\mathfrak{F}(f_\mathbf{X}) = \{[5,6], [2,6], [2,8], [0,8]\}.$$

We can represent the map $\lambda \mapsto LS_\lambda(f_\mathbf{X})$ from $[\lambda_{\min}, \lambda_{\max}]$ to $\mathfrak{F}(f_\mathbf{X})$ as shown in Figure 7.

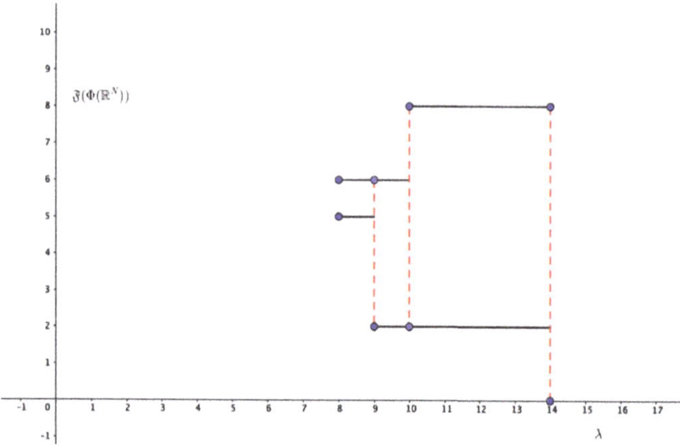

Figure 7. The map $\lambda \mapsto LS_\lambda(f_\mathbf{X})$ for $\mathbf{X} = (11, 14, 9, 7, 9, 7, 8, 10, 9)$.

Let $\mathbb{I} \in \mathfrak{F}(f_\mathbf{X})$; in order to quantify the persistence of this particular feature for the map $f_\mathbf{X}$, we use the map $\lambda \mapsto LS_\lambda(f_\mathbf{X})$ from $[\lambda_{\min}, \lambda_{\max}]$ to $\mathfrak{F}(f_\mathbf{X})$. To this end, we introduce the following definition: the birth point of the feature $\mathbb{I}$ is defined by

$$a(\mathbb{I}) = \inf\{\lambda : \mathbb{I} \in \mathfrak{F}(LS_\lambda(f_\mathbf{X}))\}$$

and the corresponding death point by

$$b(\mathbb{I}) = \sup\{\lambda : \mathbb{I} \in \mathfrak{F}(LS_\lambda(f_{\mathbf{X}}))\}.$$

In particular, we note that $a([0, N-1]) = \lambda_{\max}$ (see Figure 7). Since $a(\mathbb{I}) \leq b(\mathbb{I}) < \infty$ holds for all $\mathbb{I} \in \mathfrak{F}(f_{\mathbf{X}})$, $\mathbb{I} \neq [0, N-1]$, we call the finite interval $[a(\mathbb{I}), b(\mathbb{I})]$ *the barcode of the feature* $\mathbb{I} \in \mathfrak{F}(f_{\mathbf{X}}) \setminus \{[0, N-1]\}$.

Example 3. *From Example 1 we consider the features* $[5, 6] \in LS_{\lambda=8}(f_{\mathbf{X}})$, $[2, 6] \in LS_{\lambda=9}(f_{\mathbf{X}})$, *and* $[2, 8] \in LS_{\lambda=10}(f_{\mathbf{X}})$. *Then, the feature* $[5, 6]$ *has its birth point at* $a([5, 6]) = 8$ *and its death point at* $b([5, 6]) = 9$; *the feature* $[2, 6]$ *has its birth point at* $a([2, 6]) = 9$ *and its death point at* $b([2, 6]) = 10$. *Finally, the feature* $[2, 8]$ *has its birth point at* $a([2, 8]) = 10$ *and its death point at* $b([2, 8]) = 14$. *As a consequence, the set*

$$\mathcal{B}(f_{\mathbf{X}}) := \{([5, 6]; 8, 9), ([2, 6]; 9, 10), ([2, 8]; 10, 14)\}$$

of features and its corresponding barcodes contain the relevant information of the shape of $f_{\mathbf{X}}$ *(see Figure 7).*

Thus, we define the set of barcodes for $f_{\mathbf{X}}$ by

$$\mathcal{B}(f_{\mathbf{X}}) = \{(\mathbb{I}; a(\mathbb{I}), b(\mathbb{I})) : \mathbb{I} \in \mathfrak{F}(f_{\mathbf{X}}) \setminus \{[0, N-1]\}\}$$

and its persistence diagram as

$$\mathcal{PD}(f_{\mathbf{X}}) := \left\{(a(\mathbb{I}), b(\mathbb{I})) \in \mathbb{R}^2 : \mathbb{I} \in \mathfrak{F}(f_{\mathbf{X}}) \setminus \{[0, N-1]\}\right\}$$

(see Figure 8). An equivalent representation of the persistence diagram is the life-time diagram for $f_{\mathbf{X}}$, which is constructed by means of a bijective transformation $T(a, b) = (a, b - a)$, acting over $\mathcal{PD}(f_{\mathbf{X}})$, that is,

$$\mathcal{LT}(f_{\mathbf{X}}) := \left\{(a(\mathbb{I}), b(\mathbb{I}) - a(\mathbb{I})) \in \mathbb{R}^2 : \mathbb{I} \in \mathfrak{F}(f_{\mathbf{X}})) \setminus \{[0, N-1]\}\right\};$$

see Figure 9.

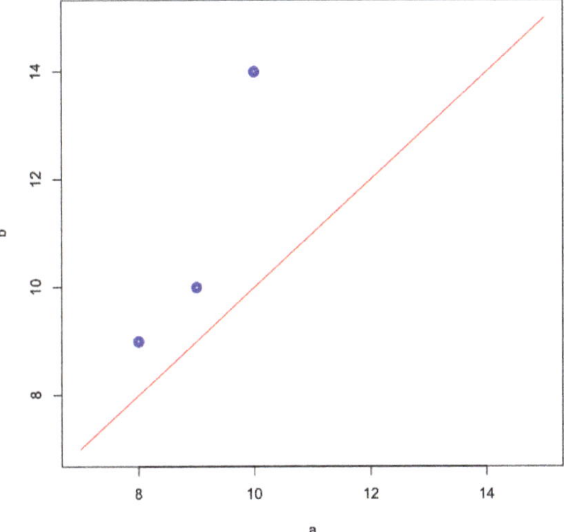

Figure 8. Persistence diagram for the map $f_{\mathbf{X}}$ when $\mathbf{X} = (11, 14, 9, 7, 9, 7, 8, 10, 9)$.

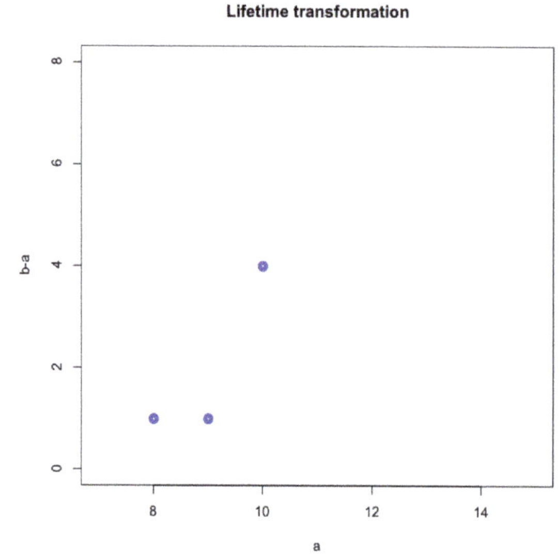

Figure 9. Life-time diagram for the map $f_\mathbf{X}$ when $\mathbf{X} = (11, 14, 9, 7, 9, 7, 8, 10, 9)$.

In order to determine the grade of similarity between two barcodes from two different time series, we need to set a similarity metric. To this end, we construct the persistent image for $f_\mathbf{X}$ as follows: we observe that $\mathcal{LT}(f_\mathbf{X})$ is a finite set of points, namely,

$$\mathcal{LT}(f_\mathbf{X}) = \{(a_1, b_1 - a_1), \ldots, (a_k, b_k - a_k)\}$$

for some natural numbers $k \geq 1$ and such that $b_1 - a_1 \leq b_2 - a_2 \ldots \leq b_k - a_k$. Then, we consider a non-negative weighting function $w : \mathcal{LT}(f_\mathbf{X}) \longrightarrow [0, 1]$ given by

$$w(a_i, b_i - a_i) = \frac{b_i - a_i}{b_k - a_k} \text{ for } 1 \leq i \leq k.$$

Finally, we fix M, a natural number, and take a bivariate normal distribution $g_u(x, y)$ centered at each point $u \in \mathcal{LT}(f_\mathbf{X})$ and with its variance $\sigma\, id = \frac{1}{M} \max_{1 \leq i \leq k}(b_i - a_i)\, id$, where id is the 2×2 identity matrix. A persistence kernel is then defined by means of a function $\rho_\mathbf{X} : \mathbb{R}^2 \to \mathbb{R}$, where

$$\rho_\mathbf{X}(x, y) = \sum_{u \in \mathcal{LT}(f_\mathbf{X})} w(u) g_u(x, y). \tag{1}$$

We associate with each $\mathbf{X} \in \mathbb{R}$ a matrix in $\mathbb{R}^{M \times M}$ as follows: let $\varepsilon > 0$ be a non-negative real number that is sufficiently small, and then consider a square region $\Omega_{\mathbf{X}, \varepsilon} = [\alpha, \beta] \times [\alpha^*, \beta^*] \subset \mathbb{R}^2$, covering the support of $\rho_\mathbf{X}(x, y)$ (up to a certain precision), such that

$$\iint_{\Omega_{\mathbf{X}, \varepsilon}} \rho_\mathbf{X}(x, y)\, dx\, dy \geq 1 - \varepsilon$$

holds. Next, we consider two equispaced partitions of the intervals

$$\alpha = p_0 \leq p_1 \ldots \leq p_M = \beta \text{ and } \alpha^* = p_0^* \leq p_1^* \ldots \leq p_M^* = \beta^*.$$

Now, we put

$$\Omega_{\mathbf{X},\varepsilon} = \bigcup_{i=0}^{M-1} \bigcup_{j=0}^{M-1} [p_i, p_{i+1}] \times [p_j^*, p_{j+1}^*] = \bigcup_{i=0}^{M-1} \bigcup_{j=0}^{M-1} P_{i,j}$$

The persistence image of $\mathbf{X}$ associated with the partition $\mathcal{P} = \{P_{i,j}\}$ is then described by the matrix given by the following equation:

$$PI(\mathbf{X}, M, \mathcal{P}, \varepsilon) = \left(\iint_{P_{i,j}} \rho_{\mathbf{X}}(x,y)dxdy \right)_{i=0,j=0}^{i=M-1,j=M-1} \in \mathbb{R}^{M \times M}. \tag{2}$$

2.5. Classification

Image classification is a procedure that is used to automatically categorize images into classes by assigning to each image a label representative of its class. A supervised classification algorithm requires a training sample for each class, that is, a collection of data points whose class of interest is known. Labels are assigned to each class of interest. The classification problem applied to a new observation is thus based on how close a new point is to each training sample. The Euclidean distance is the most common distance metric used in low-dimensional datasets. The training samples are representative of the known classes of interest to the analyst. In order to classify the persistence diagrams, we can use any state-of-the-art technique. In our case, we considered the random forest classification.

Recall that we conducted 9 different experiments, with 24 samples associated with each one of them corresponding to 3 samples for each of the different experimental conditions: relaxed rigid driver, relaxed rigid passenger, relaxed SAV driver, relaxed SAV passenger, tense rigid driver, tense rigid passenger, tense SAV driver, and tense SAV passenger. Their respective labels are $\{0,0,0,0,1,1,1,1\}$. Therefore, we designed the following training validation process: The model is trained over 144 samples and evaluated over the remaining unseen 72 experiments (two-to-one training-to-testing ratio). The split between training and sampling is achieved using random shuffling and stratification to ensure balance between the classes. In order to improve the evaluation of the model generalizability, we also performed a cross-validation procedure following a *leave-one-out* strategy, consisting of iteratively training over the full dataset except one sample that was left out and used to test and score the model. We used the *accuracy* metric to evaluate the classification model. We can represent the performance of the model using the so-called confusion matrix: a 2D entries table where elements account for the number of samples in each category, with the first axis representing the true labels and the second axis the predicted labels. We also computed the different classification metrics to obtain a more detailed reporting of the model performances.

3. Results

The trained random forest classifier model for the persistence images has a notably high accuracy score on the training dataset (144) for both approaches and high accuracy for the testing dataset (72 samples). This suggests strong differentiation of the images with the respect to their generating signals, see Figure 10. The scores on the training and testing are 93 and 83%, respectively. The leave-one-out cross-validation achieved a score of 81%, indicating a good variance–bias trade-off and good generalization potential of the model.

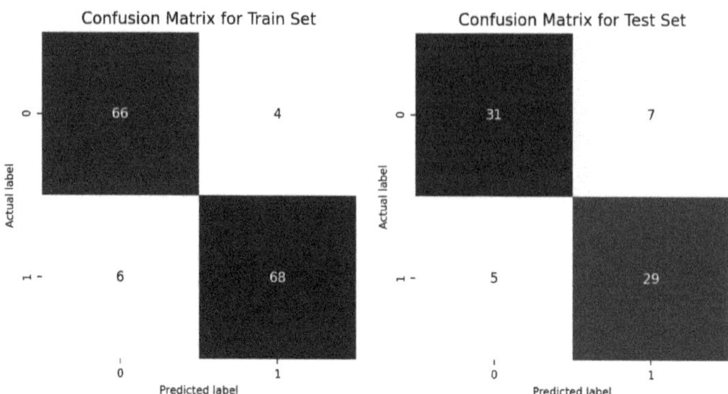

Figure 10. Model performance for prediciting the attention state.

4. Discussion

The combination of Morse theory and topological data analysis allows us to extract information from real data without the need for smoothness or regularity assumption on the time series. In our case, input data for each experiment were reduced from six-sensor time series of measurements to one single image containing the persistent pattern for attention to the road. Using the obtained persistence images as the new inputs, supervised learning proved to successfully predict the attention state of the driver or passenger.

The procedure used and described in this paper does not involve any additional pre-processing of the sensor data; is robust to noise and degraded signals; and supports large quantities of data, which makes it efficient and scalable.

It is important to highlight the fact that while the proposed methodology based on the TDA (successfully applied in large datasets [9]) seems general and powerful and it was able to extract the main data features, the validity of the driver behaviors observed in the analyzed dataset should be carefully checked due to the overly reduced dataset employed (limited to nine individuals) that does not allow for the full validation of prediction robustness.

Author Contributions: Conceptualization, F.C., H.Y.C. and M.H.; data curation, T.F., A.B. and M.H.; formal analysis, F.C., A.F., A.B., E.C., H.Y.C. and J.-L.D.; funding acquisition, F.C.; investigation, T.F., A.F. and M.H.; methodology, F.C., A.F., A.B. and H.Y.C.; software, A.B.; supervision, F.C., E.C. and H.Y.C.; writing—original draft, T.F.; writing—review editing, A.B. and E.C. All authors have read and agreed to the published version of the manuscript.

Funding: This research was funded by ESI Group, Contract 2019-0060 with the University of Zaragoza.

Institutional Review Board Statement: Ethical review and approval were waived for this study due to the research presents no risk of harm to subjects and only collects non-personalized anonymized data.

Informed Consent Statement: Informed consent was obtained from all subjects involved in the study.

Data Availability Statement: Data are available under request.

Conflicts of Interest: The authors declare no conflict of interest. The funders had no role in the design of the study; in the collection, analyses, or interpretation of data; in the writing of the manuscript, or in the decision to publish the results.

Appendix A

We can illustrate the process of computing the persistence diagrams, the lifetime diagrams, and the persistence images for the driver time series for each experimental setup:

1. Relaxed driver with SAV seat;
2. Relaxed driver with rigid seat;
3. Relaxed passenger with SAV seat;
4. Relaxed passenger with rigid seat;
5. Tense driver with SAV seat;
6. Tense driver with rigid seat;
7. Tense passenger with SAV seat;
8. Tense passenger with rigid seat.

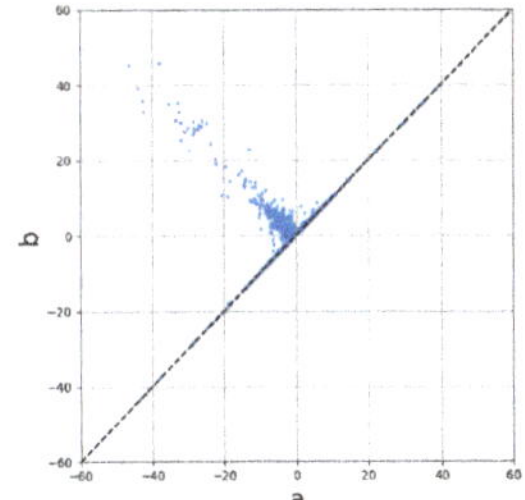 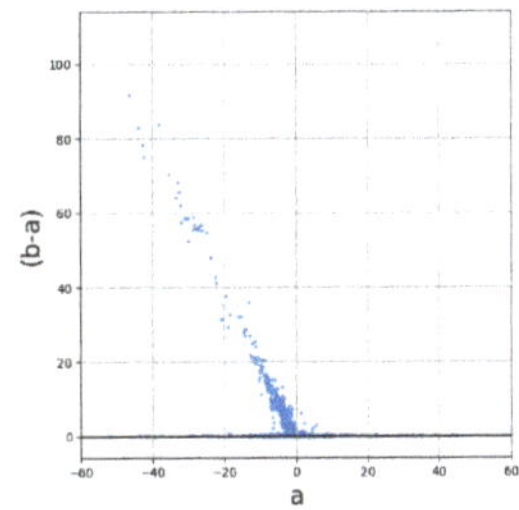 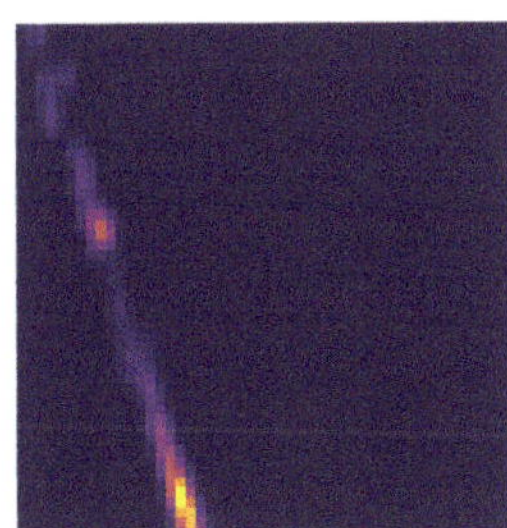

Figure A1. Relaxed driver with SAV seat.

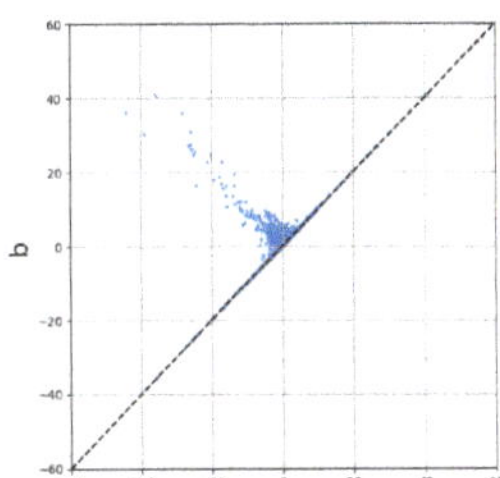 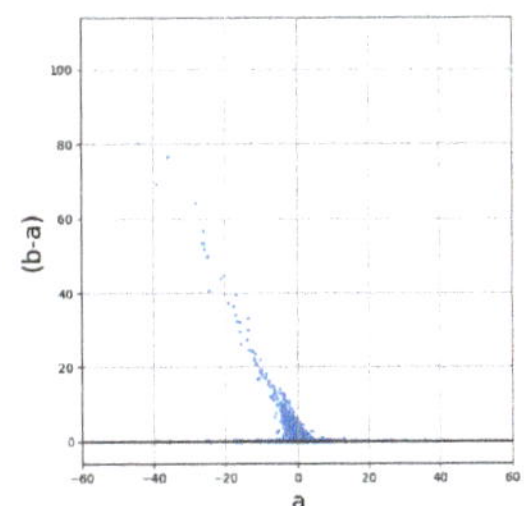 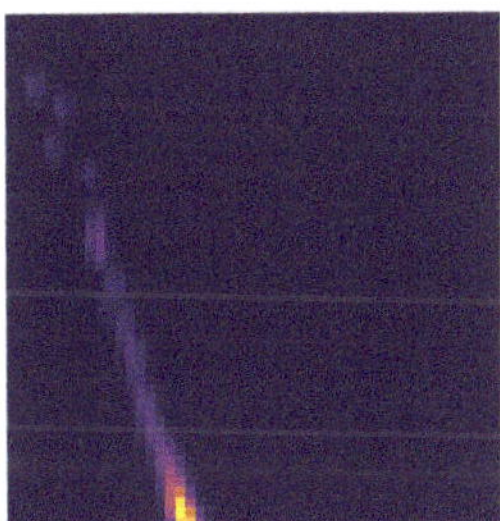

Figure A2. Relaxed driver with rigid seat.

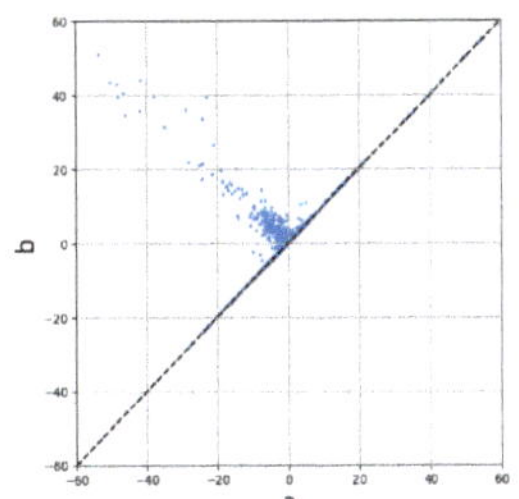 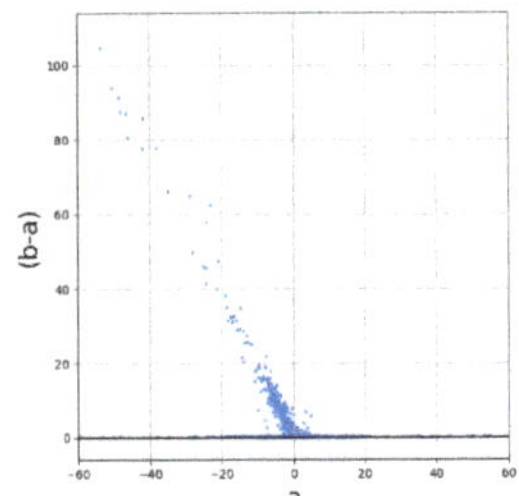 

Figure A3. Relaxed passenger with SAV seat.

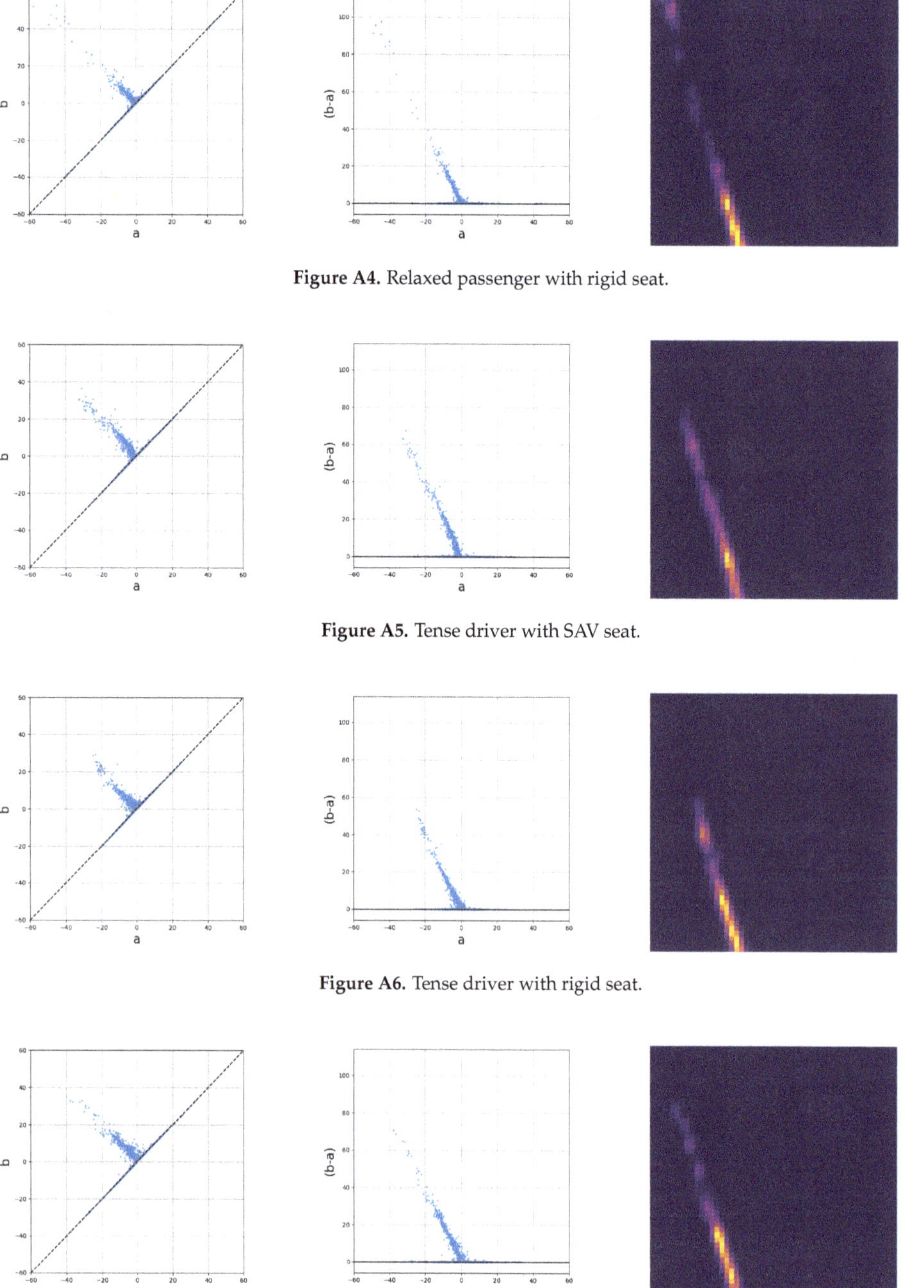

Figure A4. Relaxed passenger with rigid seat.

Figure A5. Tense driver with SAV seat.

Figure A6. Tense driver with rigid seat.

Figure A7. Tense passenger with SAV seat.

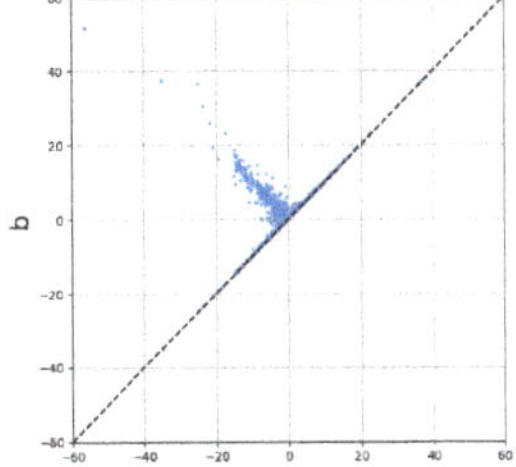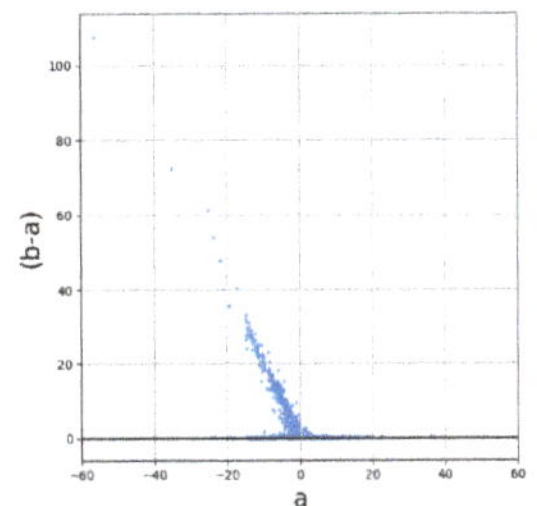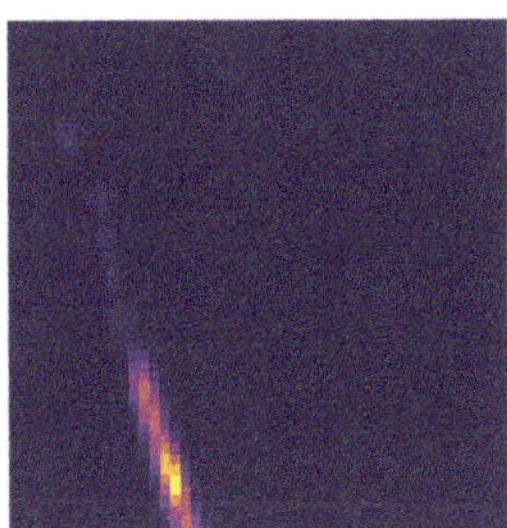

Figure A8. Tense passenger with rigid seat.

Appendix B

To better evaluate a classification model, we are interested in quantities that express how often a sample is correctly or wrongly labelled into a particular class over all the samples and all the classes:

- A *True positive* (TP): the correct prediction of a sample into a class;
- A *True negative* (TN): the correct prediction of a sample out of a class;
- A *False positive* (FP): the incorrect prediction of a sample into a class;
- A *False negative* (FN): the incorrect prediction of a sample out of class.

Therefore, we can examine in more detail the classification model performance using the following metrics:

- The *precision* P is the number of correct positive results divided by the number of all positive results.

$$P = \frac{TP}{TP + FP} \tag{A1}$$

- The *recall* R is the number of correct positive results divided by the number of all relevant samples.

$$R = \frac{TP}{TP + FN} \tag{A2}$$

- The F-1 score is the harmonic mean of precision and recall.

$$F1 = 2 \times \frac{P \times R}{P + R} \tag{A3}$$

- The *accuracy* A is the number of correct predictions over the number of all samples.

$$A = \frac{TP + TN}{TP + TN + FP + FN} \tag{A4}$$

We can summarize the presented metrics for our model in the following two reports:

	precision	recall	f1-score	support
0	0.94	0.92	0.93	72
1	0.92	0.94	0.93	72
accuracy			0.93	144
macro avg	0.93	0.93	0.93	144
weighted avg	0.93	0.93	0.93	144

(a) Training set.

	precision	recall	f1-score	support
0	0.82	0.86	0.84	36
1	0.85	0.81	0.83	36
accuracy			0.83	72
macro avg	0.83	0.83	0.83	72
weighted avg	0.83	0.83	0.83	72

(b) Testing set.

Figure A9. Classification report.

References

1. Paul, A.; Chauhan, R.; Srivastava, R.; Baruah, M. *Advanced Driver Assistance Systems (No. 2016-28-0223)*; SAE Technical Paper; SAE International: Warrendale, PA, USA, 2016.
2. Shaout, A.; Colella, D.; Awad, S.S. Advanced Driver Assistance Systems-Past, Present and Future. In Proceedings of the 2011 Seventh International Computer Engineering Conference (ICENCO'2011), Cairo, Egypt, 27–28 December 2011; pp. 72–82.
3. Geronimo, D.; Lopez, A.M.; Sappa, A.D.; Graf, T. Survey of pedestrian detection for advanced driver assistance systems. *IEEE Trans. Pattern Anal. Mach. Intell.* **2009**, *32*, 1239–1258. [CrossRef]
4. Lindgren, A.; Chen, F. State of the art analysis: An overview of advanced driver assistance systems (adas) and possible human factors issues. *Hum. Factors Econ. Asp. Saf.* **2006**, *38*, 50.
5. Izquierdo-Reyes, J.; Ramirez-Mendoza, R.A.; Bustamante-Bello, M.R. A study of the effects of advanced driver assistance systems alerts on driver performance. *Int. J. Interact. Des. Manuf.* **2018**, *12*, 263–272. [CrossRef]
6. Sikander, G.; Anwar, S. Driver fatigue detection systems: A review. *IEEE Trans. Intell. Transp. Syst.* **2018**, *20*, 2339–2352. [CrossRef]
7. Milnor, J. *Morse Theory*; Princeton University Press: Princeton, NJ, USA, 1963.
8. Epstein, C.; Carlsson, G.; Edelsbrunner, H. Topological data analysis. *Inverse Probl.* **2011**, *27*, 120201.
9. Frahi, T.; Argerich, C.; Youn, M.; Chinesta, F.; Falco, A. Tape surfaces characterization with persistence images. *AIMS Mater. Sci.* **2020**, *7*, 364–380. [CrossRef]
10. Milnor, J.W.; Thurston, W. On iterated maps of the interval, Dynamical systems (College Park, MD, 1986–87). In *Lecture Notes in Mathematics*; Springer: Berlin/Heidelberg, Germany, 1988; pp. 465–563.
11. Reeb, G. Sur les points singuliers d'une forme de Pfaff complètement intégrable ou d'une fonction numérique. *C. R. Acad. Sci. Paris* **1946**, *222*, 847–849.

 *mathematics*

Article

Impact of Malapposed and Overlapping Stents on Hemodynamics: A 2D Parametric Computational Fluid Dynamics Study

Manuel Lagache [1,2,3], **Ricardo Coppel** [1,2], **Gérard Finet** [4], **François Derimay** [4], **Roderic I. Pettigrew** [5], **Jacques Ohayon** [1,3] and **Mauro Malvè** [6,7,*]

1   Laboratory SYMME, Savoie Mont-Blanc University, 73000 Chambéry, France; manuel.lagache@univ-smb.fr (M.L.); ricardocvmty@gmail.com (R.C.); jacques.ohayon@univ-smb.fr (J.O.)
2   Savoie Mont-Blanc University, Polytech Annecy-Chambéry, 73370 Le Bourget du Lac, France
3   Laboratory TIMC-IMAG, CNRS, UMR5525, Grenoble-Alpes University, 38400 Grenoble, France
4   Department of Hemodynamics and Interventional Cardiology, Hospices Civils de Lyon and Claude Bernard University Lyon 1, INSERM Unit 886, 69622 Lyon, France; gerard.finet@univ-lyon1.fr (G.F.); Francois.derimay@chu-lyon.fr (F.D.)
5   CEO, Engineering Health (EnHealth) and Executive Dean, Engineering Medicine (EnMed), Texas A&M University and Houston Methodist Hospital, Houston, TX 77030, USA; pettigrew@tamu.edu
6   Department of Engineering, Public University of Navarra, 31006 Pamplona, Spain
7   CIBER-BBN, Research Networking in Bioengineering, Biomaterials & Nanomedicine, 50018 Zaragoza, Spain
*   Correspondence: mauro.malve@unavarra.es

Citation: Lagache, M.; Coppel, R.; Finet, G.; Derimay, F.; Pettigrew, R.I.; Ohayon, J.; Malvè, M. Impact of Malapposed and Overlapping Stents on Hemodynamics: A 2D Parametric Computational Fluid Dynamics Study. *Mathematics* **2021**, *9*, 795. https://doi.org/10.3390/math9080795

Academic Editor: Omer San

Received: 10 February 2021
Accepted: 1 April 2021
Published: 7 April 2021

Abstract: Despite significant progress, malapposed or overlapped stents are a complication that affects daily percutaneous coronary intervention (PCI) procedures. These malapposed stents affect blood flow and create a micro re-circulatory environment. These disturbances are often associated with a change in Wall Shear Stress (WSS), Time-averaged WSS (TAWSS), relative residence time (RRT) and oscillatory character of WSS and disrupt the delicate balance of vascular biology, providing a possible source of thrombosis and restenosis. In this study, 2D axisymmetric parametric computational fluid dynamics (CFD) simulations were performed to systematically analyze the hemodynamic effects of malapposition and stent overlap for two types of stents (drug-eluting stent and a bioresorbable stent). The results of the modeling are mainly analyzed using streamlines, TAWSS, oscillatory shear index (OSI) and RRT. The risks of restenosis and thrombus are evaluated according to commonly accepted thresholds for TAWSS and OSI. The small malapposition distances (MD) cause both low TAWSS and high OSI, which are potential adverse outcomes. The region of low OSI decrease with MD. Overlap configurations produce areas with low WSS and high OSI. The affected lengths are relatively insensitive to the overlap distance. The effects of strut size are even more sensitive and adverse for overlap configurations compared to a well-applied stent.

Keywords: hemodynamics; overlap; malapposition; stent; stenosis; thrombosis

1. Introduction

Percutaneous coronary intervention (PCI) with modern drug-eluting stents (DES) has revolutionized the treatment of arterial diseases. However, their benefits could be compromised by potential complications such as restenosis and thrombosis [1–4]. Complications of PCI continue to be a concern, with approximately 1–2% of stent patients dying from thrombotic occlusions and 10–15% requiring additional interventions due to restenosis [3–7]. The deployment of a coronary stent near a complex atherosclerotic lesion (i.e., located close to a bifurcation, near concomitant lesions or with eccentric plaque formation) may promote the occurrence of gaps between the vessel wall and the struts, defined as malapposition distance (MD) [8–10]. It appears in up to 33% of implanted DES and up to 75% of patients with very late (i.e., >1 year) stent thrombosis [10–12]. Furthermore, the use of two partially overlapped stents (i.e., with a certain overlap distance (OD)) may be necessary in the event of incomplete coverage of lesions with a single stent. Approximately 30% of PCI patients requiring stenting are in situations where stents overlap [7].

Stent implantation by itself generates geometric irregularities on the vascular walls that modify the hemodynamics along the entire length of the stent [13–15]. Local hemodynamic parameters, in particular abnormal wall shear stress (WSS), critically affect the evolution of atherosclerosis plaque and have been associated with an increased risk of thrombosis or restenosis [1,2,5,6,16]. Moreover, stent malapposition and overlapping have been associated with hemodynamic disturbances that may increase the risks of adverse clinical outcomes [2,8,10,17–22].

Several studies have been performed to analyze how the presence of stents perturbs the hemodynamics in a vessel. Computational fluid dynamics (CFD) calculations performed on simplified stent models inside an idealized coronary artery [5,6,8,10,22] have been used to investigate the effects of malapposed struts on the blood flow. These studies have shown that, as the MD increases, recirculation regions located near the wall tend to grow downstream from the malapposed struts until a critical MD is reached. Above this MD threshold, recirculation regions gradually reduce in size until the interaction between the wall and the misaligned struts disappears. Moreover, it has been reported that the regions near the malapposed struts (more specifically at the gaps between the wall and the struts) are subjected to high wall shear stress [5,6,10] and that the abnormal region tends to increase with MD [6]. While several studies have been conducted to highlight the hemodynamic perturbations induced by malapposed stents, few are devoted to studying overlapping configurations. An in vitro study by [18], using the particle image velocimetry technique with a vascular phantom under physiological flow conditions, showed that overlapping sections tend to disrupt the flow and create a WSS deficiency. [17,19] obtained similar conclusions after performing 3D CFD simulations based on realistic artery-stents geometries reconstructed from computed tomography images. Additionally, a 2D CFD study by [22] revealed also that strut overlapping increases the amount of flow recirculation compared to non-overlapped segments. Moreover, congruent struts (i.e., one strut on top of the other) have been identified as critical configurations due to the major flow disturbance that they produce [22] and the important drop of WSS around them [17].

Although complex 3D studies have provided promising results, identifying areas where malapposed and/or overlapping struts can lead to the development of abnormal WSS and significantly disturb blood flow on patient-specific geometries, no practical information has been given yet to clinicians to assist them in their choice when a stent is incorrectly positioned.

This lack of knowledge is mainly due to the difficulty: (1) to model complex configurations due to the computation times required (often incompatible with operational workflow) and (2) to identify the disturbance effect of each of the different parameters separately. Using static CFD models based on 2D geometries, researchers have found the location and general tendency of disturbed flow regions for different degrees of strut malapposition and overlapping [6,10,22]. However, to our knowledge, a complete parametric study designed to highlight critical malapposition and overlapping stent configurations, for which flow disturbance over a cardiac cycle becomes significant, has never been conducted.

Therefore, in this study, we designed and used two-dimensional parametric CFD models to investigate the hemodynamic effect of several stent malapposition and overlapping configurations while considering (1) pulsatile flow conditions, (2) the non-Newtonian nature of blood flow and (3) the most commonly used flow-related indices to assess the hemodynamic impact on the vessel wall over a cardiac cycle.

The use of 2D axisymmetric models requires much lower computational costs and enables a systematic evaluation of the effect of all relevant geometrical parameters (malapposition and overlapping distances) by means of several sequential simulations. The identification of specific critical configurations that can promote restenosis and thrombosis and clearly describe the temporal and spatial evolution of hemodynamic disturbances remain the main objectives of this study. The present study is complementary to the patient-specific 3D studies by providing general information and cut-off values for overlapping and malapposed distances. The purpose of this work is to provide general criteria on the effect of misalignment and overlapping distances and not patient-specific information (which must be assessed for each patient). The results could provide a new decision-making tool for cardiologists by predicting the risks of complications related to malapposition and overlapping.

2. Materials and Methods

Two distinct CFD Models of stented segments of coronary arteries were considered. The first one mimics several cases of a single malapposed stent, while the second account for overlapping stent configurations. Furthermore, all the boundary conditions and constitutive laws used in these models will be described in the following sections, as well as the chosen hemodynamic metrics to evaluate the impact of each configuration on the vessel wall.

2.1. Parametric CFD Models of Malapposed and Overlapped Stents

2.1.1. Geometries

Hemodynamic disturbances produced by the stent struts near the arterial wall were studied numerically with the CFD approach. Moreover, 2D axisymmetric geometries modeling the two "stent-artery" configurations of interest (i.e., malapposition and overlapping) were used to perform dynamic flow analyses (Figure 1A,B). The geometrical parameters used to define and design the idealized "stent-artery" were as follows: malapposition and overlapping distances (i.e., MD and OD) for the two models of interest. Realistic diversity in clinical data was investigated by varying MD $0~\mu m \leq MD \leq 450~\mu m$ and OD (in the overlapping range of 2 to 3 struts). Moreover, two strut sizes were considered in this study. The first one is a Cobalt-Chromium drug-eluting stent (CC-DES) (Synergy, Boston Scientific, Marlborough, MA, USA) with a section of $85~\mu m \times 90~\mu m$ (i.e., Height (H) $\times$ Width (W)). The second one is a bioresorbable stent (BVS, Abbot, Abbott Park, IL, USA) with a thicker section of $150~\mu m \times 215~\mu m$. The diameter of the artery lumen was taken as 3 mm. Once deployed, the inter-strut distance is 2 mm for both types of stents. Therefore, the total length of the stents is 10.54 mm and 11.29 mm for CC-DES and BVS respectively. The insertion of the apposed struts in the arterial wall has also been considered with a strut indentation of $0.15 * H + 20~\mu m$ (see Figure 1A inner box). A luminal protrusion caused by the stent struts in contact with the arterial walls is also introduced to each model with the aim of estimating with more accuracy the flow recirculation due to the presence of the struts and the associated hemodynamics variables (see Figure 1A,B). The latter could be in fact slightly underestimated, if the struts are just apposed on the arterial walls.

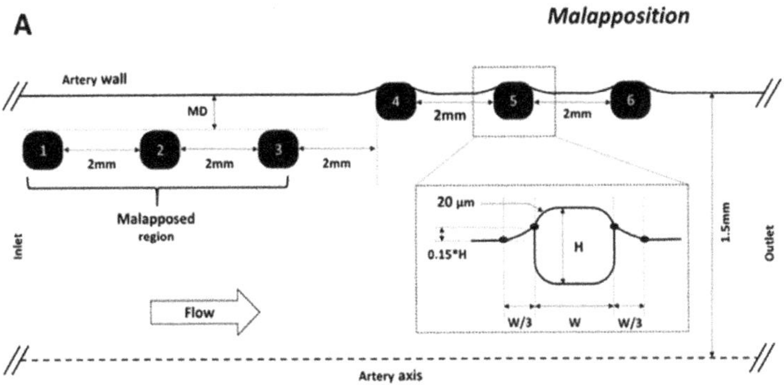

Figure 1. *Cont.*

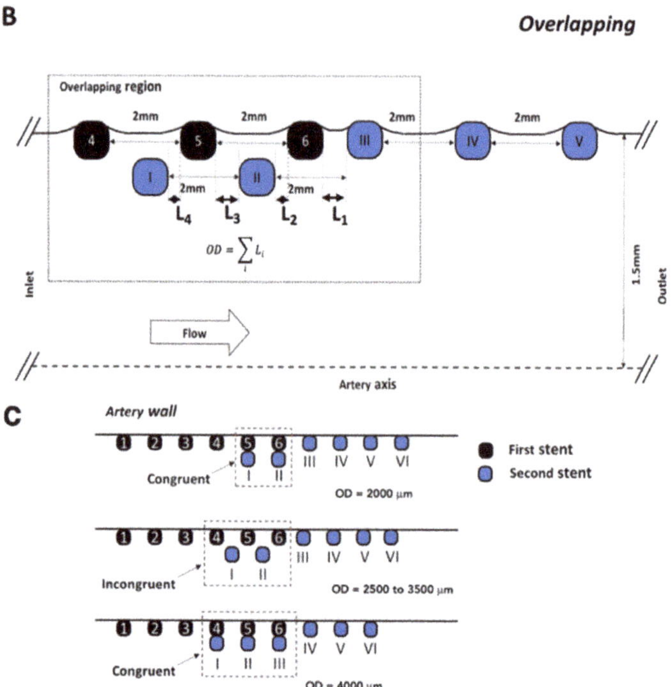

Figure 1. (**A**) Single stent geometry with three malapposed struts (# 1, 2, and 3) and three correctly apposed struts (# 4, 5, and 6) and (**B**) two overlapping stents (first stent: struts # 1 to 6, second stent: struts # I to (VI) and (**C**) 3 overlapping configurations, with two congruent struts, with incongruent struts and three congruent struts.

2.1.2. Studied Malapposition Configurations

This strut configuration is displayed in Figure 1A. It illustrates one stent with three misaligned struts and three correctly apposed ones. The MD of the first three struts was between 0 and 450 μm for both stents (i.e., CC-DES and BVS). The three apposed struts were placed downstream and remained fixed for all simulations. The following configurations were studied: for the CC-DES: MD = 0, 40, 60, 80, 115, 130, 150, 180, 225, 300 and 450 μm (i.e., $n = 11$ cases), and for the BVS: MD = 0, 40, 80, 115, 150, 180, 225, 300 and 450 μm (i.e., $n = 9$ cases). When MD = 0 μm the six struts are correctly apposed (i.e., total stented artery length of 10.54 and 11.29 mm for CC-DES and BVS, respectively), these two specific cases (one for each stent) will be considered as the optimal clinical configurations. A total of 20 distinct configurations were studied.

2.1.3. Studied Overlapping Configurations

This strut configuration is illustrated in Figure 1B,C. This configuration corresponds to the partial overlapping of 2 stents. This overlapping is for example used to treat arteries with multistenosis, bifurcations... The OD was between 2000 μm + 2W (strut width) and 4000 μm + 3W (strut width). For reasons of simplicity and homogeneity between the two stents, these distances will be named without considering the width of the struts (different for the two stents). Two types of geometric configurations were considered, congruent and incongruent struts [17,22]. In the first case, the well-apposed and overlapping struts are stacked one on top of the other, forming a higher obstacle. In the second case, both struts are separated, leaving a gap between the overlapped one and the vessel wall (see Figure 1B). For the two stent types, simulations were performed for the following configurations: OD = 2000, 2500, 3000, 3500 and 4000 μm ($n = 2 \times 5$ cases). Notice that when

OD = 2000 μm there are two pairs of congruent struts and when OD = 4000 μm there are three pairs of congruent struts and for intermediate cases (i.e., from OD = 2500 to 3500 μm) there are incongruent struts at the overlapping section. A total of 10 distinct simulations were performed.

2.1.4. Constitutive Law

Blood density was assumed to be constant with a value of 1060 kg m^{-3}. The non-Newtonian nature of blood flow was taken into account using the Carreau–Yasuda model [19]:

$$\mu = \mu_\infty + (\mu_0 - \mu_\infty)[1 + (\lambda\,\dot{S})^2]^{(m-1)/2} \tag{1}$$

where μ is the dynamic viscosity, μ_0 and μ_∞ are the viscosity values at zero and infinity shear rate, respectively, $\dot{S}$ is the shear rate, λ is the time constant, and m is the power-law index. These fluid constants values were taken from [19] and are used in [20] among other studies: μ_∞ = 0.0035 Pa s, μ_0 = 0.25 Pa s, λ = 25 s and m = 1/4.

2.1.5. Boundary Conditions

The blood flow was considered laminar and unsteady. The Reynolds number (Re) for the present simulations was 252.

The artery wall and struts were assumed to be rigid with the no-slip condition and a time-dependent velocity profile (see Figure 2) was applied at the inlet of the axisymmetric domain to mimic the pulsatile behavior of coronary blood flow (with a time period equal to 0.908 s). The physiological waveform was adapted from [23]. The same velocity profile was used for the outlet to guarantee mass conservation. The inlet and outlet regions were extended about six times the radius of the artery. These lengths were chosen with precision after the first series of simulations that proved that these extensions were sufficient to provide a fully developed flow. Additionally, in order to minimize the effect of the initial transients, two complete cardiac cycles were simulated for all the configurations studied and only the results of the second cycle were considered. A third cycle was performed as a test-case with a malapposition configuration with MD = 150 μm. However, no significant difference in velocity responses was found compared to the results from the second cycle. These different tests ensured the quality of the results.

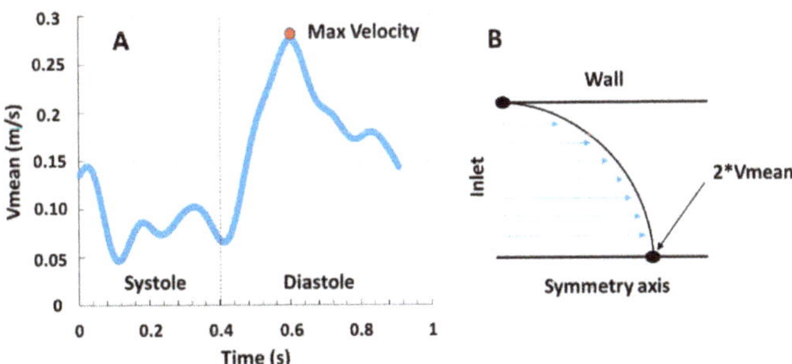

Figure 2. (**A**) Physiological velocity waveform adapted from Davies et al. (2006) and (**B**) Velocity profile imposed at the inlet.

2.1.6. Computational Fluid Dynamics Simulations

An APDL program file (ANSYS v19.2, ANSYS Inc., Canonsburg, PA, USA) was developed for the two studied configurations (malapposition and overlapping) in order to simplify the parameterization. The finite element problems are then generated and solved automatically after the selection of the different parameters. This parametric program allows easy utilization (for a stent designer, a clinician...). ANSYS FLUENT was used to mesh the fluid domain with hexahedral and triangular elements and calculate the velocity and pressure distributions. Mesh refinement was performed on the regions around the struts and vessel walls to improve the accuracy of the computations.

The malapposition configuration with MD = 150 μm was used to perform an analysis of the influence of the mesh on the convergence of the results. A baseline element size was defined for the different regions (Zone 1: central zone of the artery, Zone 2: intermediate zone of the artery and Zone 3 in the vicinity of the struts). For each of these zones, the average element sizes were: zone 1 = 5 μm, zone 2 = 3 μm, zone 3 = 1.5 μm. The mesh obtained with these values was identified as the baseline mesh. A refined mesh was obtained by dividing all element sizes by two (i.e., zone 1 = 2.5 μm, zone 2 = 1.5 μm, zone 3 = 0.75 μm). In addition, a further refinement operation was applied to all lines representing the struts and wall of the artery. Approximately 60,000 elements were obtained for the baseline mesh and 180,000 for the refined mesh. After performing a steady-state analysis, the velocity profiles obtained for the two mesh densities were compared and were found to be similar. In addition, the maximum velocity in the whole fluid domain obtained with the base mesh (0.539993 m/s) and the one obtained with the refined mesh (0.540401 m/s) showed a difference of less than 0.1%.

A similar approach was carried out for an overlapping configuration with OD = 2000 m. The conclusions were similar.

The time step was 0.001 s (908 time-steps per cardiac cycle) and convergence criteria for both pressure and velocity residuals were 10^{-6}.

2.2. Hemodynamic Metrics

WSS and its derived indexes, time-averaged WSS (TAWSS) and oscillatory shear index (OSI), are of great interest while studying the impact of stent struts on hemodynamics. The definitions of these parameters are recalled below.

TAWSS represents the average stress magnitude experienced by the vascular wall during a cardiac cycle and is derived as follows:

$$\text{TAWSS} = \frac{1}{T} \int_0^T |\text{WSS}| \, dt \tag{2}$$

where T denotes the period of the cardiac cycle and $|\text{WSS}|$ the modulus of the vector WSS. TAWSS is insensitive to the direction of the WSS vector.

OSI is a non-dimensional scalar used to evaluate the oscillatory nature of vascular flows (i.e., how much the WSS vectors change their direction over a cycle) and is calculated as follows:

$$\text{OSI} = \frac{1}{2} \left(1 - \frac{\left| \int_0^T \text{WSS} \, dt \right|}{\int_0^T |\text{WSS}| \, dt} \right) \tag{3}$$

OSI varies between 0 and 0.5 with a value of 0 when there is no oscillatory WSS and 0.5 when it is fully oscillatory.

The RRT measures how long the particles stay near the wall of the vessel. Longer time of contact between atherogenic particles and the arterial wall could cause a high risk of atherosclerosis formation [20,24,25]. High RRT (RTT > 10 Pa^{-1}) is recognized as critical for atherogenesis and in-stent restenosis [20]. Thus, RRT was defined as follows:

$$\text{RRT} = \left(\frac{1}{(1 - 2\text{OSI})\text{TAWSS}} \right) \tag{4}$$

As this previous definition shows, RRT combines the information provide by TAWSS and OSI.

2.3. Pathological OSI, TAWSS and RRT Thresholds

It is generally accepted that an abnormally low WSS increases the risk of restenosis [1,14,26,27] A TAWSS < 0.5 Pa is a common threshold value to indicate low WSS over the cardiac cycle [5,8,19]. On the other hand, high shear stresses (TAWSS > 2.5 Pa) have been associated with plaque rupture [28] that could lead to thrombosis. In addition, high shear stresses have been reported to increase the activation of platelets which are the main cellular components of a thrombus [2,6,10,20,29]. Regarding oscillatory flow, OSI > 0.1 was associated with an increased risk of arterial narrowing [16,19,30,31]. Additionally, other authors have reported that thrombus formation is enhanced at areas characterized by high OSI because slow and reversed flow promotes platelet aggregation [1,29]

In this work the following thresholds were used: TAWSS < 0.5 Pa increases the risk of restenosis, TAWSS > 2.5 Pa promotes thrombosis and OSI > 0.1 promotes both restenosis and thrombosis. RRT > 10 Pa^{-1} promotes restenosis. The threshold for RRT is variable between studies in the literature. For instance, it is 5 Pa^{-1} in [20] and 10 Pa^{-1} in [25]. In the present study, the RRT should be less than 8 Pa^{-1}. According to [24], RRT is recommended as a unique and robust measure of low and oscillating shear flow.

3. Results

3.1. Results for Malapposition

3.1.1. Effect of Malapposition on the Velocity Field

Figure 3 displays the most relevant streamlines in the vicinity of the first stent strut at the diastolic peak (See Figure 2) when the average flow velocity is the highest and recirculation regions reach their maximum extensions. For the well-apposed configuration (see MD = 0 µm for both the CC-DES and BVS stents in Figure 3), there were relatively small recirculation regions upstream and downstream from the apposed strut. It should be noticed that the recirculation zone is much larger for the BVS stent. As soon as the stent began separate from the wall (i.e., with further increments of MD), upstream recirculation disappeared but the one located downstream from the malapposed strut started growing and moving to the right until it disappeared as well (see MD = 115 µm for CC-DES stent and MD = 300 µm for BVS stent in Figure 3). As expected, flow disturbance was more enhanced for the large strut (i.e., BVS stent). Moreover, flow accelerated through the wall separation gap as MD increased giving, as a result, a larger velocity gradient.

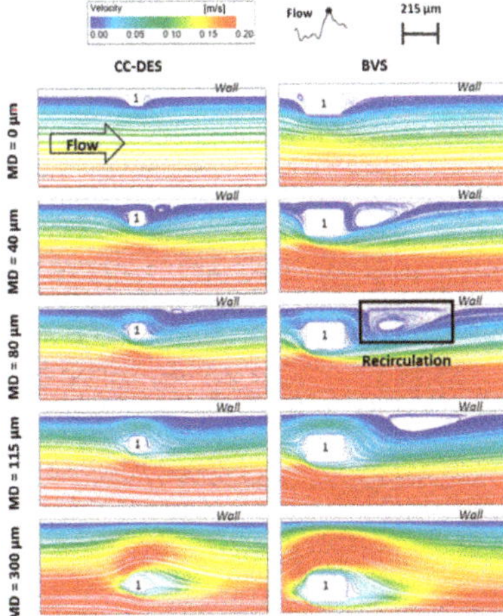

Figure 3. Most relevant streamlines in the vicinity of the first strut at the diastolic peak induced by Cobalt-Chromium drug-eluting stent (CC-DES) and bioresorbable stent (BVS) stent for MD = 0, 40, 80, 115 and 300 µm.

3.1.2. Effect of Malapposition on TAWSS

Figure 4 illustrates the TAWSS distribution along the arterial wall for the whole extension of the stent for the most pertinent MD values. First, the normal TAWSS magnitude of about 1.5 Pa (i.e., the shear stress value for a vessel without stent) was locally disturbed even for well-apposed stent. Furthermore, peaks of TAWSS, sometimes with maximum values above 2.5 Pa (see MD > 80 μm in Figure 4), developed in the malapposed region. The amplitude of these peaks increased with MD values and was higher for BVS stent (compared to CC-DES stent). In the well-apposed area, there were no TAWSS peaks for all values of MD. The TAWSS plateaus between two well-apposed struts (1.5 Pa) were slightly modified. The amplitude of the plateaus between the three well-applied struts rapidly converges to the values of the plateaus of a well-applied stent. These perturbations were more significant for BVS stents than for CC-DES stents.

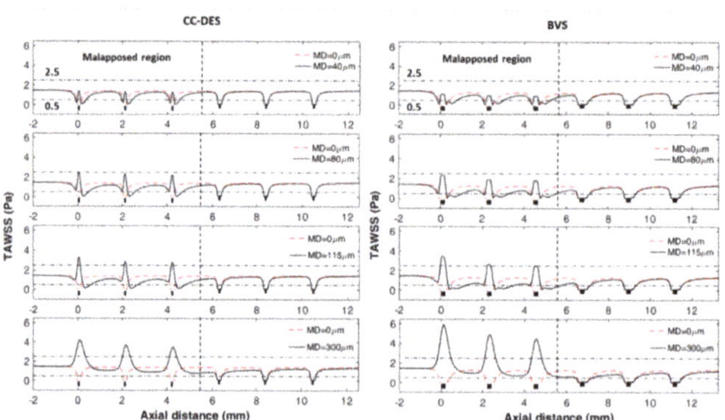

Figure 4. Time-averaged wall shear stress (TAWSS) distribution for the malapposition configuration. For simplicity reasons, only five representative configurations are displayed for each strut size (MD = 0, 40, 80, 115 and 300 μm). Strut locations are indicated with black rectangles. Notice that axial distance = 0 mm corresponds to the location of the first malapposed strut. Three TAWSS ranges can be defined: TAWSS < 0.5 Pa (Low TAWSS), 0.5 < TAWSS < 2.5 Pa (Normal TAWSS) and TAWSS > 2.5 Pa (High TAWSS).

For small malapposition distances (see MD = 40, 80 and 115 μm for both strut sizes in Figure 4), some segments of the vessel wall in the malapposed region were below 0.5 Pa, indicating an abnormally low TAWSS. These segments, with low values, vanished for MD = 300 μm (for both types of stents). These areas, with low shear stress, are located near and downstream of the struts. Additionally, BVS struts produced larger regions with low TAWSS.

3.1.3. Effect on the Oscillatory Character of WSS Caused by Malapposed Strut

Figure 5 displays the OSI distribution along the stented region of the arterial wall's five most representative MD values. First, in the case of a correctly apposed stent (see MD = 0 μm for both strut sizes in Figure 5), the effect of recirculation was similar for all the struts. However, as the stent began to separate from the wall, high peaks of OSI appeared downstream from each malapposed strut. These peaks were present for small wall separations but disappeared with further increments of malapposition distance (see Figure 5, MD = 115 μm for CC-DES and MD = 300 μm for BVS). These results suggest that malapposed struts promote significant flow recirculations during the cardiac cycle, which confirms the flow perturbation in Figure 3 by considering the temporal evolution of the flow recirculation regions.

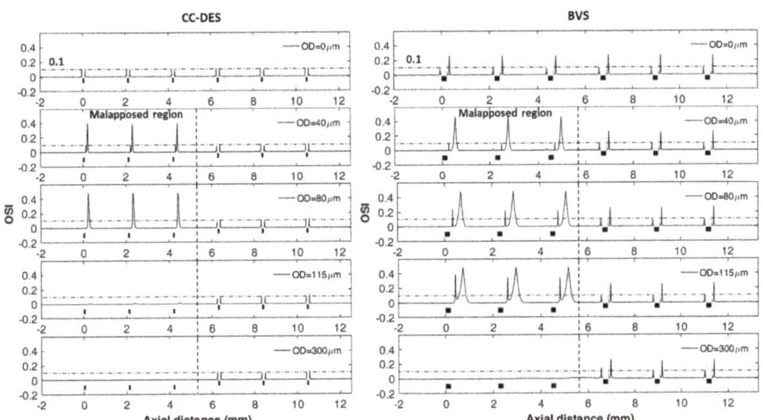

Figure 5. OSI distribution for malapposition configuration. For simplicity reasons, only five representative configurations are displayed for each strut size (MD = 0, 40, 80, 115 and 300 μm). Strut locations are indicated with black rectangles. Notice that axial distance = 0 mm corresponds to the location of the first strut. Two OSI ranges can be defined: OSI < 0.1 (Low recirculation) and OSI > 0.1 (High recirculation).

3.1.4. Effect of Malapposition on RRT

Figure 6 displays the RRT distribution along the stented region of the arterial wall across the stented region, for the five most representative MD values (MD = 0, 40, 80, 115, and 300 μm). Firstly, well-apposed struts produce peaks in the distribution of RRT located around each strut, with values significantly above the thresholds. These peaks are significantly larger for BVS stents with larger struts. They split into several peaks at low MD values. Therefore, the arterial wall affected by RRT values above the threshold is divided into several critical areas, very close to each other. This phenomenon is much more important when the dimensions of the struts increase (for BVS stent). On the other hand, all MD values do not disturb the downstream RRT distributions (for the three well-apposed struts). When MD is higher or equal to 115 μm for the CC-DES stent and 300 μm for the BVS stent, the amplitude of the peaks (for the malapposed struts) decreases drastically and is significantly below the thresholds.

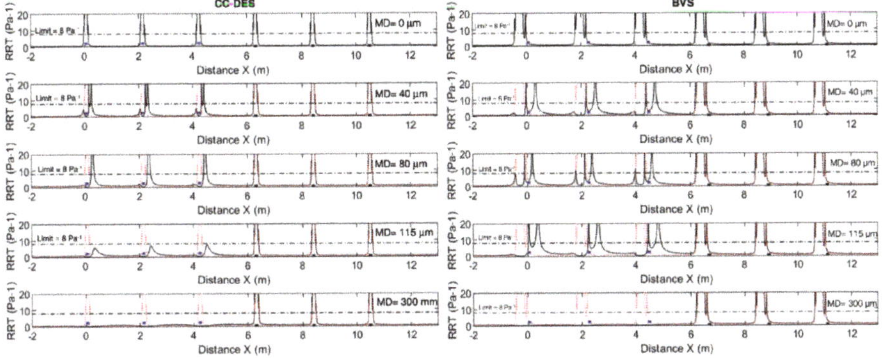

Figure 6. Relative residence time (RRT) distribution for the malapposition configuration. For sake of simplicity, only five representative configurations are displayed for each strut size (MD = 0, 40, 80, 115 and 300 μm). Strut locations are indicated with black rectangles. Notice that axial distance = 0 mm corresponds to the location of the first malapposed strut.

3.1.5. Relationship between OSI and TAWSS for Malapposion Configuration

Figure 7 plots the distribution of TAWSS versus OSI for all the nodal solutions on the arterial wall. Only the configurations including an OSI peak higher than 0.1 in the malapposed region were considered (MD = 40, 60 and 80 µm for CC-DES stent and MD = 40, 80, 115, 150 and 180 µm for BVS stent). The high OSI values were always associated with low TAWSS < 0.5 Pa.

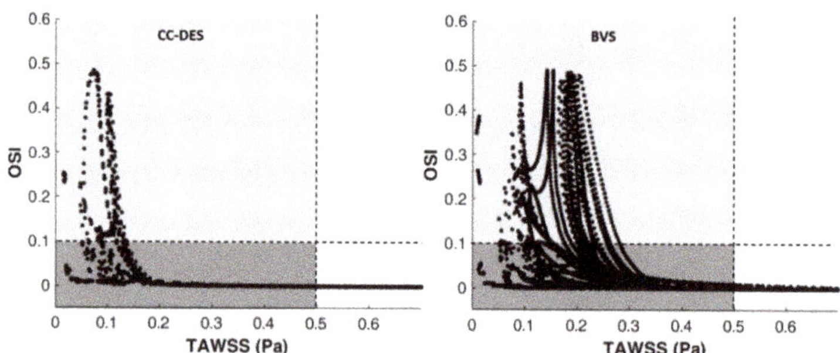

Figure 7. OSI vs. TAWSS plots. Each point represents a nodal solution of the arterial wall. The considered configurations were MD = 40, 60 and 80 µm for CC-DES stent and MD = 40, 80, 115, 150 and 180 µm for BVS stent.

3.1.6. Effect of MD Distance on Arterial Wall Extent with a Risk of Restenosis/Thrombus

Figure 8 displays the evolution of the total wall length affected by low TAWSS (<0.5 Pa), high TAWSS (>0.5 Pa) and high OSI (>0.1) versus malapposition distance. First of all, the evolution of the different affected lengths was similar for both stents, but much more significant for the BVS one. For both stents: (1) the total wall lengths affected by high OSI and low TAWSS increased until they reached a maximum value followed by a decreasing tendency that finished in a plateau (2) The wall extension affected by a high OSI reached a maximum before that corresponding to low TAWSS (i.e., at about one strut height) and (3) With regard to the total length affected by a high TAWSS, it has always tended to increase. Three zones were identified (see Figure 7), the first one prone to develop restenosis, the second one with risk of developing both restenosis and thrombosis and the last one prone to develop mainly thrombosis.

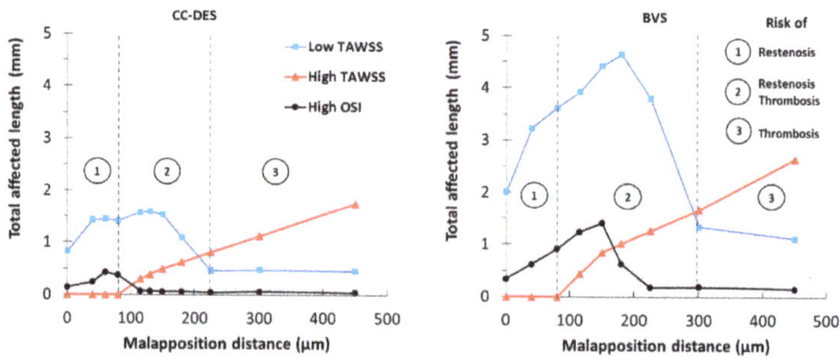

Figure 8. Evolution of affected arterial length vs. malapposition distance (MD). High risk of restenosis (zone 1), restenosis and thrombosis (zone 2) and mainly thrombosis (zone 3).

Malapposed CC-DES struts increased the total extension of the affected wall up to 1.6 mm (at MD = 130 µm) for low shear stress, 1.75 mm (MD = 450 µm) for high shear stress, and 0.42 mm (at MD = 60 µm) for high flow oscillation. For malapposed BVS struts, the total affected wall raised up to 4.64 mm (at MD = 180 µm) for low shear stress, 2.65 mm (MD = 450 µm) for high shear stress, and 1.41 mm (MD = 150 µm) for high flow oscillation.

These affected lengths were significantly larger for BVS struts (up to 2.9 times for low TAWSS, 1.51 times for high TAWSS, and 3.31 times for high OSI). The areas of the wall with high risk of restenosis/thrombus increase drastically with the dimensions of the struts.

3.2. Results for Overlapping

3.2.1. Effect of Overlapping on the Velocity Field

Figure 9 depicts the effect of different OD levels on the blood flow at the diastolic peak when recirculation regions reach their maximum extensions in the overlapping section for the diastolic peak. For congruent struts (i.e., with OD = 2000 and 4000 μm, see Figure 9), large recirculation regions appeared downstream from each pair of piled struts. With intermediate OD values (i.e., for incongruent struts with OD = 3000 μm, see Figure 9), these recirculation regions moved downstream from each overlapping strut and reduced their extensions compared to congruent configuration. Additionally, small recirculation regions reappeared downstream from each well-apposed strut. Moreover, flow acceleration occurred through the gap between the overlapping struts and the vessel wall. Finally, similar to malapposition configurations, flow disturbance was more notorious for large struts (i.e., BVS stent).

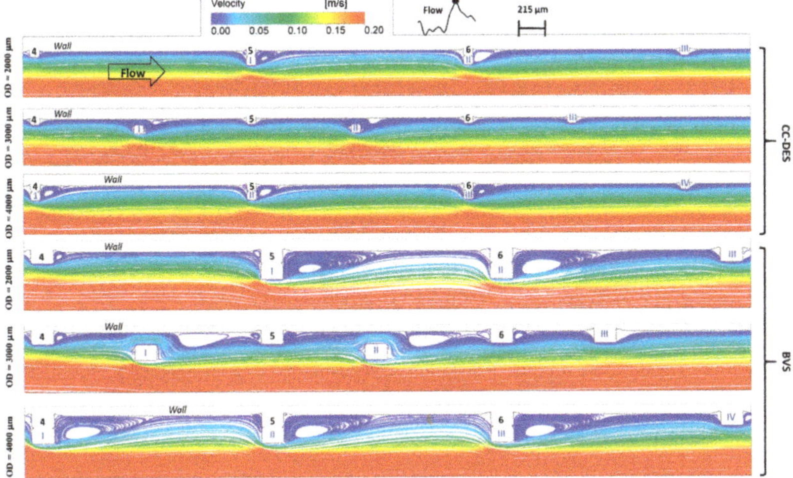

Figure 9. Streamlines at diastolic peak for overlapping stents. For simplicity reasons, just three representative configurations are presented for each strut size.

3.2.2. Effect of Overlapping on TAWSS

Figure 10 illustrates the TAWSS distribution at the overlapping region of CC-DES and BVS stents. First, the perturbation of the TAWSS distribution is localized in this overlapping zone and up to the first strut downstream. TAWSS peaks appear at the location of the overlapping struts for configuration with incongruent stents (blue rectangles in OD = 1000 μm, see Figure 10). Moreover, for all OD values, some regions with a TAWSS below 0.5 Pa (abnormally low value) appeared on the arterial wall. For congruent configurations (i.e., OD = 0 and 2000 μm, see Figure 10) and small strut size, flow reattachment downstream from each pair of piled struts allowed to recover a normal TAWSS value (1.5 Pa). However, for large struts, the perturbation was so significant that the normal TAWSS level could not be reached in the overlapping region.

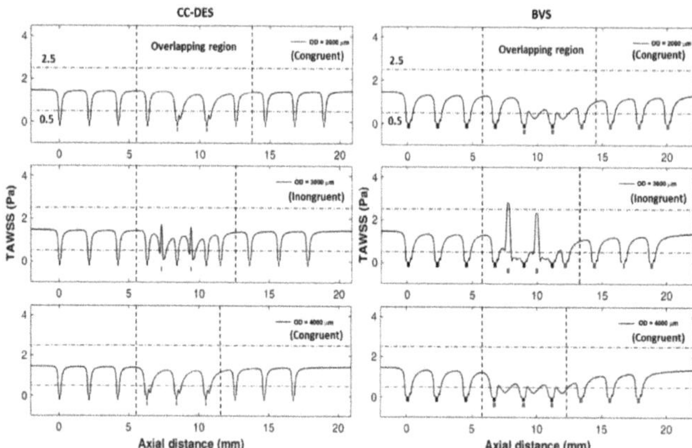

Figure 10. TAWSS distribution for overlapping configuration. For simplicity reasons, three representative configurations are displayed for each strut size. Strut locations on the X-axis are indicated with black and blue rectangles for the first and second strut respectively. Notice that axial distance = 0 mm corresponds to the location of the first stent. Three ranges can be defined: TAWSS < 0.5 Pa (Low TAWSS), 0.5 < TAWSS < 2.5 Pa (Normal TAWSS) and TAWSS > 2.5 Pa (High TAWSS).

3.2.3. Effect on the Oscillatory Character of WSS Due to Overlapping

Figure 11 displays the OSI distribution of overlapping CC-DES and BVS stents. First, the most important peaks of OSI were located at the overlapping region. Moreover, the distribution of peaks was different for congruent and incongruent cases. When the struts of two overlapping stents were piled on top of each other (see OD = 2000 and 4000 μm in Figure 11), the highest OSI peaks were located downstream from each congruent pair. On the other hand, when overlapping was incongruent, the highest peak of OSI was located downstream from each of the overlapping struts of the second stent (see Figure 11, blue rectangles in OD = 3000 μm). In general, more peaks were present for incongruent configurations. For the BVS stent, the peaks are wider and higher for all the studied configurations. Downstream of this zone, the OSI distribution rapidly reverts to that of a well-apposed single stent.

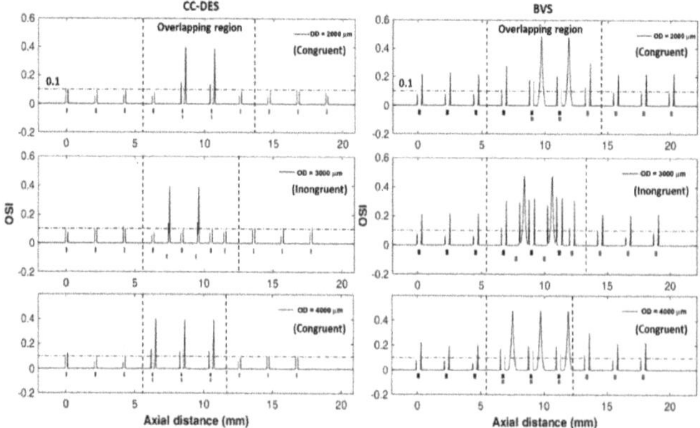

Figure 11. Oscillatory shear index (OSI) distribution for overlapping configuration. For simplicity reasons, three representative configurations are displayed for each strut size. Strut locations on the X-axis are indicated with black and blue rectangles. Notice that axial distance = 0 mm corresponds to the location of the first stent. Two ranges can be defined: OSI < 0.1 (Low recirculation) and OSI > 0.1 (High recirculation).

3.2.4. Effect of the Overlapping on the RRT

Figure 12 displays the RRT distribution of overlapping CC-DES and BVS stents. First, for the set of overlap values studied, the RRT distribution is only modified in this overlap area and not for the upstream (stent 1) and downstream (stent 2) sections. In the configurations with congruent struts (OD = 2000 µm and OD = 4000 µm), in addition to the RRT peaks located in the vicinity of the struts in contact with the arterial wall, another peak appears downstream of the congruent struts. This new peak is less wide but with values higher than the threshold chosen for this study (8 Pa^{-1}). For the configurations with non-congruent struts, the RRT distribution is significantly affected with the appearance of a weak peak upstream of the struts detached from the wall and especially a downstream zone (up to the strut in contact with the arterial wall) with several peaks clearly exceeding the RRT thresholds.

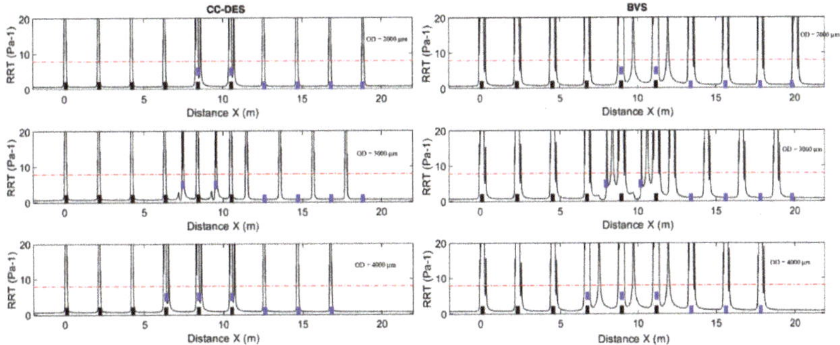

Figure 12. RRT distribution for overlapping configuration. For sake of simplicity, three representative configurations are displayed for each strut size (OD = 2000 µm, 3000 µm and 4000 µm). Strut locations on the X-axis are indicated with black (upstream stent) and blue rectangles (downstream stent). Notice that axial distance = 0 mm corresponds to the location of the first stent.

3.2.5. Relationship between OSI and TAWSS for Overlapping Configuration

Figure 13 shows the distribution of TAWSS versus OSI for all the configurations including an OSI peak higher than 0.1 in the overlapping region (OD = 2000, 2500, 3000, 3500 and 4000 µm for both the CC-DES and BVS stents). It can be seen that, similarly as found for malapposition configurations (see Figure 8), high OSI values were always associated with low TAWSS < 0.5 Pa).

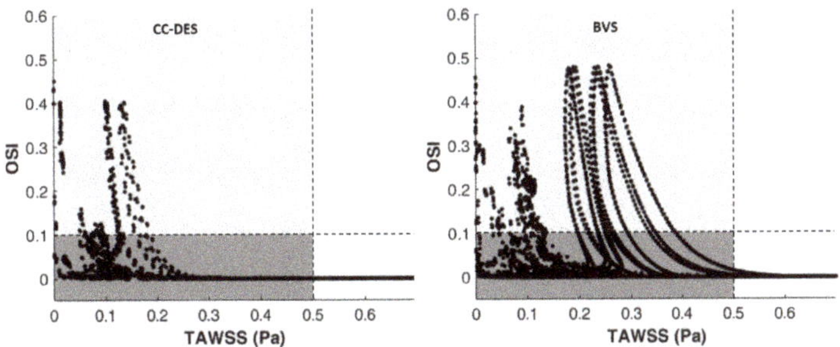

Figure 13. OSI vs. TAWSS plots. Each point represents a nodal solution of the arterial wall. The considered configurations were OD = 200, 2500, 3000, 3500 and 4000 µm for both stent types.

3.2.6. Effect of Overlapping Distance on Arterial Wall Extent with a Risk of Restenosis/Thrombus

Figure 14 displays the evolution of the total arterial extension affected by low TAWSS (<0.5 Pa), high TAWSS (>0.5 Pa) and high OSI (>0.1) versus overlapping distance (OD). In general, the effect of different OD levels on the arterial length affected by low TAWSS and high OSI was rather constant for both strut sizes but more significant for the BVS stent. Regarding high TAWSS, it did not affect significantly the vessel wall.

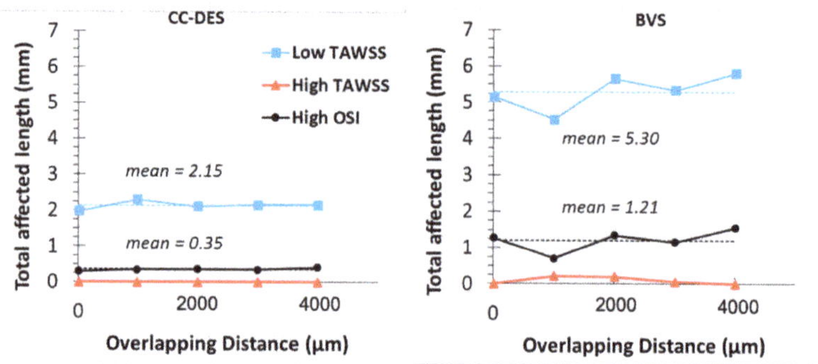

Figure 14. Evolution of affected arterial length vs. overlapping distance (OD).

Overlapping CC-DES stents increased the total arterial extension affected by low TAWSS up to 2.31 mm (at OD = 2500 µm) and the one affected by high flow oscillation up to 0.41 mm (at OD = 4000 µm). For overlapping BVS stents, the total affected wall raised up to 5.81 mm (at OD = 4000 µm) and up to 1.55 mm (at OD = 3500 µm) for low TAWSS and high flow oscillation, respectively. Moreover, the increment of wall extension affected by high TAWSS was relatively small for both stent sizes (maximum of 0 mm and 0.22 mm for CC-DES and for BVS, respectively). Finally, the increment of wall segments affected by low TAWSS and high OSI were significantly larger for BVS stents, up to 2.47 and 3.48 times respectively (i.e., taking into account the mean total affected lengths shown in Figure 14).

4. Discussion

This study investigated the hemodynamic conditions in coronary arteries with malapposed and overlapped stents while considering a pulsatile non-Newtonian blood flow and WSS-related indices computed over a cardiac cycle. The use of axisymmetric CFD models simplified the systematic analysis of each geometry by performing a parametric study with a significant number of computations ($n = 30$). The obtained results help to clarify the impact of different degrees of strut misalignment on local hemodynamics (TAWSS, OSI and RRT).

First, the regions of the vessel wall affected by high OSI were always under low TAWSS for all the studied configurations (see Figures 7 and 14). This suggests that a condition for OSI is the occurrence of low TAWSS as stated by [8,18,32].

4.1. Malapposed Configuration

The analysis of the malapposition geometries showed that regions of the arterial wall affected by both low shear stress and oscillatory flow (i.e., TAWSS < 0.5 Pa and OSI > 0.1) were present for small degrees of malapposition distance (MD). The extension of the vessel affected by high OSI reached a maximum when MD was close to one strut height (i.e., H = 85 µm and H = 150 µm for CC-DES and BVS stents, respectively). With further increments of MD, the region with high flow oscillation decreased until it vanished (see MD = 115 µm for CC-DES stent and MD = 300 µm for BVS stent in Figure 4). These results are confirmed by the RRT distribution (see Figure 6). Indeed, when the MD values increase (see MD = 115 µm for CC-DES stent and MD = 300 µm for BVS stent in Figure 6), the RRT peaks strongly decrease and fall below the threshold. The areas, in the overlap section, affected by the adverse effect of RRT tend to vanish. In addition, low shear stresses continued to develop on the arterial wall even when the OSI canceled out, indicating that the velocity gradient near the wall was low but the flow was no more oscillatory. These tendencies agree with the conclusions of previous works [8,10,22,33] and can be seen in Figure 8.

With the increment of MD and the shift of recirculation regions downstream from the malapposed struts, the free space near the arterial wall increased and the local resistance to flow decreased. Consequently, fluid accelerated through the gap between the malapposed struts and the vessel wall (see Figure 3) and caused localized regions with high shear stress (i.e., TAWSS > 2.5 Pa, see Figure 4). Moreover, the magnitude of the high shear stress and the size of the affected wall extensions increased gradually with the degree of wall separation (see Figures 4 and 8, respectively), which is consistent with the conclusions of [6,10]. Additionally, the presence of consecutive misaligned struts produced a decreasing effect on TAWSS values. It was always particularly enhanced between the first and the second strut (see Figure 4).

The configuration that led to the higher risk of potential restenosis occurred when the malapposed struts were separated from the wall of approximately one strut height (see zone 1 in Figure 8). This configuration promotes the formation of large recirculation zones downstream from each malapposed strut, resulting in abnormally low TAWSS. On the other hand, the risk of potential thrombosis was more significant for configurations with large wall separations due to the occurrence of larger segments with high TAWSS (see zone 3 in Figure 8). In such configurations, the risk of thrombus development had previously been noted by [34]. In the intermediate zone (see zone 2 in Figure 8), the risk of restenosis and thrombosis coexisted. The conclusions were similar for the two studied stents. However, the concerned length is much more important for BVS stent.

4.2. Overlapping Configuration

The analysis of the overlapping geometry revealed an important deficit of shear stress (TAWSS < 0.5 Pa) compared to non-overlapping segments of the stented artery (see Figure 10), which was in agreement with the results of [17–19]. In general, two congruent struts (i.e., with OD = 2000 μm and 4000 μm) were found to act as a single apposed strut with double height. Consequently, congruent struts produced similar TAWSS distributions than single apposed struts but with more significant hemodynamic disturbances (see OD = 2000 μm and 4000 μm in Figure 10). Moreover, the configurations with congruent struts were found to produce a large recirculation area downstream from the stacked struts at the diastolic peak (see Figure 9). As the RRT is a function of the OSI and TAWSS, the RRT provides general information combining the two previous information (see Figure 12). A similar disturbed flow region was identified by [17,22,35] after performing steady-state analyses on realistic and idealized CFD models, respectively. However, our transient studies revealed that, in terms of the hemodynamic effect on the vascular wall over the cardiac cycle, congruent struts configuration was not necessarily worse than incongruent struts configurations. As seen in Figure 14, the total arterial lengths affected by low shear stress (TAWSS < 0.5 Pa) and high oscillation (OSI > 0.1) are rather constant for all the studied range of overlapping distance. It should be noted that the total stented length varies significantly with the OD value. In fact, this length varies from 18.9 mm to 16.81 mm for the CC-DES stent and from 19.50 to 17.35 mm for the BVS one.

For overlapping configurations, the wall lengths affected by high shear stresses (TAWSS > 2.5 Pa) were relatively smaller than for malapposition configurations. Regarding incongruent strut configurations, TAWSS peaks appeared in regions with significant gaps between the vessel wall and overlapping struts (see OD = 3000 μm in Figure 10). However, these peaks were considerably lower than those caused by malapposition configurations (see Figure 4). Since the gap between the wall and the overlapping struts is less significant (i.e., 52 μm and 107 μm for CC-DES and BVS struts, respectively) compared to the cases with large MDs (i.e., up to 450 μm), the TAWSS peaks were relatively small and in the order of magnitude of small malapposition distances.

The potential risks of restenosis are relatively similar for all the overlapping struts configurations studied, as highlighted by [36]. As OSI and TAWSS are fluctuating in the non-congruent cases, the distribution of RRT is highly variable and shows several peaks in the overlap area. As seen in Figure 14, low shear stress and flow oscillation always affected the vessel wall for all the studied cases (i.e., OD from 2000 to 4000 μm). On the other hand, the risk of thrombosis seems to be reduced for these configurations.

4.3. Effect of the Strut Dimensions

It is obvious that thinner struts (i.e., CC-DES in this study) represent smaller obstacles to blood flow. So, this improves the shear stress distribution and allows a faster flow reattachment between strut cells [5,37]. In this work, BVS struts were associated with larger hemodynamic disturbances for all the studied configurations. The RRT plots (See Figures 6 and 12) for malapposition and overlapping stents show the adverse effect of the strut dimensions.

In the case of the correctly apposed stents cases (see MD = 0 μm in Figure 4), the BVS struts increased the extension of regions with low shear stress and high oscillatory flow 2.43 times and 2.37 times, respectively. However, both strut sizes allowed flow reattachment to reach normal shear stress values (TAWSS around 1.5 Pa).

Regarding malapposition and overlapping configurations, the performance difference between CC-DES and BVS struts was more notorious. For malapposition cases, the use of BVS struts increased up to 2.9, 1.51 and 3.31 times the wall segments affected by low TAWSS, high TAWSS and high OSI, respectively. Regarding overlapping configurations, the use of BVS struts increased up to 2.47 and 3.48 times the wall segments affected by low TAWSS and high OSI, respectively. These results suggest that, in the case of equal strut misalignment degrees (i.e., malapposition or overlapping), thicker struts will always induce significantly larger hemodynamic disturbances than smaller struts and will increase the risk of restenosis and/or thrombosis.

5. Study Limitations

First of all, the use of idealized axisymmetric models disregards the 3D effect that coronary stents could have on the blood flow. The models in the present study are two-dimensional, while real blood vessels are three-dimensional. The 2D models assume rotational symmetry and no tangential flow component. However, this component exists in the reality but it is neglected in the study. Additionally, stents are not axisymmetric. For these reasons, the present work is useful to show tendencies of the hemodynamic variables on the malapposition and overlapping rather than provide detailed information on the flow structures and WSS patterns.

Moreover, our model does not consider any arterial curvature (i.e., we considered straight arteries) or residual stenosis that may remain after PCI. These geometric simplifications affect the hemodynamic results. However, the use of realistic 3D models requires significant computational costs, which are not compatible with parametric studies. Furthermore, a 3D model is especially justified for the analysis of a patient-specific configuration. The use of 2D axisymmetric models let us fulfill the objectives of this work which were: (1) to clarify the hemodynamic evolution for different degrees of strut misalignment, and (2) to identify critical configurations that may be associated with restenosis and thrombosis. Such goals can be only be reached with systematic parametric analysis.

In addition, the compliance of the arterial wall was also neglected. However, it is known that stent deployment and atherosclerotic plaque reduce the compliance of the artery wall [38]. Additionally, as demonstrated in the literature [39], the WSS and its related indices are not affected by the vessel compliance for straight arteries.

Finally, the biological response of the vascular wall was not considered, and only the hemodynamic effects were investigated. Incorporating more complex models to predict drug deposition or thrombus formation [10] could give a deeper insight into this subject.

6. Clinical Application

Stent deployment is a challenging task, especially for stenoses with complex configurations (i.e., with excessive lengths, close to bifurcations, concomitant lesions, etc.). Therefore, the ideal stent implantation is difficult to achieve in clinical practice. The fact is that interventional cardiologists frequently encounter incomplete strut apposition and overlapping. The main conclusions found in this study may provide interesting information for cardiologists and stent designers to know: (1) how different degrees of malapposition and overlapping disturb blood flow and (2) which configurations are the most critical ones and their potential link to poor clinical outcomes.

First, this study highlights that malapposed struts will produce the maximal flow recirculation near the artery wall when malapposition distance is close to one strut height (i.e., critical point for restenosis). With further increments of wall separation, recirculation regions will disappear but the artery wall will be subjected to high shear stresses (critical point for thrombosis). Since there is a decreasing effect on shear stress for consecutive struts, the risk of plaque rupture and platelet activation is higher for regions close to the first group of misaligned struts. Second, stent overlapping was more prone to increase the risk of restenosis due to the appearance of segments of the artery wall subjected to low shear stress and flow recirculation. In terms of critical configurations, the risk seems to be comparable for all of them (i.e., incongruent and congruent struts). From a hemodynamic point of view, the best is to avoid overlapping if possible. Indeed, for all overlapping configuration, the extent of the zones where risks of stenosis/thrombus is significantly greater than for malapposed configuration. Finally, thicker struts are more sensitive to strut misalignment problems.

7. Conclusions

This axisymmetric numerical study allows evaluation of the risks related to a malapposition or an overlapping stent. The numerical models show that the relative extent of the areas with high risk (restenosis/thrombus) is considerably increased in regions with overlapped stent compared to regions without overlapped stent and even compared to areas of malapposed stent. Since it is generally accepted that low TAWSS (TAWSS < 0.5 Pa), high TAWSS (TAWSS < 2.5 Pa), high OSI (OSI < 0.1) and RRT > 8 Pa^{-1} are important factors for atherogenesis and thrombogenesis, the results indicate that adverse hemodynamics caused by overlapping stents may be partly responsible for adverse clinical outcomes in patients treated with overlapping stents. The development of risk areas for malapposition is significantly lower than for overlap. In addition, it was shown that the size of the struts has a very negative effect on the development of risk areas. In cases where stent overlap cannot be avoided, deployment strategies should be optimized or new stent designs should be considered to reduce the risk of restenosis.

Author Contributions: Simulations, postprocessing and writing original draft preparation, R.C.; writing original draft preparation, postprocessing, supervision, funding acquisition, project administration, M.L.; simulations, supervision, conceptualization, funding acquisition, manuscript review and editing, M.M.; Conceptualization, data curation, supervision, R.I.P.; Conceptualization, data curation, supervision, G.F., F.D.; Conceptualization, data curation, supervision, resources, project administration, writing, funding acquisition and reviewing original draft, J.O. All authors have read and agreed to the published version of the manuscript.

Funding: This research Project was supported by grants from Labex CAMI-France (project SIMPLE) and the Mexican Council of Science and Technology (CONACYT). M. Malvè was supported by the Spanish Ministry of Economy, Industry and Competitiveness through research project DPI2017-83259-R (AEI/FEDER, UE) and by the Department of Economic Development of the Navarra Government through research project PC086-087-088 CONDE.

Institutional Review Board Statement: Not applicable.

Informed Consent Statement: Not applicable.

Data Availability Statement: Data are available under request to the authors.

Acknowledgments: The support of the Instituto de Salud Carlos III (ISCIII) through the CIBER-BBN initiative is highly appreciated.

Conflicts of Interest: The authors declare no conflict of interest. The funders had no role in the design of the study; in the collection, analyses, or interpretation of data; in the writing of the manuscript, or in the decision to publish the results.

References

1. Koskinas, K.C.; Chatzizisis, Y.S.; Antoniadis, A.P.; Giannoglou, G.D. Role of Endothelial Shear Stress in Stent Restenosis and Thrombosis. *J. Am. Coll. Cardiol.* **2012**, *59*, 1337–1349. [CrossRef]
2. Jaryl, N.; Bourantas, C.V.; Torii, R.; Ang, H.Y.; Tenekecioglu, E.; Serruys, P.W.; Foin, N. Local Hemodynamic Forces After Stenting. *Arterioscler. Thromb. Vasc. Biol.* **2017**, *37*, 2231–2242.
3. Byrne, R.A.; Joner, M.; Kastrati, A. Stent Thrombosis and Restenosis: What Have We Learned and Where Are We Going? The Andreas Grüntzig Lecture ESC 2014. *Eur. Heart J.* **2015**, *36*, 3320–3331. [CrossRef] [PubMed]
4. Lewis, G. Materials, Fluid Dynamics, and Solid Mechanics Aspects of Coronary Artery Stents: A State-of-the-art Review. *J. Biomed. Mater. Res. Part B Appl. Biomater.* **2008**, *86*, 569–590. [CrossRef]
5. Beier, S.; Ormiston, J.; Webster, M.; Cater, J.; Norris, S.; Medrano-Gracia, P.; Young, A.; Cowan, B. Hemodynamics in Idealized Stented Coronary Arteries: Important Stent Design Considerations. *Ann. Biomed. Eng.* **2016**, *44*, 315–329. [CrossRef]
6. Foin, N.; Gutiérrez-Chico, J.L.; Nakatani, S.; Torii, R.; Bourantas, C.V.; Sen, S.; Nijjer, S. Incomplete Stent Apposition Causes High Shear Flow Disturbances and Delay in Neointimal Coverage as a Function of Strut to Wall Detachment Distance: Implications for the Management of Incomplete Stent Apposition. *Circ. Cardiovasc. Interv.* **2014**, *7*, 180–189. [CrossRef]
7. Räber, L.; Jüni, P.; Löffel, L.; Wandel, S.; Cook, S.; Wenaweser, P.; Togni, M. Impact of Stent Overlap on Angiographic and Long-Term Clinical Outcome in Patients Undergoing Drug-Eluting Stent Implantation. *J. Am. Coll. Cardiol.* **2010**, *55*, 1178–1188. [CrossRef] [PubMed]
8. Poon, E.K.W.; Barlis, P.; Moore, S.; Pan, W.H.; Liu, Y.; Ye, Y.; Yuan, X.; Zhu, S.J.; Andrew, S.H. Numerical Investigations of the Haemodynamic Changes Associated with Stent Malapposition in an Idealised Coronary Artery. *J. Biomech.* **2014**, *47*, 2843–2851. [CrossRef] [PubMed]
9. Yoon, H.-J.; Hur, S.-H. Optimization of Stent Deployment by Intravascular Ultrasound. *Korean J. Intern. Med.* **2012**, *27*, 30. [CrossRef]
10. Chesnutt, J.K.W.; Han, H.C. Computational Simulation of Platelet Interactions in the Initiation of Stent Thrombosis Due to Stent Malapposition. *Phys. Biol.* **2016**, *13*, 016001. [CrossRef]
11. Otsuka, F.; Nakano, M.; Ladich, E.; Kolodgie, F.D.; Virmani, R. Pathologic Etiologies of Late and Very Late Stent Thrombosis Following First-Generation Drug-Eluting Stent Placement. *Thrombosis* **2012**, 1–16. [CrossRef] [PubMed]
12. Cook, S.; Eshtehardi, P.; Kalesan, B.; Raber, L.; Wenaweser, P.; Togni, M.; Moschovitis, A. Impact of Incomplete Stent Apposition on Long-Term Clinical Outcome after Drug-Eluting Stent Implantation. *Eur. Heart J.* **2012**, *33*, 1334–1343. [CrossRef]
13. Wentzel, J.J.; Whelan, D.M.; van der Giessen, W.J.; van Beusekom, H.M.M.; Andhyiswara, I.; Serruys, P.W.; Slager, C.J.; Krams, R. Coronary Stent Implantation Changes 3-D Vessel Geometry and 3-D Shear Stress Distribution. *J. Biomech.* **2000**, *33*, 1287–1295. [CrossRef]
14. LaDisa, J.F.; Guler, I.; Olson, L.E.; Hettrick, D.A.; Kersten, J.R.; Warltier, D.C.; Pagel, P.S. Three-Dimensional Computational Fluid Dynamics Modeling of Alterations in Coronary Wall Shear Stress Produced by Stent Implantation. *Ann. Biomed. Eng.* **2003**, *31*, 972–980. [CrossRef]
15. Martin, D.M.; Murphy, E.A.; Boyle, F.J. Computational Fluid Dynamics Analysis of Balloon-Expandable Coronary Stents: Influence of Stent and Vessel Deformation. *Med. Eng. Phys.* **2014**, *36*, 1047–1056. [CrossRef] [PubMed]
16. Brindise, M.C.; Chiastra, C.; Burzotta, F.; Migliavacca, F.; Vlachos, P.P. Hemodynamics of Stent Implantation Procedures in Coronary Bifurcations: An In Vitro Study. *Ann. Biomed. Eng.* **2017**, *45*, 542–553. [CrossRef] [PubMed]
17. Rikhtegar, F.; Wyss, C.; Stok, K.S.; Poulikakos, D.; Müller, R.; Kurtcuoglu, V. Hemodynamics in Coronary Arteries with Overlapping Stents. *J. Biomech.* **2014**, *47*, 505–511. [CrossRef]
18. Charonko, J.; Karri, S.; Schmieg, J.; Prabhu, S.; Vlachos, P. In Vitro Comparison of the Effect of Stent Configuration on Wall Shear Stress Using Time-Resolved Particle Image Velocimetry. *Ann. Biomed. Eng.* **2010**, *38*, 889–902. [CrossRef]
19. Chiastra, C.; Morlacchi, S.; Gallo, D.; Morbiducci, U.; Cardenes, R.; Larrabide, I.; Migliavacca, F. Computational Fluid Dynamic Simulations of Image-Based Stented Coronary Bifurcation Models. *J. R. Soc. Interface* **2013**, *10*, 20130193. [CrossRef]
20. Foin, N.; Lu, S.; Jaryl, N.; Bulluck, H.; Hausenloy, D.; Wong, P.; Virmani, R.; Joner, R. Stent Malapposition and the Risk of Stent Thrombosis: Mechanistic Insights from an in Vitro Model. *EuroIntervention* **2017**, *13*, e1096–e1098. [CrossRef]

21. Chiastra, C.; Wu, W.; Dickerhoff, B.; Aleiou, A.; Dubini, G.; Otake, H.; Migliavacca, F.; LaDisa, J.F. Computational Replication of the Patient-Specific Stenting Procedure for Coronary Artery Bifurcations: From OCT and CT Imaging to Structural and Hemodynamics Analyses. *J. Biomech.* **2016**, *49*, 2102–2111. [CrossRef]
22. Kolandaivelu, K.; Swaminathan, R.; Gibson, W.J.; Kolachalama, V.B.; Nguyen-Ehrenreich, K.L.; Giddings, V.L.; Coleman, L.; Wong, G.K.; Edelman, E.R. Stent Thrombogenicity Early in High-Risk Interventional Settings Is Driven by Stent Design and Deployment and Protected by Polymer-Drug Coatings. *Circulation* **2011**, *123*, 1400–1409. [CrossRef] [PubMed]
23. Davies, J.E.; Whinnett, Z.I.; Francis, D.P.; Manisty, C.H.; Aguado-Sierra, J.; Willson, K.; Foale, R.A. Evidence of a Dominant Backward-Propagating 'Suction' Wave Responsible for Diastolic Coronary Filling in Humans, Attenuated in Left Ventricular Hypertrophy. *Circulation* **2006**, *113*, 1768–1778. [CrossRef] [PubMed]
24. Pinto, S.I.; Campos, J.B. Numerical study of wall shear stress-based descriptors in the human left coronary artery. *Comput. Methods Biomech. Biomed. Eng.* **2016**, *19*, 1443–1455. [CrossRef] [PubMed]
25. Paisal, M.S.A.; Taib, I.; Ismail, A.E.; Tajul Arifin, A.M.; Mahmod, M.F. Evaluation System on Haemodynamic Parameters for Stented Carotid Artery: Stent Pictorial Selection Method. *Int. J. Integr. Eng.* **2019**, *11*. Available online: https://publisher.uthm.edu.my/ojs/index.php/ijie/article/view/3117 (accessed on 10 February 2021).
26. Malek, A.M.; Alper, S.L.; Izumo, S. Hemodynamic Shear Stress and Its Role in Atherosclerosis. *JAMA* **1999**, *282*, 2035. [CrossRef]
27. Murphy, J.; Boyle, F. Predicting Neointimal Hyperplasia in Stented Arteries Using Time-Dependant Computational Fluid Dynamics: A Review. *Comput. Biol. Med.* **2010**, *40*, 408–418. [CrossRef]
28. Samady, H.; Eshtehardi, P.; McDaniel, M.C.; Suo, J.; Dhawan, S.S.; Maynard, C.; Timmins, L.H.; Quyyumi, A.A.; Giddens, D.P. Coronary Artery Wall Shear Stress Is Associated With Progression and Transformation of Atherosclerotic Plaque and Arterial Remodeling in Patients With Coronary Artery Disease. *Circulation* **2011**, *124*, 779–788. [CrossRef]
29. Bluestein, D.; Gutierrez, C.; Londono, M.; Schoephoerster, R.T. Vortex Shedding in Steady Flow through a Model of an Arterial Stenosis and Its Relevance to Mural Platelet Deposition. *Ann. Biomed. Eng.* **1999**, *27*, 763–773. [CrossRef]
30. Zarins, C.K.; Don, P.; Giddens, B.K.; Bharadvaj, V.S.; Robert, F.M.; Seymour, G. Carotid Bifurcation Atherosclerosis. *Circ. Res.* **1983**, *53*, 14.
31. Williams, A.R.; Koo, B.K.; Gundert, T.J.; Fitzgerald, P.J.; LaDisa, J.F. Local Hemodynamic Changes Caused by Main Branch Stent Implantation and Subsequent Virtual Side Branch Balloon Angioplasty in a Representative Coronary Bifurcation. *J. Appl. Physiol.* **2010**, *109*, 532–540. [CrossRef] [PubMed]
32. Katritsis, D.G.; Theodorakakos, A.; Pantos, I.; Gavaises, M.; Karcanias, N.; Efstathopoulos, E.P. Flow Patterns at Stented Coronary Bifurcations: Computational Fluid Dynamics Analysis. *Circ. Cardiovasc. Interv.* **2012**, *5*, 530–539. [CrossRef]
33. Chen, W.X.; Poon, E.K.W.; Thondapu, V.; Hutchins, N.; Barlis, P.; Ooi, A. Haemodynamic Effects of Incomplete Stent Apposition in Curved Coronary Arteries. *J. Biomech.* **2017**, *63*, 164–173. [CrossRef]
34. Brown, J.; O'Brien, C.C.; Lopes, A.C.; Kolandaivelu, K.; Edelman, E.R. Quantification of thrombus formation in malapposed coronary stents deployed in vitro through imaging analysis. *J. Biomech.* **2018**, *71*, 296–301. [CrossRef]
35. Rikhtegar, F.; Edelman, E.R.; Olgac, U.; Poulikakos, D.; Kurtcuoglu, V. Drug deposition in coronary arteries with overlapping drug-eluting stents. *J. Control.* **2016**, *238*, 1–9. [CrossRef]
36. Antoniadis, A.P. Biomechanical Modeling to Improve Coronary Artery Bifurcation Stenting: Expert Review Document on Techniques and Clinical Implementation. *JACC Cardiovasc. Interv.* **2015**, *8*, 1281–1296. [CrossRef] [PubMed]
37. Jiménez, J.M.; Davies, P.F. Hemodynamically Driven Stent Strut Design. *Ann. Biomed. Eng.* **2009**, *37*, 1483–1494. [CrossRef]
38. Yong, A.S.C.; Javadzadegan, A.; Fearon, W.F.; Moshfegh, A.; Lau, J.K.; Nicholls, S.; Martin, K.C.; Kritharides, L. The Relationship between Coronary Artery Distensibility and Fractional Flow Reserve. Edited by Michael J Lipinski. *PLoS ONE* **2017**, *12*, e0181824. [CrossRef] [PubMed]
39. Chiastra, C.; Migliavacca, F.; Martínez, M.A.; Malvè, M. On the Necessity of Modelling Fluid–Structure Interaction for Stented Coronary Arteries. *J. Mech. Behav. Biomed. Mater.* **2014**, *34*, 217–230. [CrossRef]

Article

CFD Simulations of Radioembolization: A Proof-of-Concept Study on the Impact of the Hepatic Artery Tree Truncation

Unai Lertxundi [1], Jorge Aramburu [1,*], Julio Ortega [2], Macarena Rodríguez-Fraile [3,4], Bruno Sangro [3,5], José Ignacio Bilbao [3,6] and Raúl Antón [1,3]

[1] Mechanical Engineering Department, TECNUN Escuela de Ingeniería, Universidad de Navarra, 20018 Donostia-San Sebastián, Spain; ulertxundi@tecnun.es (U.L.); ranton@tecnun.es (R.A.)

[2] Escuela de Ingeniería Mecánica, Pontificia Universidad Católica de Valparaíso, Quilpué 01567, Chile; julio.ortega@pucv.cl

[3] IdiSNA, Instituto de Investigación Sanitaria de Navarra, 31008 Pamplona, Spain; mrodriguez@unav.es (M.R.-F.); bsangro@unav.es (B.S.); jibilbao@unav.es (J.I.B.)

[4] Department of Nuclear Medicine, Clínica Universidad de Navarra, 31008 Pamplona, Spain

[5] Liver Unit and CIBEREHD, Clínica Universidad de Navarra, 31008 Pamplona, Spain

[6] Department of Radiology, Clínica Universidad de Navarra, 31008 Pamplona, Spain

* Correspondence: jaramburu@tecnun.es

Abstract: Radioembolization (RE) is a treatment for patients with liver cancer, one of the leading cause of cancer-related deaths worldwide. RE consists of the transcatheter intraarterial infusion of radioactive microspheres, which are injected at the hepatic artery level and are transported in the bloodstream, aiming to target tumors and spare healthy liver parenchyma. In paving the way towards a computer platform that allows for a treatment planning based on computational fluid dynamics (CFD) simulations, the current simulation (model preprocess, model solving, model postprocess) times (of the order of days) make the CFD-based assessment non-viable. One of the approaches to reduce the simulation time includes the reduction in size of the simulated truncated hepatic artery. In this study, we analyze for three patient-specific hepatic arteries the impact of reducing the geometry of the hepatic artery on the simulation time. Results show that geometries can be efficiently shortened without impacting greatly on the microsphere distribution.

Keywords: computational fluid dynamics; radioembolization; hemodynamics; liver cancer; hepatic artery; computational cost analysis; personalized medicine; patient specific

Citation: Lertxundi, U.; Aramburu, J.; Ortega, J.; Rodríguez-Fraile, M.; Sangro, B.; Bilbao, J.I.; Antón, R. CFD Simulations of Radioembolization: A Proof-of-Concept Study on the Impact of the Hepatic Artery Tree Truncation. *Mathematics* **2021**, *9*, 839. https://doi.org/10.3390/math9080839

Academic Editors: Mauro Malvè and Alexander Zeifman

Received: 10 February 2021
Accepted: 8 April 2021
Published: 12 April 2021

Publisher's Note: MDPI stays neutral with regard to jurisdictional claims in published maps and institutional affiliations.

1. Introduction

Liver cancer is one of the leading types of cancer in incidence and mortality rates worldwide [1]. Radioembolization (RE) is a safe and effective intraarterial targeted therapy for unresectable primary and secondary liver tumors and it consists in the microcatheter-based infusion of yttrium-90 (Y-90) radiolabeled microspheres that are transported in the bloodstream until they get lodged in the tumoral tissue, where they deliver high tumoricidal doses of radiation to cancer cells, while ideally sparing healthy liver parenchyma [2].

In the last decade, a number of studies have been published on the computational fluid dynamics-based (CFD) simulation of the hepatic artery hemodynamics and microsphere transport during RE [3]. Some studies have focused on the type of microcatheter (e.g., standard end-hole microcatheter, antireflux catheter, angled-tip microcatheter) [4–6], and others have focused on the influence of various treatment and patient parameters (e.g., injection velocity, microcatheter location, hepatic artery geometry, etc.) on the microsphere distribution [7–9]. The ultimate goal of these studies is to provide the multidisciplinary team (interventional radiologists, hepatologists, nuclear oncologists, nuclear medicine physicians, etc.) that plans the treatment with additional information that can be of interest when planning the treatment. Moreover, CFD-based computer platforms have been presented in the literature for RE planning, such as the Computational Medical

119

Management Program, by which the optimal temporal and spatial points for microsphere infusion are determined [10] or CFDose, a simulation-based tool to calculate the patient-specific dosimetry to be infused to the patient [11]. However, the former can be used with a smart microcatheter (not commercially available yet) whose tip can be placed at a specified location within the infusion cross-sectional plane and a microsphere delivery system (not commercially available yet) that infuses at a specified temporal point in the cardiac cycle.

These simulation-based tools must be fast enough in providing results to be of use in the clinical setting. The phenomena involving the microsphere–hemodynamics have been proved to be dependent on local effects near the microcatheter tip, therefore three-dimensional (3D) and transient models are needed [3]. One could use CFD simulations or fluid–structure interaction (FSI) simulations, which consider the interaction between the fluid and the (rigid or deformable) solid. However, FSI simulations are far more expensive computationally than CFD simulations. A study by Childress et al. [12] showed that using CFD simulations instead of FSI simulations resulted in an important simulation-time reduction (3.5 days vs. 11–14 h [in a 10-processor 6-core 40-GB-RAM 3.33-GHz CPU]), with minor influence in the results, so CFD simulations could suffice.

One approach to reduce current CFD simulations' computational time would be the reduction of the size of the geometry where the CFD simulation is carried out. This geometry simplification should not result in marked differences in the calculated segment-to-segment microsphere distribution. Therefore, the hypothesis behind this study is that simulation times can be reduced considerably if the arterial geometry is effectively shortened, whether downstream or upstream (or both) from the microcatheter-tip location, obtaining a segment-to-segment microsphere distribution similar to that of the baseline geometry simulation. To do so, a geometry-reduction strategy is developed with one patient-specific case and this strategy is applied to two other patient-specific cases to assess the impact of the reduction on the simulation results, in terms of downstream microsphere distribution, and simulation times were analyzed.

2. Materials and Methods

In this section, we first introduce the three patients that are modeled in this study (one for developing the geometry-reduction strategy and two additional cases where the strategy is applied and assessed). Second, we show the baseline and simplified versions of the three 3D patient-specific hepatic artery geometries used in this study. For the first geometry, Patient 1, we conducted a step-by-step upstream and downstream simplification process to analyze to what extent each simplification influences the segment-to-segment microsphere distribution. Patient 2 and Patient 3's hepatic arteries were later simplified accordingly. Finally, the CFD model is presented.

2.1. Patients: Hepatic State and Radioembolization

This study was done using the patient-specific geometry of three patients: hereafter (Patient 1, Patient 2, and Patient 3, Tables 1–3). Regarding the geometries, these were reconstructed with MeVis (MeVis Medical Solutions AG, Bremen, Germany). Regarding the liver segment volumes, with segments defined as proposed by Couinaud [13], these were either obtained from the report provided by MeVis (Patients 1 and 3, Tables 1 and 3) or they were taken from the literature to be physiologically realistic (total volume according to reference [14] and fractional segmental volume according to reference [15]) (Patient 2, Table 2). As for the cancer scenarios, the same fictional cancer scenario was posited in the three patients. The scenario consists of a hepatocellular carcinoma (HCC) in segment 8. The tumor volume is equal to 20% of the healthy tissue volume of segment 8.

Table 1. Patient 1 liver mass volumes and blood flow rates per segment.

Segment	Healthy Tissue Volume (mL)	Tumor Volume (mL)	Volumetric Flow Rate (mL/min)
S1	-	-	-
S2	241	-	24.1
S3	96	-	9.6
S4a	24	-	2.4
S4b	48	-	4.8
S5	330	-	33
S6	183	-	18.3
S7	228	-	22.8
S8	340	68	68
Total	1490	68	183

Table 2. Patient 2 liver mass volumes and blood flow rates per segment.

Segment	Healthy Tissue Volume (mL)	Tumor Volume (mL)	Volumetric Flow Rate (mL/min)
S1	-	-	-
S2	157.8	-	15.8
S3	157.4	-	15.7
S4	309.4	-	30.9
S5	151.7	-	15.2
S6	193.1	-	19.3
S7	155.3	-	15.5
S8	385.4	77.1	77.1
Total	1510	77.1	189.5

Table 3. Patient 3 liver mass volumes and blood flow rates per segment.

Segment	Healthy Tissue Volume (mL)	Tumor Volume (mL)	Volumetric Flow Rate (mL/min)
S1	62	-	6.2
S2	128	-	12.8
S3	181	-	18.1
S4a	73	-	7.3
S4b	11	-	1.1
S5	124	-	12.4
S6	169	-	16.9
S7	373	-	37.3
S8	204	39.8	40.3
Total	1325	39.8	152.4

The perfusion model developed by Aramburu et al. was used to determine the segmental arterial blood flow rates [16]. A normal/healthy tissue perfusion of $k_1 = 0.1$ mL min^{-1} mL^{-1} was adopted for all segments, with a tumor tissue perfusion of $k_2 = 0.5$ mL min^{-1} mL^{-1} [17,18]. In this model, the average blood flow rate flowing towards a segment s, i.e., q_s, is calculated with Equation (1):

$$q_s = V_{0,s}k_1 + V_{c,s}k_2,\tag{1}$$

where q_s is the volumetric flow rate to segment s (with s from segment 1 [S1] to segment 8 [S8]), $V_{0,s}$ is the volume of healthy tissue in segment s, $V_{c,s}$ is the volume of the tumor tissue in segment s, k_1 is the healthy tissue arterial perfusion, and k_2 is the tumor tissue arterial perfusion. Tables 1–3 collect the liver volumes and flow rates of each patient analyzed per segment.

The RE treatment was computer-simulated with Y-90 resin SIR-Spheres® (Sirtex Medical Limited, Australia). The activity to be delivered was calculated with the body surface area method [19], assuming a 1.76 m 74 kg male in all cases. According to this method, 1.7 GBq must be administered in the analyzed cases, that is, approximately 34 million microspheres.

The infusion device modeled was a 2.7 F end-hole microcatheter with inner and outer diameters of D_{in} = 0.65 mm and D_{out} = 0.9 mm, respectively, modeling a Progreat® microcatheter (Terumo, Tokyo, Japan). A selective catheter location was modeled and located before the first branch that irrigates segment 8, approximately in the initial 1/3 of the branch. The tip of the microcatheter was radially centered in the lumen of the artery. An additional microcatheter location was assumed for Patient 2, explained later.

2.2. Baseline and Simplified Hepatic Artery Geometries

As previously said, the geometry of Patient 1's hepatic artery was modified step by step, with the aim of generating a rule that ensures that the segment-to-segment microsphere distributions calculated from the simulations with the baseline and simplified geometries are similar. Once the simplification with the desired characteristics (i.e., minor impact on segment-to-segment microsphere distribution) were obtained, hepatic arteries for Patient 2 and Patient 3 were likewise simplified. For Patient 1 (*Patient 1-Baseline*), four simplifications were made to the baseline geometry. Figure 1 shows the geometries obtained from the simplifications. The arrowhead illustrates the microcatheter-tip position in each case, arrows indicate the branches feeding the tumor-bearing segment 8, and labels S1–S8 indicate the segment(s) that each outlet irrigate(s). First, the upstream branches were removed, giving as a result two simplifications. The first one consists of a simplified geometry where the branches irrigating the segments with no tumors are truncated (*Patient 1-Reduc1*, see Figure 1b). In the second (upstream) simplification, in addition to the simplifications made in the first (upstream) simplification, all the upstream branches that are farther than 3 cm from the microcatheter-tip are removed (*Patient 1-Reduc2*, see Figure 1c). Then, the downstream branches were simplified, truncating at locations where the bifurcation gives rise to two daughter vessels that irrigate the same segment (*Patient 1-Truncated-3cm*, see Figure 1d). Finally, the geometry has been truncated before the first bifurcation, adding a branch in the perpendicular direction to the inlet boundary, with the same diameter and a length of 1 cm (*Patient 1-Reduc3*, see Figure 1e). This final simplification is for obtaining a fully developed-like flow on original inlet section.

The criterion that we are going to establish to deem a simplification as valid is a maximum of 10 percent points of difference at a given segment between the segment-to-segment microsphere distributions of the simulations of the baseline and simplified geometries. In the clinical application of these simulations, we are interested in predicting the segment-to-segment microsphere distribution for a potential improvement in the treatment planning, so the validity criterion of the simplification is based on these results. Qualitative assessment of velocity contours and vectors is used to analyze if the geometry reduction-related changes in blood-flow patterns are excessive.

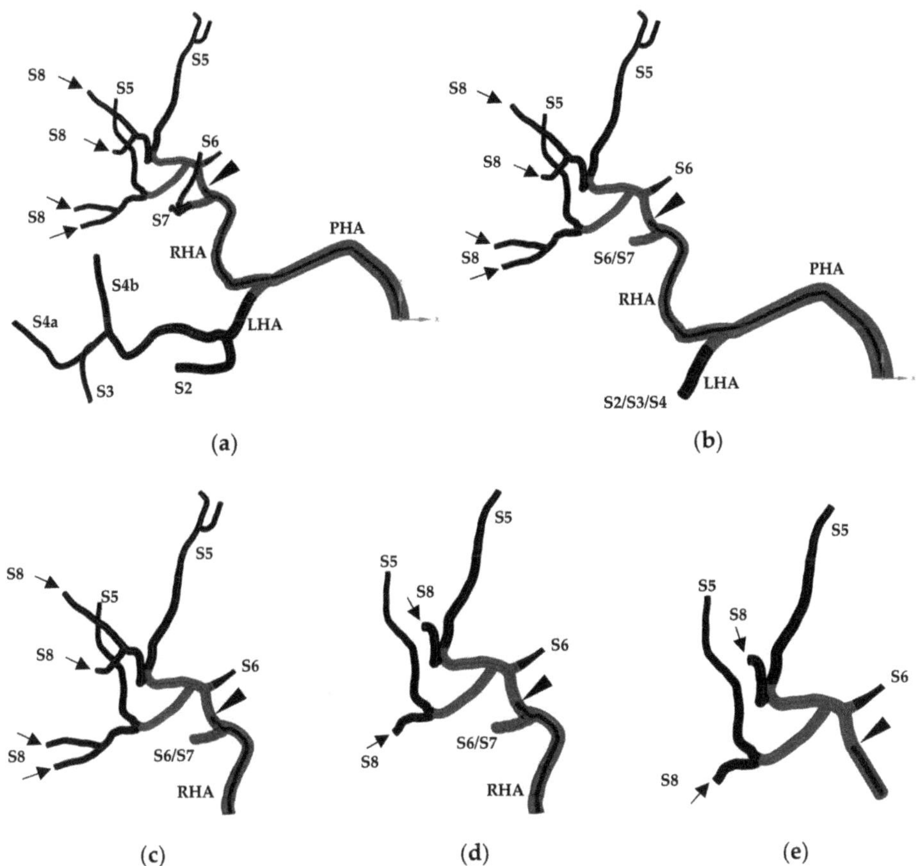

Figure 1. Patient 1 hepatic artery tree simplifications: (**a**) Baseline geometry (i.e., *Patient 1-Baseline*); (**b**) Upstream branches truncation (i.e., *Patient 1-Reduc1*); (**c**) Upstream truncation, 3 cm from the microcatheter position (i.e., *Patient 1-Reduc2*); (**d**) Intra-segmental branches truncation (i.e., *Patient 1-Truncated-3cm*); (**e**) Truncation before the first upstream branch (i.e., *Patient 1-Reduc3*). PHA: proper hepatic artery; LHA: left hepatic artery; RHA: right hepatic artery. Arrowheads indicate microcatheter-tip location. Arrows indicate branches feeding the tumor-bearing segment. S1–S8 in each outlet indicates that all the downstream branches arising from that outlet feed that segment.

After analyzing (quantitatively) the segment-to-segment microsphere distributions and (qualitatively) the flow characteristics (velocity magnitude contours and vectors) near the microcatheter tip for the truncated geometries of Patient 1, the selected truncated geometry is the geometry shown in Figure 1d, that is, the geometry with the following characteristics: all the upstream branches that are farther than 3 cm from the microcatheter tip are truncated and all the downstream branches that, after a bifurcation feed, had the same segment truncated as well. This is because this geometry is the shortest geometry that fulfills the imposed 10 percent point difference limit in microspheres segment-to-segment distribution for a given segment. The same rule followed to generate *Patient 1-Truncated-3cm* was used to define the truncated geometries of Patient 2 (*Patient 2a-Truncated-3cm*) and Patient 3 (*Patient 3-Truncated-3cm*), shown in Figure 2.

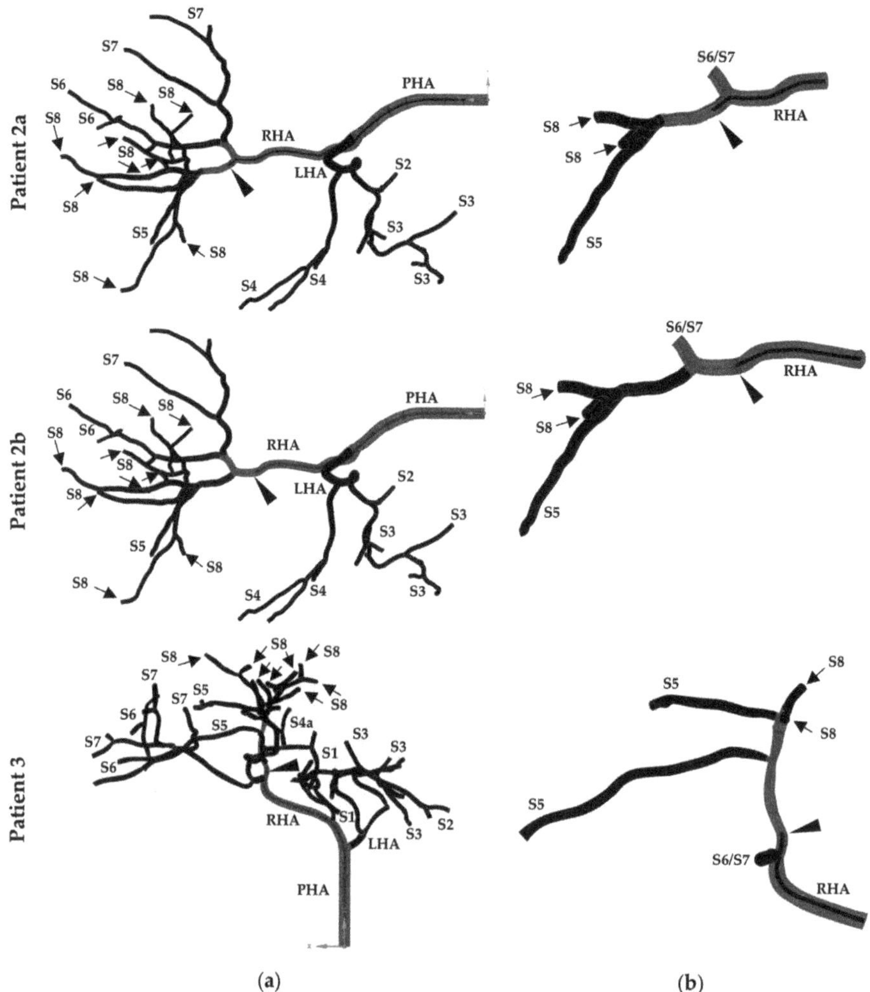

(a)   (b)

Figure 2. Patient 2 and Patient 3 hepatic artery trees: (**a**) Baseline geometries (i.e., *Patient 2a-Baseline*, *Patient 2b-Baseline*, and *Patient 3-Baseline*); (**b**) Truncated geometries (i.e., *Patient 2a-Truncated-3cm*, *Patient 2b-Truncated-3cm*, and *Patient 3-Truncated-3cm*). PHA: proper hepatic artery; LHA: left hepatic artery; RHA: right hepatic artery. Arrowheads indicate microcatheter-tip location. Arrows indicate branches feeding the tumor-bearing segment. S1–S8 in each outlet indicates that all the downstream branches arising from that outlet feed that segment.

The three cases reported have a bifurcation between the microcatheter tip and the 3-cm upstream truncation. A second microcatheter-tip location was chosen for Patient 2, to study an extra case that has no bifurcation in this inlet-to-microcatheter-tip zone, hereafter Patient 2b, being Patient 2a the case with the original microcatheter-tip location (see Figure 2 top and middle panels).

2.3. Preprocessing

2.3.1. Spatial Domain Discretization

All the geometries described in the Section 2.2 were discretized following the same procedure and Fluent Meshing 2020R1 (ANSYS Inc., Canonsburg, PA, USA) software package. The type of elements used was poly-hexahedral. The general settings for generating the mesh include a maximum element size of 0.2 mm, a minimum element size of 0.02 mm and a growth rate of 1.1. A local body sizing was also used for the microcatheter with an element size of 0.05 mm. Finally, inflation layers were used in all the walls to capture the boundary layer in sufficient detail. This refinement has 4 layers, the first layer with 0.01 mm and increases with a growth rate of 1.1. Table 4 shows the quantity of elements of the baseline and truncated geometries.

Table 4. Element quantity comparison (in millions).

Patient	Baseline Geometry	Truncated Geometry
Patient 1	3	0.8
Patient 2a	5	0.8
Patient 2b	4.7	0.95
Patient 3	3.8	0.6

2.3.2. Mathematical Modeling

In modeling the RE treatment, the hemodynamics and the Y-90 resin microsphere transport were modeled. The mathematical model used in this work is the one presented in Aramburu et al. [8]. Regarding the hemodynamics, blood was assumed an isothermal, incompressible, non-Newtonian fluid flowing in laminar regime. The governing equations are the conservation of mass (Equation (2)) and the conservation of the linear momentum (per unit blood volume) (Equation (3)) read:

$$\nabla \cdot \mathbf{u} = 0, \tag{2}$$

$$\rho\left(\frac{\partial \mathbf{u}}{\partial t} + \nabla \cdot (\mathbf{uu})\right) = -\nabla p + \nabla \cdot \boldsymbol{\tau} + \mathbf{F}, \tag{3}$$

where $\mathbf{u}$ is the fluid velocity vector, ρ is the fluid density (1050 kg/m^3 [20]), p is the fluid pressure, $\mathbf{F}$ is the body force per unit volume, and $\boldsymbol{\tau}$ is the stress tensor, which is defined as in Equation (4) for blood:

$$\boldsymbol{\tau} = \mu_{\text{app}}(\dot{\gamma})[\nabla \mathbf{u} + (\nabla \mathbf{u})^{\text{T}}], \tag{4}$$

where $\mu_{\text{app}}(\dot{\gamma})$ is the apparent blood viscosity, which depends on the shear rate ($\dot{\gamma}$) as indicated in Equations (5) and (6) [7]:

$$\mu(\dot{\gamma}) = \max\left\{\mu_0, \left(\sqrt{\mu_\infty} + \frac{\sqrt{\tau_0}}{\sqrt{\lambda} + \sqrt{\dot{\gamma}}}\right)^2\right\}, \tag{5}$$

$$\dot{\gamma} = \sqrt{\nabla \mathbf{u}[\nabla \mathbf{u} + (\nabla \mathbf{u})^{\text{T}}]}, \tag{6}$$

where $\mu_0 = 0.00309$ Pa·s, $\mu_\infty = 0.002654$ Pa·s, $\tau_0 = 0.004360$ Pa, and $\lambda = 0.02181$ s^{-1} are the minimum viscosity, asymptotic viscosity, the apparent yield stress, and the shear stress modifier, respectively [7,12,21–25].

Microspheres were modeled as 32 µm 1600 kg/m^3 spheres. The Discrete Phase Model of Fluent 2020R1 (ANSYS® Inc.) was used to calculate the microsphere trajectory in a Lagrangian reference frame, using a two-way blood–microsphere interaction coupling, and considering no interaction between microspheres. Microsphere motion is governed by the Newton's Second Law of Motion, in which the virtual mass force, pressure gradient force, gravitational force, and drag forces are considered as in reference [8]. This governing equation expressed per unit microsphere mass reads (Equation (7)):

$$\frac{d\mathbf{u}_{\text{p}}}{dt} = \mathbf{f}_{\text{V}} + \mathbf{f}_{\text{P}} + \mathbf{f}_{\text{G}} + \mathbf{f}_{\text{D}}, \tag{7}$$

where $\mathbf{u}_p$ is the velocity of the particle (i.e., the microsphere), and $\mathbf{f}_V$, $\mathbf{f}_P$, $\mathbf{f}_G$, and $\mathbf{f}_D$, are the virtual mass (Equation (8)), pressure gradient (Equation (9)), gravitational (Equation (10)), and drag (Equation (11)) forces, respectively:

$$\mathbf{f}_V = C_V \frac{\rho}{\rho_P}\left[(\mathbf{u}_p\cdot\nabla)\mathbf{u} - \frac{d\mathbf{u}_p}{dt}\right], \tag{8}$$

where C_V is the virtual mass force coefficient and ρ_P is the particle (i.e., microsphere) density,

$$\mathbf{f}_P = \frac{\rho}{\rho_P}(\mathbf{u}_p\cdot\nabla)\mathbf{u}, \tag{9}$$

$$\mathbf{f}_G = \frac{\rho_P - \rho}{\rho_P}\mathbf{g}, \tag{10}$$

where $\mathbf{g}$ is the vector of the acceleration of gravity, 9.81 m/s^2 in magnitude and considering patient's recumbent position during injection,

$$\mathbf{f}_D = \frac{18\mu_{app}}{\rho_P d_p^2}\frac{C_D Re_p}{24}(\mathbf{u} - \mathbf{u}_p), \tag{11}$$

where d_p is the particle (i.e., microsphere) diameter, C_D is the drag coefficient, and Re_p is the particle (i.e., microsphere) Reynolds number (Equation (12)):

$$Re_p = \frac{\rho d_p \left|\mathbf{u}_p - \mathbf{u}\right|}{\mu_{app}}. \tag{12}$$

2.3.3. Boundary Conditions

The boundary conditions prescribed are similar to those presented in Aramburu et al. [16], i.e., inflows at the inlets and inflow fractions (percentages of the total inflow) at the outlets. This model considers the arterial blood flow need of each segment and distributes the blood in the computational domain according to these needs. Regarding the inlet section of the artery, an inflow waveform was prescribed. Figure 3 shows the six periodic (period = 1 s) waveforms used, two per patient. To calculate the mean values of the waveforms, the perfusion model developed by Aramburu et al. [16] and applied to these three patients in Section 2.1 was used (see row "Total" in Tables 1–3) for the baseline geometries. For the truncated geometries, the segments irrigated from the inlet section were considered. These waveforms were translated into uniform velocities at the inlet section. At the inlet section of the microcatheter, a uniform constant velocity equal to the systolic peak velocity was prescribed. With regard to the outlet sections, the flow fractions were prescribed, i.e., the percentage of the total inflow flowing through each outlet. The flow split was defined as explained in Aramburu et al. [16], where the flow fraction at an outlet depends on the flow toward the segment that the outlet is feeding (calculated with Equation (1)) and the level of generation the outlet is located at. For example, if a segment is irrigated by a tree consisting of an artery that bifurcates into branch 1 and branch 2, which bifurcates into branch 3, and branch 4 and outlets are at branches 1, 3, and 4, then 50% of the blood flow will be provided by the outlet at branch 1, and 25% will be provided by each outlet at branches 3 and 4. Walls were assumed as rigid, impermeable, with no relative velocity between the wall velocity and fluid velocity (no-slip condition) and the impact between walls and microspheres was assumed as elastic.

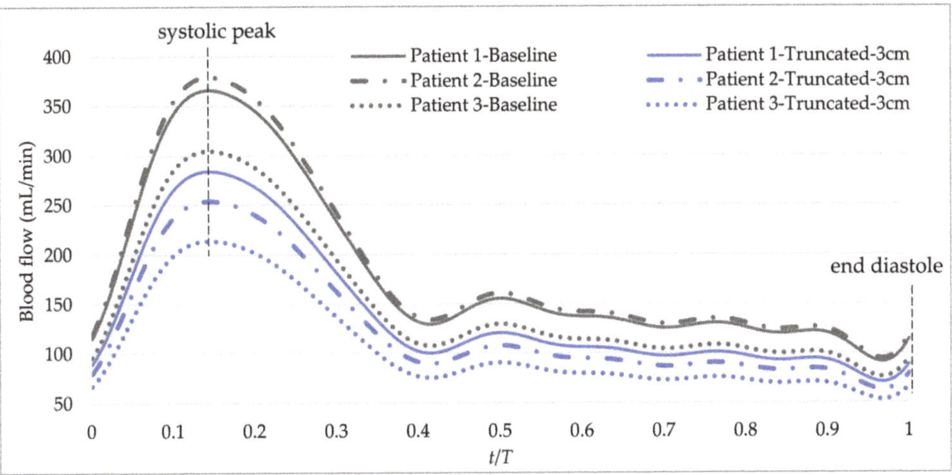

Figure 3. Inlet volumetric blood flows. Systolic peak (t/T = 0.15, with T = 1 s) and end diastole (t/T = 1, with T = 1 s) are the points where the postprocessing is done.

For the mass flow rate of microspheres to be injected, it was assumed that all the microspheres were diluted in a 5-mL vial. Microspheres were assumed to be injected with the initial concentration in the vial. Hence, the prescribed microsphere mass flow rates for Patient 1, Patient 2, and Patient 3 were 2.040×10^{-5} kg/s, 2.119×10^{-5} kg/s, and 2.096×10^{-5} kg/s, respectively.

2.4. Solver Settings

The governing equations of hemodynamics and microsphere transport were solved numerically using the finite volume method-based Fluent 2020R1 (ANSYS Inc.). The pressure and velocity were solved in a segregated way using the SIMPLE scheme for pressure–velocity coupling. Gradients were computed using a least squares cell-based algorithm, and the pressure and momentum interpolations were done with a second order algorithm. Finally, the transient formulation was defined as a second order implicit, with a fixed time step of 2×10^{-3} s. The maximum iterations per time step was limited to 80.

Regarding the convergence criteria used, minimum residuals of 1×10^{-5} were fixed for continuity and the three components of the velocity.

Regarding the number of cardiac cycles simulated, RE was simulated as follows: a first cardiac cycle was simulated ($t = -1$ s to $t = 0$ s) to eliminate the influence of the initial value (i.e., the convergence cycle). Then the actual simulation began. Microspheres were injected during the first cardiac cycle ($t = 0$ s to $t = 1$ s) (i.e., the injection cycle). Two extra cardiac cycles were simulated without injecting more microspheres ($t = 1$ s to $t = 3$ s) (i.e., extra cycle 1 and extra cycle 2). These two extra cardiac cycles are simulated to ensure that the fraction of exiting microspheres reach a steady value over time. Therefore, four cardiac cycles are simulated per simulation. With these results, we could extrapolate the final distribution of all the microsphere of the vial, assuming that the microsphere distribution will remain unchanged throughout the treatment. This model provides a prediction of the outlet-to-outlet microsphere distribution (of the microspheres that have been injected in one second). The simulation tracks the travel of microspheres since they are injected until they exit the computational domain. With these outlet-to-outlet microsphere distribution results, the segment-to-segment microsphere distribution can be predicted, although no extra information can be given about dynamics of the microspheres once they leave the computational domain.

The CFD model (type of geometry plus governing equations plus type of boundary conditions) and the simulation strategy (simulation of one cardiac cycle where microspheres are injected plus sufficient additional cardiac cycles to ensure that most of the injected microspheres exit the computational domain) used in this study have been recently validated in vivo in a proof-of-concept study, where the CFD model was defined using pre-RE imaging techniques and the simulated segment-to-segment microsphere distribution was compared with the measured segment-to-segment activity distribution taken from post-RE imaging techniques [26].

2.5. Postprocessing

The main objective of this research is to analyze the possibility of reducing the computational time of RE simulations by reducing the size of the geometry, without influencing much in the segment-to-segment microsphere distribution. A threshold of 10 percent points was established as a maximum difference to be accepted for a given segment in the simulation results of the baseline and truncated geometries. To analyze the effects of geometry truncation on simulation time reduction and on microsphere distribution, the following postprocessing of the results was done:

- Cycle-to-cycle computational time: to quantitatively assess the cost (in time) of each of the cycles.
- Segment-to-segment microsphere distribution: to qualitatively assess the segment-to-segment microsphere distribution at the end of the treatment. This is calculated from the cumulative number of microspheres exiting through each outlet. The final outlet-to-outlet microsphere distribution is translated into the segment-to-segment microsphere distribution, which is the important parameters due to its clinical implication.
- Velocity magnitude contours and velocity vectors at the systolic peak and end diastole (see Figure 3) at the cross-section of the microcatheter tip: to qualitatively analyze important changes in the blood-flow pattern near the microcatheter tip. The important postprocessing is the segment-to-segment microsphere distribution for its clinical implications, but blood-flow patterns should be similar between the baseline and truncated geometries so that the microsphere distribution is also similar.

3. Results

For each patient, blood flow velocity magnitude contours and vectors at the cross-sectional plane of the microcatheter tip of baseline and truncated geometries are presented, together with segment-to-segment prescribed blood flow split and calculated microsphere distributions. These results are in Figures 4 and 5 for Patient 1, Figures 6 and 7 for Patient 2a, Figures 8 and 9 for Patient 2b, and Figures 10 and 11 for Patient 3. Figures 5, 7, 9 and 11 show the distribution of the microspheres that exit the computational domain toward the segments. These values are normalized with respect to the total number of microspheres that exit the computational domain. In order to be able to compare the blood flow split and the microsphere distribution, blood flow values have been normalized with the blood flow through the artery branch where the microcatheter tip is located. It is important to note that microsphere and blood flow distributions are given in percentage values. When reporting differences between these magnitudes at a given segment, the absolute difference is reported, with units being percent points (%). The computational cost is also studied, focusing on the number of cardiac cycles required for microspheres to exit the computational domain and on the time reduction due to geometry simplification.

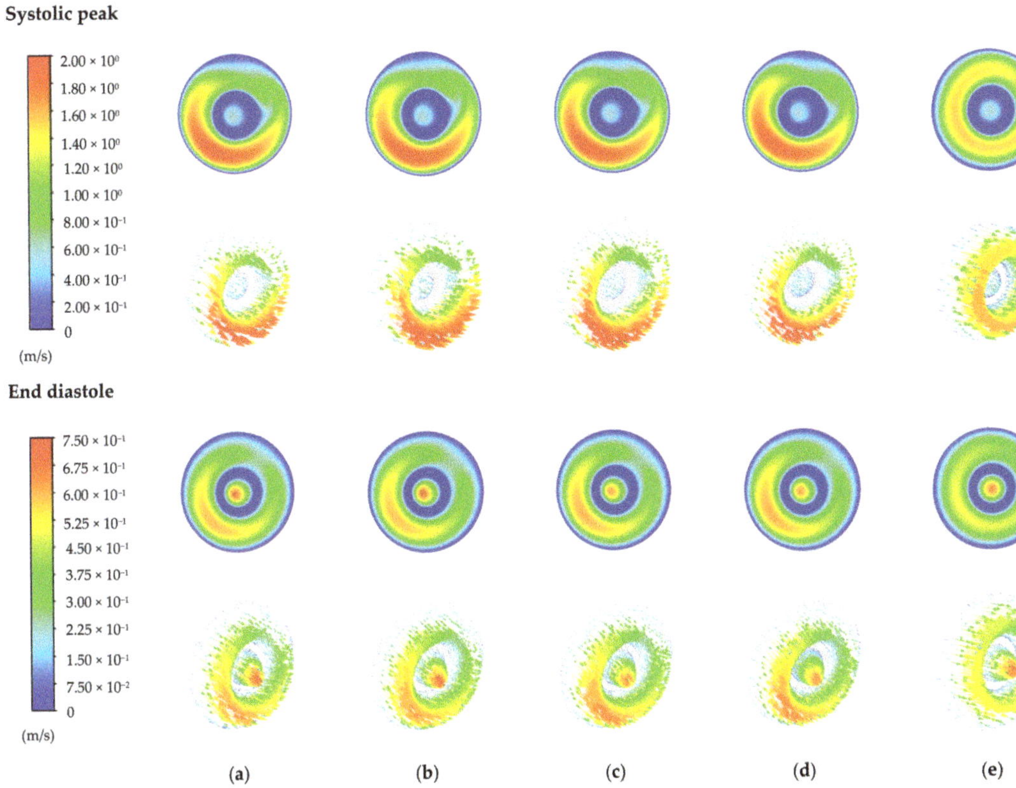

Systolic peak

End diastole

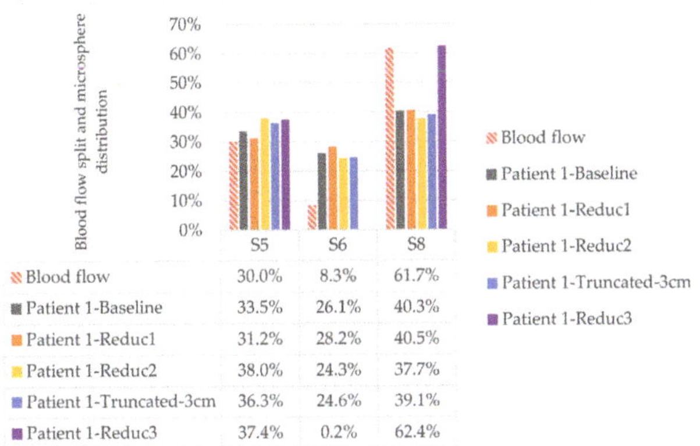

Figure 4. Patient 1 velocity magnitude contours and vectors: (**a**) Baseline geometry (i.e., *Patient 1-Baseline*); (**b**) Upstream branches truncation (i.e., *Patient 1-Reduc1*); (**c**) Upstream truncation, 3 cm from the microcatheter position (i.e., *Patient 1-Reduc2*); (**d**) Intra-segmental branches truncation (i.e., *Patient 1-Truncated-3cm*); (**e**) Truncation before the first upstream branch (i.e., *Patient 1-Reduc3*).

	S5	S6	S8
Blood flow	30.0%	8.3%	61.7%
Patient 1-Baseline	33.5%	26.1%	40.3%
Patient 1-Reduc1	31.2%	28.2%	40.5%
Patient 1-Reduc2	38.0%	24.3%	37.7%
Patient 1-Truncated-3cm	36.3%	24.6%	39.1%
Patient 1-Reduc3	37.4%	0.2%	62.4%

Figure 5. Patient 1, segment-to-segment blood flow distribution (normalized with the blood flow through the artery branch where the microcatheter tip is located) and microsphere distributions in the segments downstream from the injection plane (percentage of microspheres that have reached each segment, normalized with the total number of microspheres that exited the computational domain).

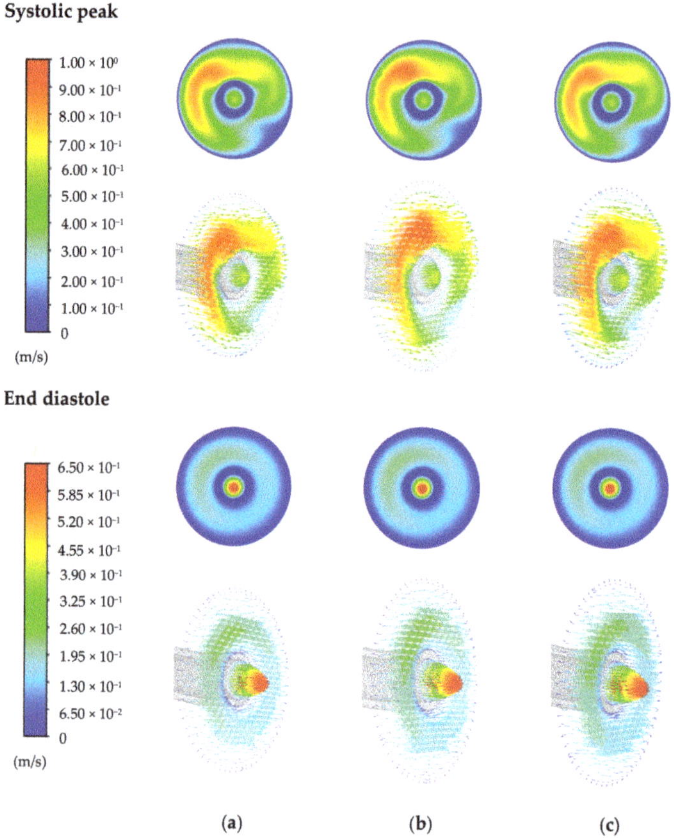

Figure 6. Patient 2a velocity magnitude contours and vectors: (**a**) Baseline geometry (i.e., *Patient 2a-Baseline*); (**b**) Upstream truncation, 3 cm from the microcatheter-tip location, and downstream truncation, intra-segmental branches (i.e., *Patient 2a-Truncated-3cm*); (**c**) Upstream truncation, 4 cm from the microcatheter-tip location, and downstream truncation, intra-segmental branches (i.e., *Patient 2a-Truncated-4cm*).

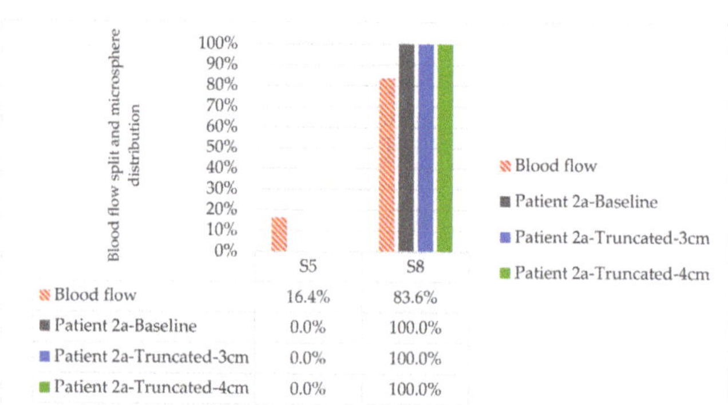

	S5	S8
Blood flow	16.4%	83.6%
Patient 2a-Baseline	0.0%	100.0%
Patient 2a-Truncated-3cm	0.0%	100.0%
Patient 2a-Truncated-4cm	0.0%	100.0%

Figure 7. Patient 2a, segment-to-segment blood flow distribution (normalized with the blood flow through the artery branch where the microcatheter tip is located) and microsphere distributions in the segments downstream from the injection plane (percentage of microspheres that have reached each segment, normalized with the total number of microspheres that exited the computational domain).

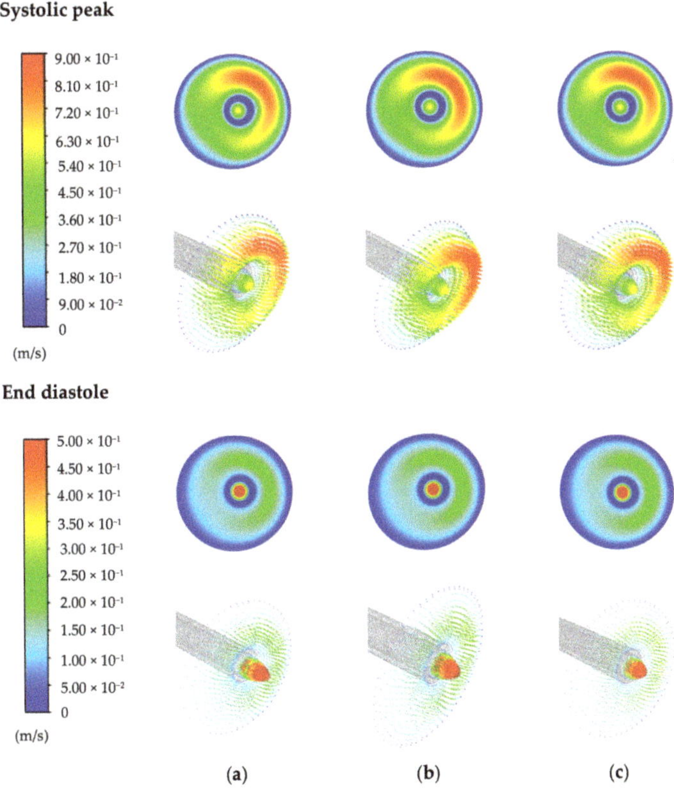

Figure 8. Patient 2b velocity magnitude contours and vectors: (**a**) Baseline geometry (i.e., *Patient 2b-Baseline*); (**b**) Upstream truncation, 3 cm from the microcatheter-tip location and downstream truncation, intrasegmental branches (i.e., *Patient 2b-Truncated-3cm*); (**c**) Upstream truncation, 4 cm from the microcatheter-tip location, and downstream truncation, intrasegmental branches (i.e., *Patient 2b-Truncated-4cm*).

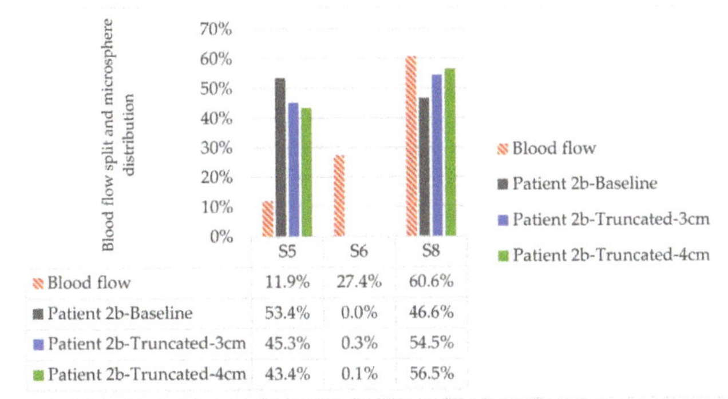

Figure 9. Patient 2b, segment-to-segment blood flow distribution (normalized with the blood flow through the artery branch where the microcatheter tip is located) and microsphere distribution in the segments downstream from the injection plane (percentage of microspheres that have reached each segment, normalized with the total number of microspheres that exited the computational domain).

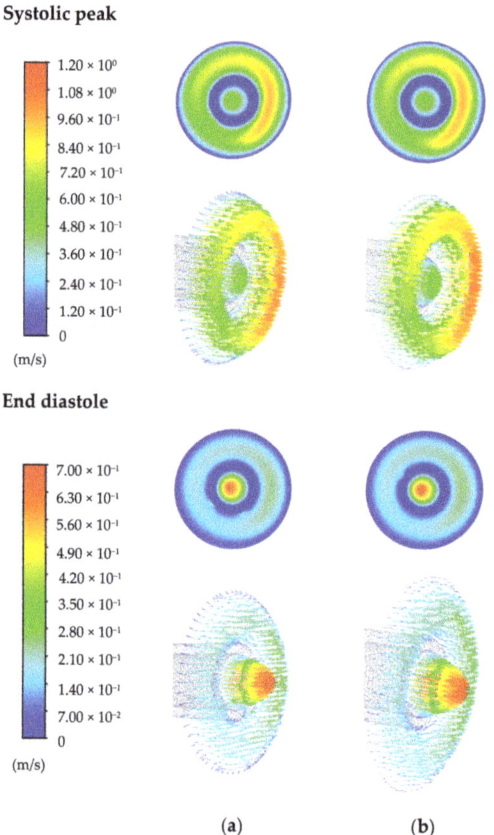

Figure 10. Patient 3 velocity magnitude contours and vectors: (**a**) Baseline geometry (i.e., *Patient 3-Baseline*); (**b**) Upstream truncation, 3 cm from the microcatheter-tip location, and downstream truncation, intra-segmental branches (i.e., *Patient 3-Truncated-3cm*).

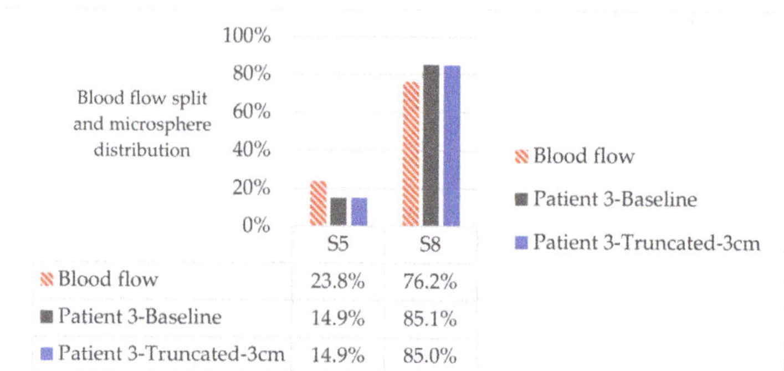

	S5	S8
Blood flow	23.8%	76.2%
Patient 3-Baseline	14.9%	85.1%
Patient 3-Truncated-3cm	14.9%	85.0%

Figure 11. Patient 3, segment-to-segment blood flow distribution (normalized with the blood flow through the artery branch where the microcatheter tip is located) and microsphere distributions in the segments downstream from the injection plane (percentage of microspheres that have reached each segment, normalized with the total number of microspheres that exited the computational domain).

3.1. Patient 1

Figure 4 shows minor differences in velocity magnitude contours and velocity vectors between *Patient 1-Baseline* and almost all simplified cases except for *Patient 1-Reduc3*. Regarding this last case, *Patient 1-Reduc3*, blood flow patterns near the microsphere injection location have changed considerably because of an excessive geometry reduction. This change in blood flow pattern translates into differences in microsphere distributions. In fact, differences of 3.9%, 25.9%, and 33.9% are observed in S5, S6, and S8, respectively, between *Patient 1-Baseline* and *Patient 1-Reduc3* (see Figure 5). With the defined criterion where a threshold of 10 percent points is taken as the maximum difference we can accept at a segment, then the simplification done for *Patient 1-Reduc3* is unacceptable.

However, if we compare *Patient 1-Baseline* and *Patient 1-Truncated-3cm*, where no considerable differences are seen in velocity contours and vectors (see Figure 4), the differences in microsphere distributions are 2.8% in S5, 1.5% in S6, and 1.2% in S8 (see Figure 5). The maximum difference in percent points is below 3%, so this case is regarded as a case with a valid geometry truncation. As explained before, the same simplification criteria used to create *Patient 1-Truncated-3cm* was used in Patients 2 and 3, to analyze the usefulness of the criteria.

3.2. Patient 2

Additional geometries were generated for Patient 2a and 2b, after observing the differences in velocity contours and vectors in Patient 2a (see the high-velocity regions in the systolic peak contours in Figure 6a,b) and in microsphere distributions for Patient 2b (see Figure 9). For the additional truncated geometries, an extra centimeter was left in the geometries, resulting in a truncation of all the upstream branches farther than 4 cm from the microcatheter tip.

3.2.1. Patient 2a

Figure 6 shows differences between *Patient 2a-Baseline* and *Patient 2a-Truncated-3cm* velocity magnitude contours and vectors in the systolic peak (see the area of the high-velocity regions Figure 6a,b). This could potentially result in differences at the injection conditions and therefore in the microsphere distribution. As explained before, an extra case with an extra upstream centimeter was built to assess this difference. Indeed, Figure 6c shows that the results of *Patient 2a-Truncated-4cm* improve compared to *Patient 2a-Truncated-3cm*. Regarding the segment-to-segment microsphere distributions, all the cases show a S8-directed treatment (i.e., 100% microspheres targeting S8). Thus, even though some minor qualitative differences can be seen between *Patient 2a-Baseline* and *Patient 2a-Truncated-3cm*, the latter can be taken as valid because no differences in microsphere distribution are observed.

3.2.2. Patient 2b

Figure 8 shows that velocity magnitude contours and vectors are similar in all cases. However, the segment-to-segment microsphere distributions differ (see Figure 9). Segment to segment microsphere distribution differences between *Patient 2b-Baseline* and *Patient 2b-Truncated-3cm* are 8.1% in S5, of 0.3% in S6, and 7.9% in S8.

As explained before, an extra case was generated due to these differences. An extra centimeter was left upstream, giving as a result of the case *Patient 2b-Truncated-4cm*. The differences between *Patient 2b-Baseline* and *Patient 2b-Truncated-4cm* are 10% in S5, of 0.1% in S6, and 9.9% in S8. It is important to note that even though smaller truncation is applied in *Patient 2b-Truncated-4cm*, the differences with respect to the baseline case increased, compared to the differences between *Patient 2b-Truncated-3cm* and the baseline case. Thus, the case *Patient 2b-Truncated-3cm* has been considered a valid simplification.

3.3. Patient 3

Results show similar qualitative results of velocity contours and vectors between *Patient 3-Baseline* and *Patient 3-Truncated-3cm* (see Figure 9), and segment-to-segment microsphere distributions match—a difference of 0.1 percent points is seen in the microspheres targeting S8, see Figure 10.

3.4. Computational Times

Reducing computational times by effectively reducing the hepatic artery geometry is the main goal of this study. In order to reduce the computational time, the size of the computational domain can be reduced. This reduction potentially allows for an additional reduction in simulation times because a smaller number of cardiac cycles are necessary to ensure that most of the injected microspheres exit the computational domain. Before carrying out the present study, a preliminary study was conducted to analyze the influence of the number of cardiac cycles simulated on the fraction of injected microspheres that exited the computational domain. The computational domains used in this preliminary study were not the same as the ones reported in the present study. Results can be regarded as representative because the baseline geometries are the same in the preliminary and actual studies and the reduced geometries are similar in the preliminary and actual studies. Figure 12 shows the fraction of microspheres that have exited the computational domain over time since the start of the injection—microspheres are injected between 0 s and 1 s, $t = 1$ s is the end of the first cycle (injection cycle) without microsphere injection $t = 2$ s is the end of the second cycle (extra cycle 1), etc.

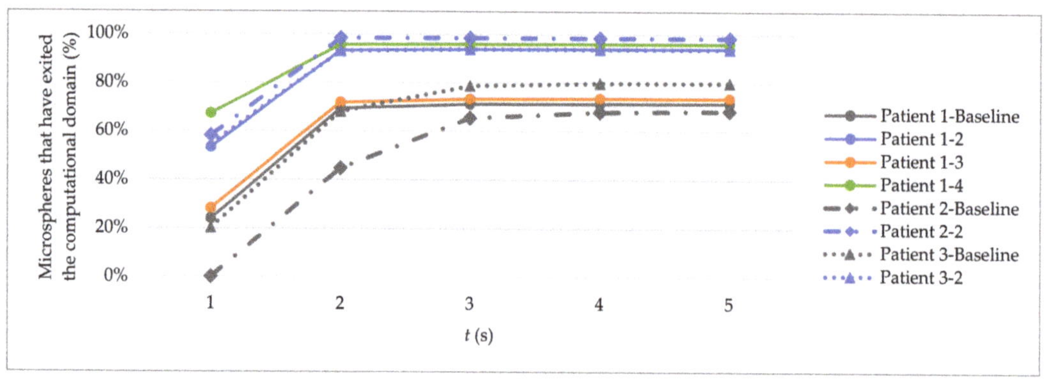

Figure 12. Fraction of microspheres that exited the computational domain over time.

Results show that in most cases a single extra cycle, i.e., extra cycle 1, is sufficient ($t = 1$–2 s) to achieve a steady number of the fraction of exited microspheres, but there are some baseline cases where a second extra cycle, i.e., extra cycle 2, is necessary ($t = 2$–3 s) (see Figure 12, *Patient 2-Baseline* and *Patient 3-Baseline*). The number of microspheres that exit the domain in extra cycles 3 and 4 ($t = 3$–5 s) are negligible. Accordingly, two extra cycles were considered in the simulations of the actual study. Hence, four cycles were necessary per simulation: an initial cardiac cycle for the blood flow convergence ($t = -1$ s to $t = 0$ s), another cycle to inject the microspheres ($t = 0$ s to $t = 1$ s), and two extra cycles ($t = 1$ s to $t = 3$ s).

For the simulations reported in the actual study, the computational time (i.e., cost) per cycle was analyzed and the cost of each cycle was assessed. Table 5 collects the fraction of time needed for each cycle, averaged for the simulations presented in this article. Around 10% of the time is needed for the convergence cycle, a third of the time for the injection cycle, and half of the time for the extra cycles.

Table 5. Average relative computational time of each cardiac cycle of the simulation.

Cycle	Relative Computational Time
Cycle 1: Convergence	11.6%
Cycle 2: Injection	34.1%
Cycle 3: Extra cycle 1	28.0%
Cycle 4: Extra cycle 2	26.3%

Finally, the relation between the number of mesh elements and the simulation duration was studied for the *"Baseline"* and *"Truncated-3cm"* geometries of the actual study, shown in Figure 13. The first observation is that simplifying the geometry results in a reduction in the computational time in all cases. This trend was expected due to the reduction in the number of mesh elements, but it can be quantified. The reduction in Patient 1 is from 23 h and 13 min to 16 h and 38 min (i.e., 28.4%); in Patient 2a, from 29 h and 11 min to 8 h and 03 min (i.e., 72.4%); in Patient 2b, from 47 h and 48 min to 8 h and 52 min (i.e., 81.5%); and in Patient 3, from 25 h and 35 min to 11 h and 00 min (i.e., 57%). Another observation is the fact that all the *"Truncated-3cm"* computational domains consist of a similar number of elements (near one million elements), even though the *"Baseline"* geometries differ in the number of elements. This comes as a result of applying the same geometry reduction strategy. It is also important to note that other factors that influence the simulation time include the quality of the elements of the mesh and the complexity of the studied blood flow.

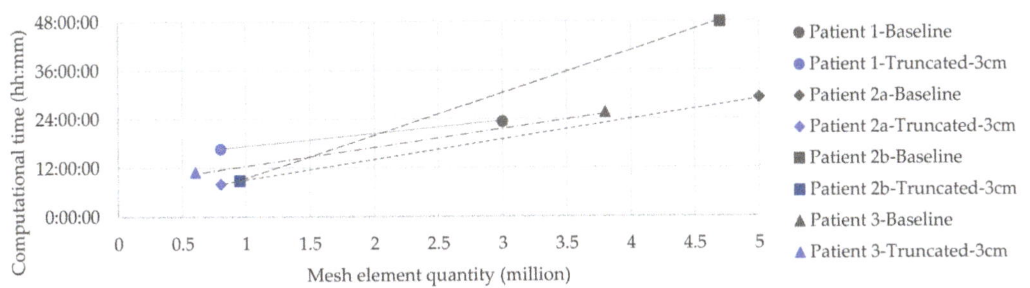

Figure 13. Computational time vs. mesh size.

4. Discussion

In this study, we analyzed the possibility of reducing the computational time of RE simulations by reducing the size of the geometry, without losing much accuracy in the segment-to-segment microsphere distribution prediction. For that purpose, three patient-specific hepatic artery trees were used. The geometry of the first patient was used as a case study to define a simplification strategy. This strategy consisted of removing from the computational domain branches that were upstream and downstream from the microcatheter tip. For upstream branches, all the branches that were farther than 3 cm were removed; for downstream branches, branches were removed at bifurcations where both daughter vessels fed the same liver segment.

The first observation is that the imposed blood flow split differs from the predicted or calculated microsphere distribution (see Figures 5, 7, 9 and 11). This finding is not novel, it is indeed observed in previous studies that used a microcatheter to inject the microspheres [8] or the studies that used the particle release maps [7,9,26] as a research tool—particle release maps correlate each point in the injection cross-sectional plane with the computational domain outlet from which microspheres would exit. Even if a blood flow split-matching microsphere distribution can be achieved [5], this cannot be assumed as a general rule. Therefore, unless a blood flow split-matching microsphere distribution is promoted, CFD simulations are necessary to account for the influence of the local hemodynamic phenomena taking place during RE [3].

With regard to the geometry truncations, the four cases (Patient 1, Patient 2a, Patient 2b, and Patient 3) meet the criterion we established. It can be seen that the more tortuous and more intricate the geometry, the easier it is to obtain a good match between baseline and truncated simulations' results. For Patient 1 and Patient 3, the biggest differences in microsphere distributions at a given segment are 2.8 and 0.1 percent points, respectively. These geometries are very tortuous, having bends close to 90° and bifurcations in the 3 cm of artery before the microcatheter tip. Regarding Patient 2a, the microsphere distribution in the truncated geometry is the same as the one obtained in the baseline geometry, even though minor differences in blood flow conditions near the injection location are qualitatively seen. Patient 2b gives the worst results when the microsphere distributions are compared in the baseline and truncated cases. However, the established criterion is still met. In these Patient 2 cases, the geometry has neither bifurcations nor large tortuosity in its initial part, supporting the hypothesis that the more the intricacies, the easier to replicate the baseline flow patterns in the simplified geometries. Therefore, it could be concluded that if the blood flow is intricate inside the 3 cm between the microcatheter tip and the upstream truncation, then the results of the truncated geometry will match better to the baseline geometry than if it is not.

Regarding the computational time analysis, it has been first concluded that two extra cycles are enough to ensure the fraction of exited microspheres reach a steady value over time (see Figure 12), and a four-cycle simulation is proposed: the first cycle for blood flow convergence, the second cycle for microsphere injection, and the third and fourth cycles for microspheres to exit the computational domain. Regarding the fraction of injected microspheres that do not exit the baseline computational domain, these values are 13.8%, 18.4%, 17.3%, and 9.7% for Patient 1, Patient 2a, Patient 2b, and Patient 3, respectively. These values are similar to those reported by Bomberna et al. [9]. In the truncated geometries, the fraction of non-exiting microspheres reduces to 13.8%, 0%, 6.2%, and 7.6% for Patient 1, Patient 2a, Patient 2b, and Patient 3, respectively. The simulation time of the baseline cases was 31 h 26 min on average, and that of the truncated cases 11 h 08 min on average (see Figure 13), meaning an average reduction of 64.6%. The comparison with other studies in the literature is tricky because the simulation time depends on the modeling approach and on the computer used to solve the model. Among the models used in the literature to study RE, the most similar one to the present model is that of Aramburu [27], with simulations taking 100 h approximately. Other similar models took 11–14 h [12] and around 38 h [7]. Other models have considered reduced order 1D and 0D models, resulting in simulations times of 1 h [28] and 5 s [29], respectively. However, these last models cannot capture local effects, which are essential in the microsphere distribution unless blood flow split-matching microsphere distribution is promoted. In the present case, the simulations have been carried out on a server with 80 cores and 125 GB of RAM. Although with this reduction it is possible to simulate the CFD model in half a day, it is still considered an excessive time for a feasible CFD-based computer platform. It is necessary to reduce more the computational time to offer to the treatment planning multidisciplinary team a versatile CFD-based computer platform to plan the intervention based on a patient-specific basis, being able to adjust the dosimetry with greater accuracy, potentially increasing the effectiveness of the intervention.

From the clinical point of view, we have posited a cancer scenario and we have placed the microcatheter at a reasonable location to target the tumor-bearing segment 8. Once we have performed the analysis with the aim of reducing the simulation time, we should analyze the results from the medical point of view. Results show that in the reduced geometries, the fraction of exiting microspheres targeting segment 8 are 39.1%, 100%, 54.5%, and 85%, for Patient 1, Patient 2a, Patient 2b, and Patient 3, respectively. When seeking to preserve as much healthy liver as possible, the aim is to advance the microcatheter as distally as possible. This allows to deliver high radiation doses via the microspheres released in the tumor-feeding arteries [2]. In this study, the site of injection of microspheres is optimal in Patient 2a, with 100% of microspheres reaching segment 8, and acceptable in the case of Patient 3 with 85% of microspheres reaching the tumoral segment (S8). This is not the case for the remaining cases. In Patient 1 and Patient 2b most of the radioactive load is directed to healthy tissue, with only 40% and 50% of the microspheres reaching segment 8, respectively. Patient 2 shows the importance of a correct microcatheter longitudinal location. In the case Patient 2a, 100% of the microspheres reach the tumor-bearing segment, whereas in the case of Patient 2b only 50% reach the target segment. However, the aim of the study was not to optimize the microcatheter location for an optimal microsphere delivery, but to assess the possibility of reducing the computational time by reducing the size of the simulated geometry. Moreover, results show that there is room for improvement and the microcatheter location could be modified in the simulations to improve the segment 8 targeting via simulation-based assessment. Again, the local effects would probably play an important role in the microsphere distribution and it would be very difficult to mimic the exact microcatheter location during the actual treatment.

Study Limitations

There are some limitations to this study. Regarding the modeling approach, patient-specific hepatic arteries have been used, but the hepatic and cancer states have been posited, with tumors localized in segment 8. Moreover, the first seconds have been taken as representative of the treatment that usually lasts half an hour—computationally unaffordable—without considering the microembolic effect of microspheres.

Regarding the methodology, only three patients—which can be seen as four cases—have been tested. Further research is needed to make sure that the rule works in most cases. Moreover, how to further reduce the computational time should be explored in future studies.

5. Conclusions

In the numerical simulation of RE, a patient-specific CFD analysis is necessary to capture in detail the influence of local effects near the injection location in order to predict the segment-to-segment microsphere distribution. Besides, the rule created to shorten the geometry of the hepatic artery for simulating the microsphere–hemodynamics during RE has resulted in simulations with differences smaller than 10 percent points in a given segment between segment-to-segment microsphere distribution results of the baseline and truncated geometries, as a result of minor changes in blood flow patterns, allowing for an important reduction in simulation time—in this study, an average time reduction of 62%. This simulation time reduction could be a step forward in the development of a simulation-based tool to be used in the clinical setting for personalized RE therapy planning to optimize the treatment.

Author Contributions: Conceptualization, J.A. and R.A.; formal analysis, U.L.; funding acquisition, M.R.-F. and J.I.B.; investigation, U.L. and J.A.; methodology, J.A., J.O. and R.A.; supervision, J.A. and R.A.; writing—original draft, U.L. and J.A.; writing—review & editing, J.O., M.R.-F., B.S., J.I.B. and R.A. All authors have read and agreed to the published version of the manuscript.

Funding: This work has been financed by the PI18/00692 project, integrated in the 2013–2016 National R&D Plan and co-financed by the ISCIII- General Division for Research Evaluation and Promotion and the European Regional Development Fund. The authors acknowledge the support of Cátedra Fundación Antonio Aranzábal-Universidad de Navarra and U.L. gratefully acknowledges the financial support of Eusko Jaurlaritzako Hezkuntza Saila (Basque Government Department of Education) through a Non-Doctor Research Personnel Predoctoral Training Program. The APC was funded by the PI18/00692 project, integrated in the 2013–2016 National R&D Plan and co-financed by the ISCIII- General Division for Research Evaluation and Promotion and the European Regional Development Fund.

Institutional Review Board Statement: For this retrospective study, the University of Navarra ethics committee approved the protocol (186/2018) for this study and it was performed in accordance with the ethical standards laid down in the 1964 Declaration of Helsinki and all subsequent revisions. Informed consent was signed by each patient.

Informed Consent Statement: Not applicable.

Data Availability Statement: Data and codes available under request to the authors.

Conflicts of Interest: The authors declare no conflict of interest. The funders had no role in the design of the study; in the collection, analyses, or interpretation of data; in the writing of the manuscript, or in the decision to publish the results.

References

1. Bray, F.; Ferlay, J.; Soerjomataram, I.; Siegel, R.L.; Torre, L.A.; Jemal, A. Global cancer statistics 2018: GLOBOCAN estimates of incidence and mortality worldwide for 36 cancers in 185 countries. *CA Cancer J. Clin.* **2018**, *68*, 394–424. [CrossRef]
2. Bajwa, R.; Madoff, D.C.; Kishore, S.A. Embolotherapy for Hepatic Oncology: Current Perspectives and Future Directions. *Dig. Dis. Interv.* **2020**, *4*, 134–147. [CrossRef]
3. Aramburo, J.; Antón, R.; Rivas, A.; Ramos, J.C.; Sangro, B.; Bilbao, J.I. Liver Radioembolization: An Analysis of Parameters that Influence the Catheter-Based Particle-Delivery via CFD. *Curr. Med. Chem.* **2020**, *27*, 1600–1615. [CrossRef]
4. Xu, Z.; Jernigan, S.; Kleinstreuer, C.; Buckner, G.D. Solid Tumor Embolotherapy in Hepatic Arteries with an Anti-reflux Catheter System. *Ann. Biomed. Eng.* **2015**, *44*, 1036–1046. [CrossRef] [PubMed]
5. Aramburu, J.; Antón, R.; Rivas, A.; Ramos, J.C.; Sangro, B.; Bilbao, J.I. Computational assessment of the effects of the catheter type on particle–hemodynamics during liver radioembolization. *J. Biomech.* **2016**, *49*, 3705–3713. [CrossRef]
6. Aramburu, J.; Antón, R.; Rivas, A.; Ramos, J.C.; Sangro, B.; Bilbao, J.I. The role of angled-tip microcatheter and microsphere injection velocity in liver radioembolization: A computational particle–hemodynamics study. *Int. J. Numer. Methods Biomed. Eng.* **2017**, *33*. [CrossRef] [PubMed]
7. Basciano, C.A.; Kleinstreuer, C.; Kennedy, A.S.; Dezarn, W.A.; Childress, E. Computer Modeling of Controlled Microsphere Release and Targeting in a Representative Hepatic Artery System. *Ann. Biomed. Eng.* **2010**, *38*, 1862–1879. [CrossRef]
8. Aramburu, J.; Antón, R.; Rivas, A.; Ramos, J.C.; Sangro, B.; Bilbao, J.I. Computational particle-haemodynamics analysis of liver radioembolization pretreatment as an actual treatment surrogate. *Int. J. Numer. Methods Biomed. Eng.* **2016**, *33*, e02791. [CrossRef] [PubMed]
9. Bomberna, T.; Koudehi, G.A.; Claerebout, C.; Verslype, C.; Maleux, G.; Debbaut, C. Transarterial drug delivery for liver cancer: Numerical simulations and experimental validation of particle distribution in patient-specific livers. *Expert Opin. Drug Deliv.* **2020**, 1–14. [CrossRef] [PubMed]
10. Kleinstreuer, C. Drug-targeting methodologies with applications: A review. *World J. Clin. Cases* **2014**, *2*, 742–756. [CrossRef]
11. Roncali, E.; Taebi, A.; Foster, C.; Vu, C.T. Personalized Dosimetry for Liver Cancer Y-90 Radioembolization Using Computational Fluid Dynamics and Monte Carlo Simulation. *Ann. Biomed. Eng.* **2020**, *48*, 1499–1510. [CrossRef] [PubMed]
12. Childress, E.M.; Kleinstreuer, C. Impact of Fluid–Structure Interaction on Direct Tumor-Targeting in a Representative Hepatic Artery System. *Ann. Biomed. Eng.* **2013**, *42*, 461–474. [CrossRef] [PubMed]
13. Couinaud, C. The anatomy of the liver. *Ann. Ital. Chir.* **1992**, *63*, 693–697. [PubMed]
14. Oktar, S.O.; Yücel, C.; Demirogullari, T.; Uner, A.; Benekli, M.; Erbas, G.; Ozdemir, H. Doppler Sonographic Evaluation of Hemodynamic Changes in Colorectal Liver Metastases Relative to Liver Size. *J. Ultrasound Med.* **2006**, *25*, 575–582. [CrossRef] [PubMed]
15. Mise, Y.; Satou, S.; Shindoh, J.; Conrad, C.; Aoki, T.; Hasegawa, K.; Sugawara, Y.; Kokudo, N. Three-dimensional volumetry in 107 normal livers reveals clinically relevant inter-segment variation in size. *HPB* **2014**, *16*, 439–447. [CrossRef]
16. Aramburu, J.; Antón, R.; Rivas, A.; Ramos, J.C.; Sangro, B.; Bilbao, J.I. Liver cancer arterial perfusion modelling and CFD boundary conditions methodology: A case study of the haemodynamics of a patient-specific hepatic artery in literature-based healthy and tumour-bearing liver scenarios. *Int. J. Numer. Methods Biomed. Eng.* **2016**, *32*, e02764. [CrossRef]
17. Yang, H.F.; Du, Y.; Ni, J.X.; Zhou, X.P.; Li, J.D.; Zhang, Q.; Xu, X.X.; Li, Y. Perfusion computed tomography evaluation of angiogenesis in liver cancer. *Eur. Radiol.* **2010**, *20*, 1424–1430. [CrossRef]
18. Tsushima, Y.; Funabasama, S.; Aoki, J.; Sanada, S.; Endo, K. Quantitative perfusion map of malignant liver tumors, created from dynamic computed tomography data. *Acad. Radiol.* **2004**, *11*, 215–223. [CrossRef]
19. Sirtex Medical Limited. SIR-Spheres ®Y-90 Resin Microspheres (Yttrium-90 Microspheres). Available online: https://www.google.com/url?sa=t&rct=j&q=&esrc=s&source=web&cd=&ved=2ahUKEwjVj_uy4dzuAhVxRhUIHe0VCrUQFjAAegQI-Ax-AC&url=https%3A%2F%2Fwww.sirtex.com%2Fmedia%2F169247%2Fssl-us-14-sir-spheres-microspheres-ifu-us.pdf&usg=AOvVaw1W0waSBfewI6noitPu_xzW (accessed on 9 February 2021).
20. Kenner, T. The measurement of blood density and its meaning. *Basic Res. Cardiol.* **1989**, *84*, 111–124. [CrossRef]
21. Basciano, C.A. Computational Particle-Hemodynamics Analysis Applied to an Abdominal Aortic Aneurysm with Thrombus and Microsphere-Targeting of Liver Tumors. Ph.D. Thesis, North Carolina State University, Raleigh, NC, USA, 2010.

22. Basciano, C.A.; Kleinstreuer, C.; Kennedy, A.S. Computational Fluid Dynamics Modeling of 90Y Microspheres in Human Hepatic Tumors. *J. Nucl. Med. Radiat. Ther.* **2011**, *1*. [CrossRef]
23. Kleinstreuer, C.; Basciano, C.A.; Childress, E.M.; Kennedy, A.S. A New Catheter for Tumor Targeting With Radioactive Microspheres in Representative Hepatic Artery Systems. Part I: Impact of Catheter Presence on Local Blood Flow and Microsphere Delivery. *J. Biomech. Eng.* **2012**, *134*. [CrossRef] [PubMed]
24. Childress, E.M.; Kleinstreuer, C.; Kennedy, A.S. A New Catheter for Tumor-Targeting with Radioactive Microspheres in Representative Hepatic Artery Systems—Part II: Solid Tumor-Targeting in a Patient-Inspired Hepatic Artery System. *J. Biomech. Eng.* **2012**, *134*. [CrossRef]
25. Childress, E.M.; Kleinstreuer, C. Computationally Efficient Particle Release Map Determination for Direct Tumor-Targeting in a Representative Hepatic Artery System. *J. Biomech. Eng.* **2013**, *136*. [CrossRef]
26. Antón, R.; Antoñana, J.; Aramburu, J.; Ezponda, A.; Prieto, E.; Andonegui, A.; Ortega, J.; Vivas, I.; Sancho, L.; Sangro, B.; et al. A proof-of-concept study of the in-vivo validation of a computational fluid dynamics model of personalized radioembolization. *Sci. Rep.* **2021**, *11*, 1–12. [CrossRef] [PubMed]
27. Aramburu, J. Liver Radioembolization: Computational Particle-Hemodynamics Studies in a Patient-Specific Hepatic Artery under Literature-Based Cancer Scenarios. Ph.D. Thesis, Universidad de Navarra, Pamplona, Spain, 2016.
28. Umbarkar, T.S.; Kleinstreuer, C. Computationally Efficient Fluid-Particle Dynamics Simulations of Arterial Systems. *Commun. Comput. Phys.* **2015**, *17*, 401–423. [CrossRef]
29. Aramburu, J.; Antón, R.; Rivas, A.; Ramos, J.C.; Larraona, G.S.; Sangro, B.; Bilbao, J.I. Numerical zero-dimensional hepatic artery hemodynamics model for balloon-occluded transarterial chemoembolization. *Int. J. Numer. Methods Biomed. Eng.* **2018**, *34*, e2983. [CrossRef]

 *mathematics*

Article

Simulating Extraocular Muscle Dynamics. A Comparison between Dynamic Implicit and Explicit Finite Element Methods

Jorge Grasa [1,2,*] and Begoña Calvo [1,2]

1 Aragón Institute of Engineering Research (i3A), Universidad de Zaragoza, 50018 Zaragoza, Spain; bcalvo@unizar.es
2 Centro de Investigación Biomédica en Red en Bioingeniería, Biomateriales y Nanomedicina (CIBER-BBN), 28029 Madrid, Spain
* Correspondence: jgrasa@unizar.es

Abstract: The finite element method has been widely used to investigate the mechanical behavior of biological tissues. When analyzing these particular materials subjected to dynamic requests, time integration algorithms should be considered to incorporate the inertial effects. These algorithms can be classified as implicit or explicit. Although both algorithms have been used in different scenarios, a comparative study of the outcomes of both methods is important to determine the performance of a model used to simulate the active contraction of the skeletal muscle tissue. In this work, dynamic implicit and dynamic explicit solutions are presented for the movement of the eye ball induced by the extraocular muscles. Aspects such as stability, computational time and the influence of mass-scaling for the explicit formulation were assessed using ABAQUS software. Both strategies produced similar results regarding range of movement of the eye ball, total deformation and kinetic energy. Using the implicit dynamic formulation, an important amount of computational time reduction is achieved. Although mass-scaling can reduce the simulation time, the dynamic contraction of the muscle is drastically altered.

Keywords: finite element method; implicit FEM; explicit FEM; skeletal muscle

Citation: Grasa, J.; Calvo, B. Simulating Extraocular Muscle Dynamics. A Comparison between Dynamic Implicit and Explicit Finite Element Methods. *Mathematics* **2021**, *9*, 1024. https://doi.org/10.3390/math9091024

Academic Editor: Vince Grolmusz

Received: 22 March 2021
Accepted: 29 April 2021
Published: 1 May 2021

Publisher's Note: MDPI stays neutral with regard to jurisdictional claims in published maps and institutional affiliations.

1. Introduction

The extraocular muscles (EOM) are responsible for the eye movements of the upper eyelid and the eyeball. The group that controls the eyeball contains six muscles: four muscles that run almost a straight course from origin to insertion and hence are called recti and two muscles that run a diagonal course, the oblique muscles [1]. The group that controls eye movement in the cardinal directions are the superior (responsible for elevation, incyclotorsion and adduction), inferior (responsible for depression, extorsion (outward, rotational movement) and adduction), lateral (responsible for abduction) and medial (responsible for adduction) rectus muscles. The movements of the extraocular muscles take place under the influence of a system of extraocular soft tissue pulleys in the orbit. The extraocular muscle pulley system is fundamental to the movement of the eye muscles, in particular to ensure conformity to Listing's law. Certain diseases of the pulleys (heterotopy, instability and hindrance of the pulleys) cause particular patterns of incomitant strabismus [2]. Simulating and analyzing eye movements is useful for assessing the role of these tissues and for exploring the equilibrium of the applied forces that can be impaired and lead to different pathologies [3,4].

The finite element method (FEM) has been widely used to simulate the behavior of skeletal muscle both passively and actively [5–10]. The vast majority of studies have focused on determining the essential parameters that best fit the experimental evidence [4,10,11]. Although different approximations and scenarios have been evaluated with the help of this numerical technique, the contraction of the muscle has been analyzed assuming quasi-static conditions with no inertia effects [4].

Mathematics **2021**, *9*, 1024. https://doi.org/10.3390/math9091024   https://www.mdpi.com/journal/mathematics

An important aspect to consider when reproducing the eyeball movements using FEM is the fast response of the muscles when activated and how the result of this contraction is translated to induce the system motion. This movement is achieved in a few tenths of a second [11]. When tackling these small time periods, realistic simulations should account for the inertia effect of the mass of the different elements. In such scenarios, the use of a dynamic formulation of the FEM is essential. Time integration algorithms for dynamic problems in FEM analysis can be classified as implicit or explicit. Basically, the implicit method computes the state of the model at each time increment based on the information of that same time increment and the previous time increment, while the explicit method uses the data of the previous time increment to solve the motion equations during the new time increment [12]. Implicit time integration schemes are more expensive per time step, but can obtain the solution for larger time steps and provide a control on the dynamic residual force vector, since they are usually used in conjunction with an iterative procedure within each time step [13]. The explicit algorithm can be solved directly without requiring iteration. This method is conditionally stable, and the critical time step for the operator (without damping) is a function of the material specification and the size of the smallest element of the mesh [12]. Increments larger than this critical time can cause the solution to be unstable and oscillations to occur in the model's response, which can lead to excessively distorted elements. To decrease the total computational time, a mass-scaling technique is commonly used, whereby the solver can use larger time increments by artificially increasing the density of the system. However, it is important to ensure that the added mass does not change the behavior of the model, which in simulating active muscle contraction is decisive. The choice between implicit and explicit methods with or without mass scaling has been the subject of many studies [12,14–16]

The aim of this study was therefore to compare dynamic implicit and explicit solutions using ABAQUS software [17] in the analysis of the contraction of the EOM for eyeball movements. More specifically, we compared the prediction of dynamic effects, potential convergence problems, the accuracy and stability of the calculations, the computational time between the two methods and the influence of mass-scaling in the explicit formulation.

The rest of the paper is organized as follows. Section 2 describes the formulation of the skeletal muscle behavior, the implicit and explicit algorithms, the implementation of the user subroutine and the description of the finite element model. In Section 3, selected results of the model comparing both algorithms are presented and then discussed in Section 4. Finally, in Section 5, the main conclusions are summarized.

2. Materials and Methods

2.1. Muscle Contraction Model

Let Ω_0 be a three-dimensional portion of the space representing the solid initial geometry of the muscle tissue. This region defines a set of points at a reference configuration which are identified by the position vector $\mathbf{X}$. The motion of the solid defines the current configuration at time t, Ω_t and can be described by the map $\mathbf{x} = \chi(\mathbf{X}) = \mathbf{x}(\mathbf{X})$.

Let $\mathbf{F} = \frac{\partial \mathbf{x}}{\partial \mathbf{X}}$ be the deformation gradient associated with the motion, where $J \equiv det\mathbf{F} > 0$ is the Jacobian of the transformation. A multiplicative decomposition of this deformation gradient into volume-changing and volume-preserving parts is established to handle the quasi-incompressibility constraint presented by soft tissues ($J \cong 1$) [8]

$$\mathbf{F} = J^{\frac{1}{3}}\bar{\mathbf{F}}, \qquad \bar{\mathbf{F}} = J^{-\frac{1}{3}}\mathbf{F} \tag{1}$$

$$\mathbf{C} = \mathbf{F}^T\mathbf{F}, \qquad \bar{\mathbf{C}} = J^{-\frac{2}{3}}\mathbf{C} = \bar{\mathbf{F}}^T\bar{\mathbf{F}} \tag{2}$$

$$\mathbf{b} = \mathbf{F}\mathbf{F}^T, \qquad \bar{\mathbf{b}} = J^{-\frac{2}{3}}\mathbf{b} = \bar{\mathbf{F}}\bar{\mathbf{F}}^T \tag{3}$$

where $J^{\frac{1}{3}}\mathbf{I}$ and $\bar{\mathbf{F}}$ represent the volumetric and deviatoric deformation gradients, respectively. $\mathbf{C}$ and $\mathbf{b}$ are the right and left Cauchy–Green strain tensors and $\bar{\mathbf{C}}$ and $\bar{\mathbf{b}}$ their modified counterparts.

It is assumed in this work that the contraction process can be modelled as two fictitious steps [8,18] (see Figure 1). The first step is associated with the relative motion of the protein filaments myosin and actin during the power stroke of the cross bridges, and the second step relates to the elastic deformation of cross bridges. This contraction process can be expressed as a multiplicative decomposition of the deformation gradient $\bar{\mathbf{F}}$:

$$\bar{\mathbf{F}} = \bar{\mathbf{F}}_e \bar{\mathbf{F}}_a \tag{4}$$

where $\bar{\mathbf{F}}_a$ is the deformation gradient associated with the contractile response induced by the actin and myosin translation, whereas $\bar{\mathbf{F}}_e$ defines a deformation due to the cross bridges elasticity. The gradient $\bar{\mathbf{F}}_a$ represents the active contraction so it does not need to be integrable. Thus, infinitesimal parts of the tangent space Ω_0 are deformed independently, and the configuration they form after the motion may not be compatible. The gradient $\bar{\mathbf{F}}_e$ guarantees the compatibility in the deformed configuration Ω_t. Accordingly, let $\bar{\mathbf{C}}_e = \bar{\mathbf{F}}_e^T \bar{\mathbf{F}}_e = \bar{\mathbf{F}}_a^{-T} \bar{\mathbf{C}} \bar{\mathbf{F}}_a^{-1}$ be a deformation measure due to the titin and cross bridges motion which is not a state variable since it depends on $\bar{\mathbf{C}}$ and $\bar{\mathbf{F}}_a$.

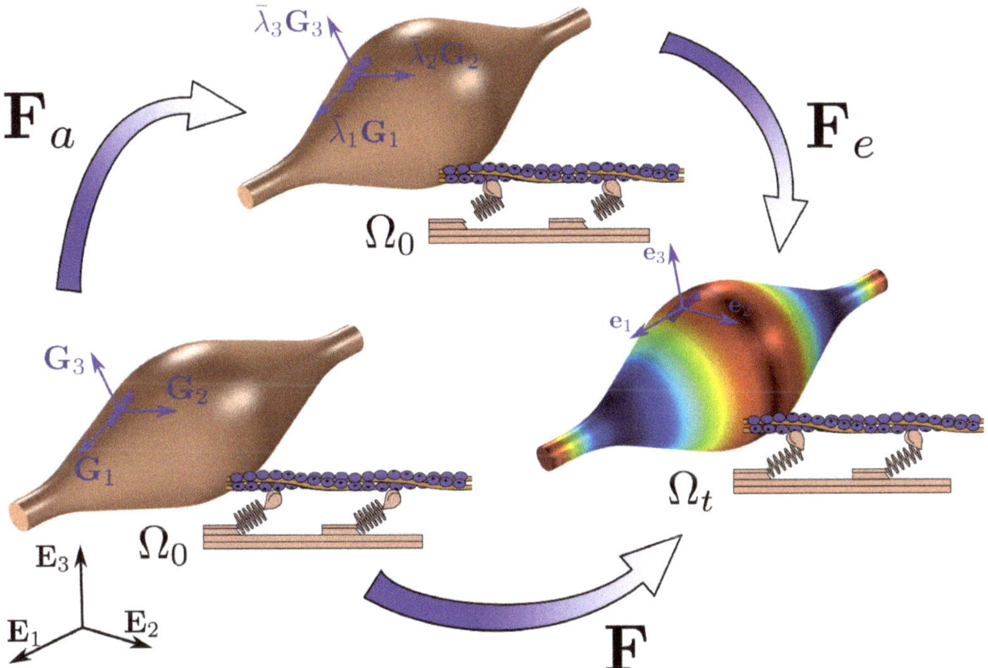

Figure 1. Illustration of the contraction process modelled as two fictitious steps. $\mathbf{G}_i$ vectors located at a point $\mathbf{X}$ in a muscle fiber in the initial configuration Ω_0 transform into new vectors $\bar{\lambda}_i \mathbf{G}_i$ by the active contraction $\mathbf{F}_a$. The intermediate step is associated with the relative motion of the protein filaments due to cross bridges power stroke. In the final step, the cross-bridges are deformed by $\mathbf{F}_e$ to restore the compatibility of the deformed configuration Ω_t.

The active contraction occurs along the direction that is defined by the muscle fibers, so let us introduce this direction as $\mathbf{n}_0$, and let $\bar{\lambda}_a$ be the active stretch. Thus, the active contractile tensor, $\bar{\mathbf{F}}_a'$, can be written in the local coordinate system, $\mathbf{G}_i$, as:

$$\bar{\mathbf{F}}_a' = \bar{\lambda}_a \mathbf{G}_1 \otimes \mathbf{G}_1 + \bar{\lambda}_a^{-1/2} \mathbf{G}_2 \otimes \mathbf{G}_2 + \bar{\lambda}_a^{-1/2} \mathbf{G}_3 \otimes \mathbf{G}_3 \tag{5}$$

where we assume the active contractile tensor $\bar{\mathbf{F}}'_a$ to be isochoric. This local coordinate system varies along the fiber length and is represented at a particular point of the tissue in Figure 1.

The components of the contractile tensor expressed in the global system of coordinates $\mathbf{E}_i$, $\bar{\mathbf{F}}_a$, can be obtained as:

$$\bar{\mathbf{F}}_a = \mathbf{R}^T \bar{\mathbf{F}}'_a \mathbf{R} \tag{6}$$

where $\mathbf{R}$ is the rotation tensor.

A strain energy density formulation decoupled into volume-changing and volume-preserving parts is commonly taken to formulate the elastic constitutive law for transversely isotropic materials such as skeletal muscle [8,18,19]. This energy is formulated in this work as:

$$\Psi = \Psi_{vol}(J) + \bar{\Psi}_p(\bar{\mathbf{C}}, \mathbf{N}) + \bar{\Psi}_a(\bar{\mathbf{C}}_e, \bar{\lambda}_a, \mathbf{N}) \tag{7}$$

where $\mathbf{N} = \mathbf{n}_0 \otimes \mathbf{n}_0$. Equivalently, Ψ can be expressed as a function of the invariants of the strain tensors:

$$\Psi = \Psi_{vol}(J) + \bar{\Psi}_p(\bar{I}_1, \bar{I}_2, \bar{I}_4) + \bar{\Psi}_a(\bar{J}_4, \bar{\lambda}_a) \tag{8}$$

where $\bar{I}_1 = tr\bar{\mathbf{C}}$ and $\bar{I}_2 = \frac{1}{2}((tr\bar{\mathbf{C}})^2 - tr\bar{\mathbf{C}}^2)$ are the first and second modified strain invariants of the symmetric modified Cauchy–Green tensor $\bar{\mathbf{C}}$, and $\bar{I}_4 = \mathbf{n}_0 \cdot \bar{\mathbf{C}}\mathbf{n}_0 = \bar{\lambda}^2$ is the pseudo-invariant related to the anisotropy of the passive response (collagen fibers). Similarly, the active contribution of the strain energy function, $\bar{\Psi}_a$, is expressed in terms of the pseudo-invariant associated to $\bar{\mathbf{C}}_e$ and the preferred direction $\mathbf{n}_0$, $\bar{J}_4 = \mathbf{n}_0 \cdot \bar{\mathbf{C}}_e\mathbf{n}_0 = \lambda_e^2$. As shown, the anisotropy in the formulation is induced by a single orientation for both the passive and the active behavior. Although in fusiform muscles such as the EOM this is commonly accepted, in other muscle architectures two families of fibers should be considered to adopt a more suitable formulation [9,20].

The third term in Equation (8) represents the strain energy associated with the active response and, consequently, with the actin-myosin interaction. This term is written here as a function $\bar{\Psi}'_a$ that relates to the energy stored in the cross-bridges, while $f_1(\bar{\lambda}_a)$ is a function that accounts for the filament overlap and $f_2(t)$ for the muscle activation level:

$$\Psi = \Psi_{vol}(J) + \bar{\Psi}_p(\bar{I}_1, \bar{I}_2, \bar{I}_4) + f_1(\bar{\lambda}_a) f_2(t) \bar{\Psi}'_a(\bar{J}_4) \tag{9}$$

The function $0 < f_1(\bar{\lambda}_a) < 1$ has been experimentally characterized in previous studies for different muscles [19,21] and fitted by a smooth exponential relationship:

$$f_1(\bar{\lambda}_a) = \exp^{\frac{-(\bar{\lambda}_a - \lambda_{opt})^2}{2\zeta^2}} \tag{10}$$

where λ_{opt} is the optimum length of the muscle at which isometric maximum stress is developed and ζ determines the curvature of the function. To formulate $f_2(t)$, we assume in this work that all muscle fibers are completely recruited in each contraction, and this function can be expressed as:

$$f_2(t) = \alpha \tanh^2(s_1 t) \tag{11}$$

where $0 < \alpha < 1$ governs the activation level, s_1 regulates the initial slope of the function and t is the time variable. Finally, the energy stored in the cross-bridges is expressed in terms of the invariant associated to $\bar{\mathbf{C}}_e$ in the direction of the muscle fibers $\mathbf{n}_0$ and a parameter P_0 related to the maximum active stress:

$$\bar{\Psi}'_a = \frac{1}{2} P_0 (\bar{J}_4 - 1)^2 \tag{12}$$

During the muscle contraction process, the second law of thermodynamics can be formulated in the shape of the Clausius–Planck inequality neglecting the thermal dissipation rate. This inequality allows us to consider that some of the power produced internally is stored while another portion is dissipated [8]:

$$\mathcal{D}_{int} = -\dot{\Psi} + \frac{1}{2}\mathbf{S} : \dot{\mathbf{C}} + \frac{1}{2}\mathbf{S}_a : \dot{\mathbf{C}}_a \geq 0 \tag{13}$$

In Equation (13), $\mathbf{S}_a$ represents active stress and $\frac{1}{2}\mathbf{S}_a : \dot{\mathbf{C}}_a$ the muscle power stroke [18]. Following the work of Hernández-Gascón et al. [8], the following constitutive relations are obtained:

$$\mathbf{S} = 2\frac{\partial \Psi}{\partial \mathbf{C}} + \mathbf{F}_a^{-1}\left(2\frac{\partial \Psi}{\partial \mathbf{C}_e}\right)\mathbf{F}_a^{-T} \tag{14}$$

$$\left(\mathbf{P}_a - 2\mathbf{F}_a\frac{\partial \Psi}{\partial \mathbf{C}_a} + 2\mathbf{C}_e\frac{\partial \Psi}{\partial \mathbf{C}_e}\mathbf{F}_a^{-T}\right) : \dot{\mathbf{F}}_a \geq 0 \tag{15}$$

where $\mathbf{P}_a$ is the first Piola–Kirchoff active stress. Since contraction occurs along the muscle fiber only, Equation (15) reduces to:

$$\left[P_a - \frac{\partial \bar{\Psi}}{\partial \bar{\lambda}_a} + \left(2\bar{\mathbf{C}}_e\frac{\partial \bar{\Psi}}{\partial \bar{\mathbf{C}}_e}\bar{\mathbf{F}}_a^{-T}\right) : \frac{\partial \bar{\mathbf{F}}_a}{\partial \bar{\lambda}_a}\right]\dot{\bar{\lambda}}_a \geq 0 \tag{16}$$

This expression leads to the following constitutive relation for the active contraction velocity $\dot{\bar{\lambda}}_a$:

$$P_a - \frac{\partial \bar{\Psi}}{\partial \bar{\lambda}_a} + \left(2\bar{\mathbf{C}}_e\frac{\partial \bar{\Psi}}{\partial \bar{\mathbf{C}}_e}\bar{\mathbf{F}}_a^{-T}\right) : \frac{\partial \bar{\mathbf{F}}_a}{\partial \bar{\lambda}_a} = C\dot{\bar{\lambda}}_a \tag{17}$$

assuming that:

$$C = \frac{1}{v_0}P_0 f_1(\bar{\lambda}_a)f_2(t) \tag{18}$$

where v_0 is associated with the initial contraction velocity. The active stress P_a from Equation (17) is defined as a function of P_0, $f_1(\lambda_a)$ and $f_2(t)$, and v is a friction parameter that takes into account the relative sliding speed between actin and myosin:

$$P_a = -vP_0 f_1(\bar{\lambda}_a)f_2(t) \tag{19}$$

Substituting Equations (18) and (19) and the last term of Equation (9) into Equation (17) leads to the expression for the contraction velocity:

$$\dot{\bar{\lambda}}_a = v_0\left[-v - \frac{1}{f_1(\bar{\lambda}_a)}\frac{\partial f_1(\bar{\lambda}_a)}{\partial \bar{\lambda}_a}\bar{\Psi}_a'(\bar{J}_4) + \left(2\bar{\mathbf{C}}_e\frac{\partial \bar{\Psi}_a'(\bar{J}_4)}{\partial \bar{\mathbf{C}}_e}\bar{\mathbf{F}}_a^{-T}\right) : \frac{\partial \bar{\mathbf{F}}_a}{\partial \bar{\lambda}_a}\right] \tag{20}$$

Since $\bar{\Psi}_a'$ depends on $\bar{J}_4$, Equation (20) reduces to:

$$\dot{\bar{\lambda}}_a = v_0\left[-v - \frac{1}{f_1(\bar{\lambda}_a)}\frac{\partial f_1(\bar{\lambda}_a)}{\partial \bar{\lambda}_a}\bar{\Psi}_a'(\bar{J}_4) + 2\frac{\bar{\lambda}_e^2}{\bar{\lambda}_a}\frac{\partial \bar{\Psi}_a'(\bar{J}_4)}{\partial \bar{J}_4}\right] \tag{21}$$

Taking the first constitutive relation (Equation (14)) and the particular form of the strain energy density function, the expressions for the Cauchy stress tensor and the elasticity tensor can be derived. Both tensors must be provided to define the mechanical constitutive model in the implicit user material subroutine in ABAQUS/Standard [17], whereas only the definition of the Cauchy stress is needed in ABAQUS/Explicit [17].

From Equations (7) and (14), the second Piola–Kirchhoff stress tensor is found to be:

$$\mathbf{S} = \mathbf{S}_{vol} + \bar{\mathbf{S}}_p + \bar{\mathbf{S}}_a \tag{22}$$

The Cauchy stress tensor is obtained by means of a weighted push-forward operation of $\mathbf{S}$, $\sigma = J^{-1}\chi_*(\mathbf{S}) = J^{-1}\mathbf{F}\mathbf{S}\mathbf{F}^T$:

$$\sigma = \sigma_{vol} + \bar{\sigma}_p + \bar{\sigma}_a = p\mathbf{1} + \frac{1}{J}dev\left[\bar{\mathbf{F}}\tilde{\mathbf{S}}_p\bar{\mathbf{F}}^{\mathbf{T}}\right] + \frac{1}{J}dev\left[\bar{\mathbf{F}}_{\mathbf{e}}\tilde{\mathbf{S}}_e\bar{\mathbf{F}}_{\mathbf{e}}^{\mathbf{T}}\right]$$
$$= p\mathbf{1} + \frac{1}{J}dev[\tilde{\sigma}_p] + \frac{1}{J}dev[\tilde{\sigma}_e] \tag{23}$$

with:

$$p = \frac{d\Psi_{vol}(J)}{dJ}, \qquad dev[\cdot] = (\cdot) - \frac{1}{3}tr[\cdot]\mathbf{1} \tag{24}$$

Differentiating Equation (22) with respect to $\mathbf{C}$ leads to the material elasticity tensor $\mathbb{C}$, which can be divided into volumetric and deviatoric parts associated with the passive and active responses as follows:

$$\mathbb{C} = \mathbb{C}_{vol} + \bar{\mathbb{C}}_p + \bar{\mathbb{C}}_a = 2\frac{\partial \mathbf{S}_{vol}}{\partial \mathbf{C}} + 2\frac{\partial \bar{\mathbf{S}}_p}{\partial \mathbf{C}} + 2\frac{\partial \bar{\mathbf{S}}_a}{\partial \mathbf{C}} \tag{25}$$

The elasticity tensor in the spatial configuration, c, is obtained by a weighted push-forward operation of $\mathbb{C}$, which can be expressed as $c = J^{-1}\chi_*(\mathbb{C})$ and results in:

$$c = \mathbb{C}_{vol} + \mathbb{C}_p + \mathbb{C}_a \tag{26}$$

For a detailed explanation about these expressions and further information, the reader is referred to the works of Weiss et al. [22] and Hernández-Gascón et al. [8].

2.2. Principle of Virtual Work and Finite Element Discretization

The principle of virtual work that allows to establish the finite element formulation is derived from the balance of momentum of a body V with boundary S that can be written as:

$$\int_S tdS + \int_V \rho gdV = \int_V \rho\ddot{a} \tag{27}$$

where t is the stress vector, ρ is the material density, g is the gravity acceleration and $\ddot{a}$ is the accelerations vector. Applying the relation between the stress vector and the stress tensor $t = n\sigma$ with n the normal surface vector and the divergence theorem [13], the following relation must be fulfilled at each material point:

$$\nabla \cdot \sigma + \rho g = \rho\ddot{a} \tag{28}$$

Multiplying this equation by a virtual displacement field δa and integrating over the domain V:

$$\int_V \delta a(\nabla \cdot \sigma + \rho g - \rho\ddot{a}) = 0 \tag{29}$$

After applying the divergence theorem and some manipulations, the weak form of the equation of motion that represents the principle of virtual work is obtained:

$$\int_V \sigma : \delta e dV = \int_S \rho(g - \ddot{a})\delta a dV + \int_S t\delta a dS \tag{30}$$

with $e = \frac{1}{2}(\mathbf{1} - F^{-T}F^{-1})$ the Euler–Almansi strain tensor.

Equation (30) is the basis for the finite element discretization where the displacements at the nodes of the mesh elements are considered as the fundamental unknowns. The continuous displacement field a can be approximated at each element as:

$$a = \sum_{k=1}^{n} \phi_k(\xi, \eta, \zeta)u_k \tag{31}$$

where ϕ_k are the interpolation functions of an element supported by n nodes and (ξ, η, ζ) the coordinates of the reference element. The proper definition of the interpolation functions using polynomials (linear in this work) and assembling element matrix contributions to the integration over all the solid volume in Equation (30) conduct to the well-known finite element equation system.

2.3. Implicit Solution Method

The dynamic response in ABAQUS/Standard for nonlinear models is obtained by direct time integration of all of the degrees of freedom of the finite element model [17]. The time step for implicit integration can be chosen automatically on the basis of the "half-increment residual" by monitoring the values of equilibrium residuals at $t + \Delta t/2$ once the solution at $t + \Delta t$ has been obtained. The accuracy of the solution can be assessed and the time step adjusted appropriately.

The equilibrium equations are written at the end of the time step (time $t + \Delta t$) and are calculated from the time integration operator. The finite element approximation is:

$$\mathbf{M}\ddot{\mathbf{u}} + \mathbf{I} = \mathbf{F} \tag{32}$$

where $\mathbf{F}$ is the vector of externally applied forces, $\mathbf{I}$ is the vector of internal element forces, $\mathbf{M}$ is the lumped mass matrix and $\ddot{\mathbf{u}}$ is the accelerations vector.

The algorithm defined by Hilber et al. [23] is:

$$\mathbf{M}\ddot{\mathbf{u}}^{(i+1)} + (1+\alpha)\mathbf{K}\mathbf{u}^{(i+1)} - \alpha\mathbf{K}\mathbf{u}^{(i)} = (1+\alpha)\mathbf{F}^{(i+1)} - \alpha\mathbf{F}^{(i)} \tag{33}$$

where $\mathbf{u}$ is the displacement vector and maintaining Newmark's assumption that the acceleration $\ddot{\mathbf{u}}$ varies linearly over the time step [13]:

$$\mathbf{u}^{(i+1)} = \mathbf{u}^{(i)} + \Delta t \dot{\mathbf{u}}^{(i)} + \Delta t^2 \left[\left(\frac{1}{2} - \beta \right) \ddot{\mathbf{u}}^{(i)} + \beta \ddot{\mathbf{u}}^{(i+1)} \right] \tag{34}$$

$$\dot{\mathbf{u}}^{(i+1)} = \dot{\mathbf{u}}^{(i)} + \Delta t \left[(1-\gamma)\ddot{\mathbf{u}}^{(i)} + \gamma \ddot{\mathbf{u}}^{(i+1)} \right] \tag{35}$$

being $\dot{\mathbf{u}}$ the velocities vector and

$$\beta = \frac{1}{4}(1-\alpha)^2; \qquad \gamma = 1/2 - \alpha; \qquad -\frac{1}{3} \leq \alpha \leq 0 \tag{36}$$

$\alpha = -0.05$ is chosen by default in ABAQUS/Standard as a small damping term to remove the high frequency noise without affecting the lower frequency response [15]. Taking $\alpha = 0$, the Newmark method is obtained. For the algorithmic implementation, it is necessary to obtain $\ddot{\mathbf{u}}^{(i+1)}$ from Equation (34):

$$\ddot{\mathbf{u}}^{(i+1)} = \frac{1}{\beta \Delta t^2}\Delta\mathbf{u} - \frac{1}{\beta \Delta t}\dot{\mathbf{u}}^{(i)} - \frac{1-2\beta}{2\beta}\ddot{\mathbf{u}}^{(i)} \tag{37}$$

where $\Delta\mathbf{u} = \mathbf{u}^{(i+1)} - \mathbf{u}^{(i)}$. Substituting this expression into Equation (33) and taking $\mathbf{I}^{(i)} = \mathbf{K}\mathbf{u}^{(i)}$ yields:

$$\Delta\mathbf{u} = (\mathbf{K}^*)^{-1}\mathbf{F}^* \tag{38}$$

with the algorithmic tangential stiffness matrix

$$\mathbf{K}^* = (1+\alpha)\mathbf{K} + \frac{1}{\beta \Delta t^2}\mathbf{M} \tag{39}$$

and the vector $\mathbf{F}^*$ defined as:

$$\mathbf{F}^* = (1+\alpha)\mathbf{F}^{(i+1)} - \alpha\mathbf{F}^{(i)} - \mathbf{I}^{(i)} + \mathbf{M}\left(\frac{1}{\beta \Delta t}\dot{\mathbf{u}}^{(i)} + \frac{1-2\beta}{2\beta}\ddot{\mathbf{u}}^{(i)} \right) \tag{40}$$

Three factors should be considered when selecting the maximum allowable time step size: the rate of variation of the applied loading, the complexity of the nonlinear damping and stiffness properties and the typical period of vibration of the structure [17]. A maximum increment versus period ratio $\Delta t / T < 1/10$ is recommended for obtaining reliable results [17], where T is the period of typical modes of response. The Hilber et al. [23] α-method time integration scheme for solving the implicit problem can be summarized as:

1. Initialize $\mathbf{u}^0$, $\dot{\mathbf{u}}^0$ and $\mathbf{I}^0$
2. Compute the mass matrix $\mathbf{M}$
3. For each time step increment:

    (a) Initialize the displacement increment $\Delta\mathbf{u}_0$ and the internal force $\mathbf{I}_0^{(i+1)} = \mathbf{I}^{(i)}$

    (b) Iterations $j = 0, \dots$ for finding "dynamic equilibrium" within the time step increment:

    - Compute tangential stiffness matrix: $\mathbf{K}_j$
    - Compute the algorithmic stiffness matrix: $\mathbf{K}_j^* = (1 + \alpha)\mathbf{K}_j + \frac{1}{\beta\Delta t^2}\mathbf{M}$
    - Compute $\mathbf{F}_j^* = (1 + \alpha)\mathbf{F}^{(i+1)} - \alpha\mathbf{F}^{(i)} - \mathbf{I}_j^{(i)} + \mathbf{M}\left(\frac{1}{\beta\Delta t}\dot{\mathbf{u}}^{(i)} + \frac{1 - 2\beta}{2\beta}\ddot{\mathbf{u}}^{(i)}\right)$
    - Solve the linear system: $d\mathbf{u}_{j+1} = (\mathbf{K}_j^*)^{-1}\mathbf{F}_j^*$
    - Update the displacement increments: $\Delta\mathbf{u}_{j+1} = \Delta\mathbf{u}_j + d\mathbf{u}_{j+1}$
    - For each integration point k:
        - Compute the strain increment: $\Delta\mathbf{u}_{j+1} \rightarrow \Delta\varepsilon_{k,j+1}$
        - Compute the stress increment: $\Delta\varepsilon_{k,j+1} \rightarrow \Delta\sigma_{k,j+1}$
        - Compute the total stress: $\sigma_{k,j+1} = \sigma_k^{(i)} + \Delta\sigma_{k,j+1}$
    - Compute internal force: $\mathbf{I}_{j+1}^{(i+1)}$
    - Compute accelerations: $\ddot{\mathbf{u}}_{j+1}^{(i+1)} = \frac{1}{\beta\Delta t^2}\Delta\mathbf{u}_{j+1} - \frac{1}{\beta\Delta t}\dot{\mathbf{u}}^{(i)} - \frac{1 - 2\beta}{2\beta}\ddot{\mathbf{u}}^{(i)}$
    - Compute residual: $\mathbf{r}_{j+1}^* = (1 + \alpha)\mathbf{F}^{(i+1)} - \alpha\mathbf{F}^{(i)} - \mathbf{I}_{j+1}^{(i+1)} - \mathbf{M}\ddot{\mathbf{u}}_{j+1}^{(i+1)}$
    - Check convergence: if $\left\|\mathbf{r}_{j+1}^*\right\| < \eta$, with η the convergence tolerance, go to Step (c).

    (c) Compute the velocities and displacements at the end of the time step:

    - Velocities: $\dot{\mathbf{u}}^{(i+1)} = \dot{\mathbf{u}}^{(i)} + \Delta t\left[(1 - \gamma)\ddot{\mathbf{u}}^{(i)} + \gamma\ddot{\mathbf{u}}^{(i+1)}\right]$
    - Displacements $\mathbf{u}^{(i+1)} = \mathbf{u}^{(i)} + \Delta\mathbf{u}$

2.4. Explicit Solution Method

The explicit dynamics analysis procedure in ABAQUS/Explicit is established by an explicit integration rule together with the use of diagonal or "lumped" element mass matrices. The equations of motion for the body are integrated using the explicit central difference integration rule as follows:

$$\mathbf{u}^{(i+1)} = \mathbf{u}^{(i)} + \Delta t^{(i+1)}\dot{\mathbf{u}}^{(i+\frac{1}{2})} \tag{41}$$

$$\dot{\mathbf{u}}^{(i+\frac{1}{2})} = \dot{\mathbf{u}}^{(i-\frac{1}{2})} + \frac{\Delta t^{(i+1)} + \Delta t^{(i)}}{2}\ddot{\mathbf{u}}^{(i)} \tag{42}$$

The superscripts i, $i - \frac{1}{2}$ and $i + \frac{1}{2}$ refer to the increment number and mid-increment numbers. The state of the analysis is advanced by assuming constant values for the velocities, $\dot{\mathbf{u}}$, and the accelerations, $\ddot{\mathbf{u}}$, across half time intervals [16]. The accelerations are computed at the start of the increment by:

$$\ddot{\mathbf{u}}^{(i)} = \mathbf{M}^{-1}(\mathbf{F}^{(i)} - \mathbf{I}^{(i)}) \tag{43}$$

As the lumped mass matrix is diagonalized, it is a trivial process to invert it, unlike the global stiffness matrix in the implicit solution method. Therefore, each time increment is computationally inexpensive to solve. Several possibilities of this lumping process are available, such as nodal quadrature, row-sum lumping, or a "special lumping technique" where only the latter method produces positive lumped masses for any element type [13]. A stability limit determines the size of the time increment:

$$\Delta t \leq \frac{2}{\omega_{max}} \tag{44}$$

where ω_{max} is the maximum element eigenvalue. A conservative and practical method of implementing the above inequality is:

$$\Delta t = min \frac{L^e}{c^d} \tag{45}$$

where L^e is the characteristic element length and c^d is the dilatational wave speed:

$$c^d = \sqrt{\frac{\lambda + 2\mu}{\rho}} \tag{46}$$

λ and μ are the Lamé elastic constants and ρ is the material density.

To maintain efficiency of the analysis, it is important to ensure that the sizes of the elements are as regular as possible since one small element reduces the time increment for the whole model [16]. If the inertia effects in the model are negligible or can be considered as quasi-static, it is not useful to maintain the stable time increment as the simulation would take too long. One method that can be used to artificially reduce the runtime of the simulation involves scaling the density of the material in the model. According to Equations (45) and (46), when the density is scaled by a factor, f^2, the runtime is reduced by a factor f. ABAQUS introduces in the explicit solver a bulk viscosity parameter, which introduces damping associated with the volumetric straining to improve the high speed dynamics simulations [17]. Two types of bulk viscosity parameter are considered: the linear set to 0.03 in this study and the quadratic set to 1.2.

The central difference time integration scheme for non-linear problems employed in most explicit computer codes [13] can be resumed in the following steps:

1. Initialize $\mathbf{u}^0$ and $\dot{\mathbf{u}}^0$
2. Compute the mass matrix $\mathbf{M}$
3. Compute $\dot{\mathbf{u}}^{\left(\frac{1}{2}\right)} = \dot{\mathbf{u}}^0 + \frac{\Delta t^{(1)}}{2}\left(\mathbf{M}^{-1}\left(\mathbf{F}^0 - \mathbf{I}^0\right)\right)$
4. For each time step increment:

    (a) Solve for total displacements: $\mathbf{u}^{(i+1)} = \mathbf{u}^{(i)} + \Delta t^{(i+1)}\dot{\mathbf{u}}^{(i+\frac{1}{2})}$

    (b) Compute the displacement increment: $\Delta\mathbf{u} = \mathbf{u}^{(i+1)} - \mathbf{u}^{(i)}$

    (c) For each integration point j:

    - Compute the strain increment: $\Delta\mathbf{u} \rightarrow \Delta\varepsilon_j$
    - Compute the stress increment: $\Delta\varepsilon_j \rightarrow \Delta\sigma_j$
    - Compute the total stress: $\sigma_j^{i+1} = \sigma_j + \Delta\sigma_j$

    (d) Compute the internal force vector: $\mathbf{I}^{(i+1)}$

    (e) Solve for the new accelerations: $\ddot{\mathbf{u}}^{(i+1)} = \mathbf{M}^{-1}\left(\mathbf{F}^{(i+1)} - \mathbf{I}^{(i+1)}\right)$

    (f) Compute the velocities at new mid-time: $\dot{\mathbf{u}}^{i+\frac{3}{2}} = \dot{\mathbf{u}}^{i+\frac{1}{2}} + \frac{\Delta t^{(i+1)} + \Delta t^{(i)}}{2}\ddot{\mathbf{u}}^{(i)}$

2.5. Development of User Material Subroutine

The active behavior of the muscle is not provided in the libraries of the commercial finite element codes. It is therefore necessary to implement the active behavior in the form of a user-defined stress update algorithm. This was implemented in the finite element code ABAQUS/Standard by means of a UMAT. Additionally, for ABAQUS/Explicit, a VUMAT was also developed. Much of the coding involved in the two algorithms is the same but there are several key issues that must be addressed to maintain consistency of the results between the two solvers. These subroutines, written in Fortran, implemented the behavior of the material in the form of a stress update algorithm that is called at each integration point for every iteration during the finite element simulation. At these integration points, it is also necessary to define the anisotropy of the material to form the different tensors and to obtain the set of invariants. The subroutines were able to read a discretized fiber orientations from an external file during the first time increment.

The most important difference between the two programmed subroutines is that the explicit one does not update the tangential stiffness matrix. Nevertheless, when writing the implicit subroutine this matrix should be accurately represented to obtain correct solutions. The initial time increment used in ABAQUS/Standard is chosen by the user, and subsequent increments are controlled by an automatic incrementation control. To determine the size of the initial time increment in ABAQUS/Explicit a bogus set of tiny strain increments are passed to the VUMAT at the start of the analysis. From the stress response of the material, a conservative value for the stable time increment is calculated [16].

2.6. Eyeball and EOM Finite Element Model

The geometrical data of the model were obtained from the database of Mitsuhashi et al. [24], which was created from a whole-body set of 2 mm interval MRI images of a male volunteer. Some adjustments to the surfaces were made by hand to remove discontinuities and increase smoothness. The contours of the right eyeball and the four recti muscles were defined. As intorsion and extorsion movements were not considered in this study, the oblique muscles were not included in the model [10] To prevent the eye muscles from slipping away while the globe rotates, connective tissue surrounds the globe and stabilizes the muscles acting as pulleys and serving as functional EOM origins [25].

Figure 2 shows the finite element mesh used in the model. Solid hexahedral finite elements were used to mesh the eyeball and the EOMs, whereas one-dimensional truss elements simulated the action of the connective tissue pulleys. In Table 1, the number of nodes and elements of the solid parts of the model are shown together with their volume and mass according to the eyeball density [26] and muscle density [27].

Table 1. Mesh size, volume and mass of the model parts.

Part	Nodes	Elements	Volume (mm^3)	Mass (g)
Eyeball	2686	2211	8134	8.134
Lateral EOM	2573	1920	488	0.517
Inferior EOM	2291	1680	576	0.611
Medial EOM	1953	1440	419	0.444
Superior EOM	1377	936	461	0.489

The set of mechanical properties used for the muscle material behavior is included in Table 2. Both passive and active parameters were adapted from a previous work [19]. The passive properties of connective tissues such as the tendon ends of the EOMS and the pulleys were taken from Calvo et al. [5]. For the latter, circular sections of 1 mm radius were applied to the truss elements and a rigid body constraint was applied to the eyeball.

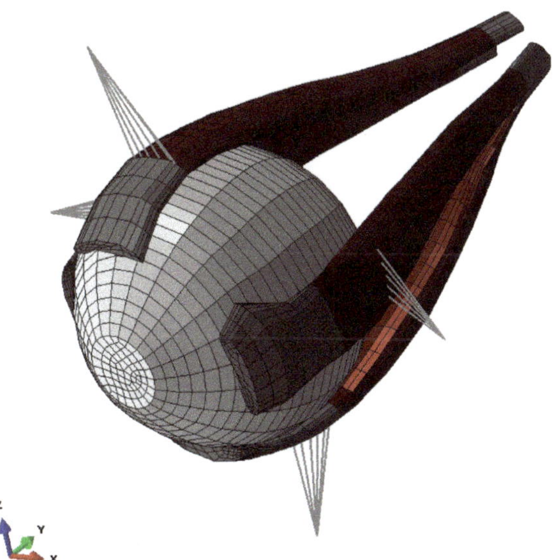

Figure 2. Finite element mesh of the right eyeball, the EOMs and the zones of tendinous insertions obtained from Mitsuhashi et al. [24] incorporating connector elements to simulate the action of soft tissue pulleys.

Table 2. Parameters considered in the model for the material behavior of the skeletal-muscle tissue [19].

	Parameters
Passive behavior	$c_1 = 0.008837$ MPa $c_3 = 0.00987$ MPa $c_4 = 2.23787$ $c_5 = 3.06367$ MPa $c_6 = -4.75963$ MPa $c_7 = -2.76353$ MPa $\bar{I}_{4_0} = 1.25638$ $\bar{I}_{4_{ref}} = 1.25638$
Maximum isometric stress	$P_0 = 0.1$ MPa
Force length relationship	$\lambda = 1$ $\lambda_{opt} = 1$ $\xi = 0.1$
Force time relationship	$\alpha = 1$ $s_1 =$ variable

To account for the anisotropy present in the muscles due to the presence of fibers, a set of directions was generated inside the volume of the EOMs (Figure 3). These directions define both the passive behavior of the tissue and the direction in which the active contraction occurs.

Each of the ends of the EOMs in contact with the eyeball was tied, fixing the three degrees of freedom of both surfaces. The nodes of the other end of the muscles were clamped, restraining all movements. No contact was considered between the surfaces of the model since no interaction of the different parts of the model was detected in the movement. The truss elements connected nodes of the EOMs to a common node that was pinned. Finally, a rigid solid constraint was applied to the eyeball fixing an instant center of rotation at the center of the sphere.

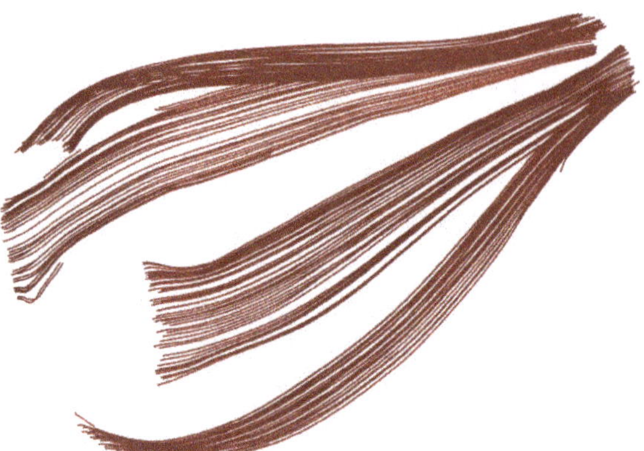

Figure 3. Representation of the muscle and collagen fiber orientations.

3. Results

As the maximum time increment in a dynamic implicit algorithm depends on the typical period of vibration of the system, the natural frequencies and mode shapes were obtained for the model of the right eyeball, the pulleys and the EOMs. The initial four mode shapes are represented in Figure 4 which correspond to bending shapes of the four different muscles at frequencies from 55.33 to 84.16 Hz. The lower natural frequency corresponds to the inferior EOM muscle which is the muscle with the largest volume. Taking the largest characteristic frequency obtained, a maximum recommended increment for an implicit dynamic simulation should be less than 0.001 s.

The model was analyzed under four different movement scenarios characterized by the evolution of the activation function in Equation (11) along time. Figure 5a shows the activation function $f_2(t)$ considering four s_1 parameters that will induce eyeball movements from a very slow one $s_1 = 10$ to an intended nearly instantaneous $s_1 = 100$. Although all the muscles contribute in a certain way to the eyeball movements (elevation–depression and abduction–adduction), the results presented in this work are those obtained for the single activation of the medial EOM (adduction), as shown in Figure 5b. In this figure, the distribution of the maximum principal stress is presented. As can be observed, the maximum values are reached in the medial EOM at the final point of the simulation at $t = 0.3$ s.

The inertia of the system provides a response slower than the activation function, as shown in Figure 6a. The displacement in the x direction is represented for a point located at the center of the pupil area in the simulated eyeball. Implicit and explicit results are presented for the four activation function parameters. Comparing both algorithms, differences under 3% at the end of the simulations were found. Figure 6b shows the x displacement field at the end of the adduction movement considering $s_1 = 10$ for the implicit simulation.

To compare the performance of both methodologies, the evolution of the total strain energy (Figure 7a) and total kinetic energy (Figure 7b) was analyzed in the model. The total strain energy accumulated by the model for both algorithms and for all the activation signals is nearly the same at the beginning of the simulation. In contrast, at the end of the simulation, the models calculated with the explicit algorithm show an extra level of internal energy compared with the implicit simulations. As can be observed, the total energy is higher when the contraction velocity increases induced by the activation function. These higher levels of total energy are a consequence of a greater amount of kinetic energy when increasing the contraction velocity. The total kinetic energy is shown in Figure 7b and again the explicit results outperform the implicit ones.

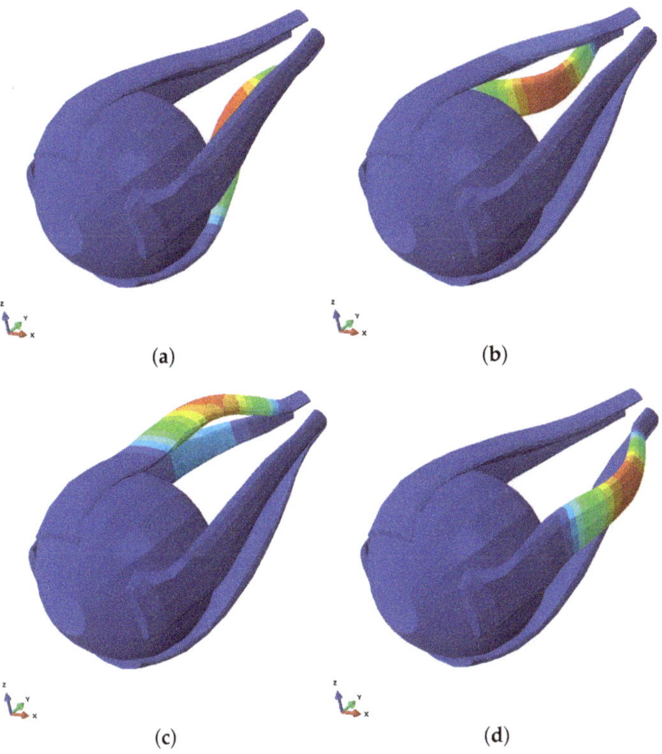

Figure 4. Initial four mode shapes obtained for the right eyeball and EOMs system. These modes correspond with bending of the EOMs at natural frequencies of: (**a**) 55.33 Hz for the inferior EOM; (**b**) 72.57 Hz for the lateral EOM; (**c**) 72.84 Hz flexion mode for the superior EOM; and (**d**) 84.16 Hz for the medial EOM.

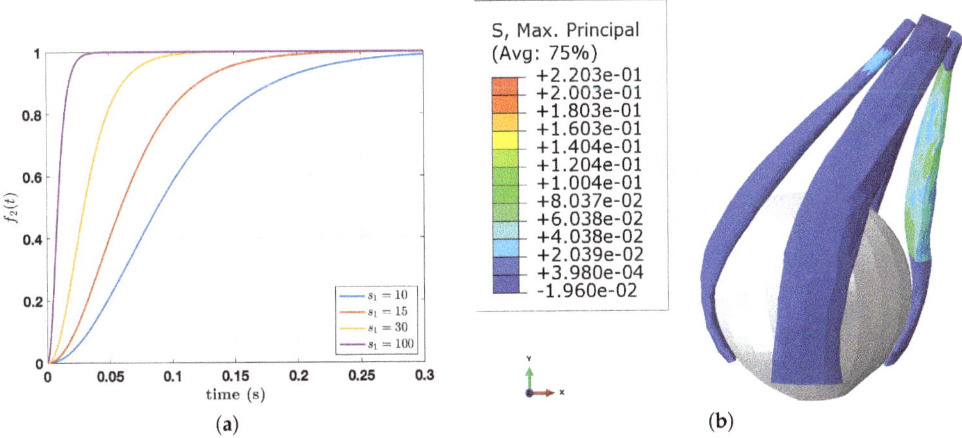

Figure 5. (**a**) Evolution of the activation function $f_2(t)$ with $\alpha = 1$ for different s_1 parameter values simulating four contraction velocities; and (**b**) maximum principal stress distribution for the medial EOM activation with $s_1 = 10$.

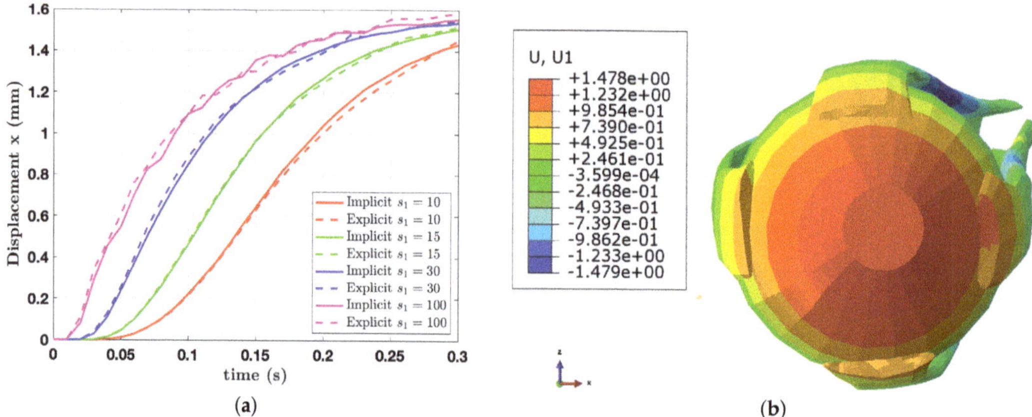

Figure 6. (**a**) Evolution of the displacement of a node located at the center of the surface of the mesh where the pupil is located for all the activation signals considered using both algorithms; and (**b**) displacement field in the x direction for an implicit simulation with $s_1 = 10$ at $t = 0.3$ s.

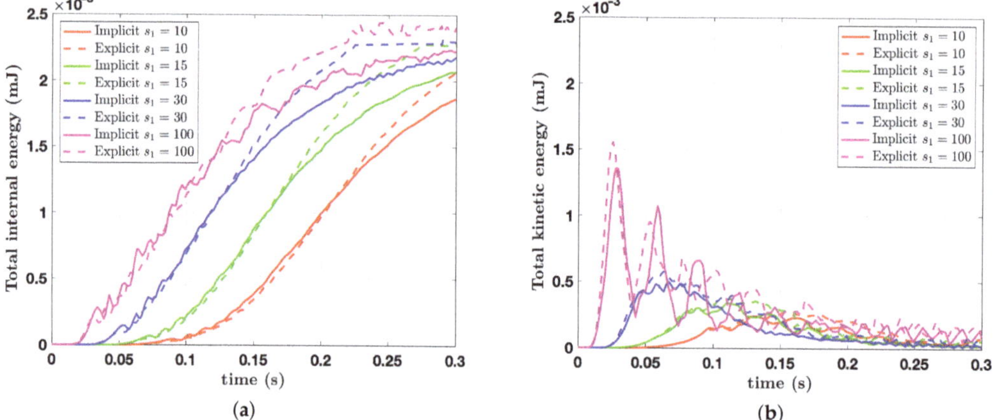

Figure 7. (**a**) Evolution of the total strain energy for all the activation signals considered using both algorithms; and (**b**) evolution of the total kinetic energy for all the activation signals considered using the implicit algorithm.

Finally, a comparative analysis of the computational time is summarized in Table 3. As can be seen, the implicit algorithm takes only a 5% of the time spent with the explicit algorithm without mass scaling. A series of global mass scaling factors was applied to the model. Increasing this factor, higher values of fictitious mass are added, penalizing the range of motion of the system. Although a factor of 100 notably reduces the simulation time to levels near that of the implicit algorithm, the maximum displacement at the end of the simulation differs by 36.5% with respect to that obtained with no mass scaling.

Table 3. Computational time in percent relative to the dynamic explicit simulation without mass scaling (m.s.) for both algorithms and incorporating different m.s. factors.

	Mass Added (%)	Calculation Time (%)	Maximum Displacement Reduction (%)
Implicit	0	5	1.13
Explicit no m.s.	0	100	0
Explicit m.s. factor 1.01	1	99	0.05
Explicit m.s. factor 1.1	10	96	0.52
Explicit m.s. factor 4	300	53	0.9
Explicit m.s. factor 6	500	35	1.6
Explicit m.s. factor 10	900	30	3.1
Explicit m.s. factor 100	9900	11	36.5

4. Discussion

When studying extraocular mechanics, muscle activation and deformation are important parameters to characterize the movement of the eyeball [11]. In this paper, the activation of the medial EOM is analyzed under four activation signals which induce increasing contraction velocities. This function was simplified unlike more realistic previous models [8,9,19] to reduce the computation of unnecessary terms. This was assumed considering that all muscle fibers are recruited during activation (tetanic contraction). The range of motion associated with the region that represents the pupil in the model is in good agreement with those reported previously in the literature [4,10], although other authors simulated even larger angles of rotation [11]. The maximum horizontal or x displacement in Figure 6a can be translated according to the position of the instant center of rotation to a rotational angle of 7.4 degrees, which is far from the 20 degrees simulated in the work of Wei et al. [11]. Differences in the maximum isometric stress and in the force length relationship could be responsible for this disagreement since the properties of the muscle active behavior incorporated in this model were taken from a previous work [19]. The use of a single model with a particular geometry also limits the comparison with previous results but, on the other hand, it proves the potential of the methodology to develop a functional model based on medical image.

The results obtained for the four activation signal paths (Figure 6a) show that the predictions of the two algorithms differ at the end of the simulations. Larger time increments in the implicit method could lead to underestimating the contraction velocity (Equation (21)) implemented in the user subroutine, and consequently the muscle contracts more slowly.

Using the explicit analyses, the mass-scaling option is available to increase the stable time increment by artificially adding mass to the system. Although mass-scaling could decrease the mean computational time in the simulation of the eyeball movement (see Table 3), the range of motion is reduced drastically. It has been suggested in the literature that mass-scaling results are acceptable when the proportion of kinetic energy to strain energy is less than 5% [12]. In this case, the dynamic effect is negligible and problems can be solved with quasi-static solutions. As can be observed in Figure 7, this ratio is not satisfied in this model at the beginning of the simulations. Future analysis should consider increasing the maximum isometric stress developed by the activated muscle to explore whether it is possible to balance the addition of mass to obtain more realistic results.

As pointed out by other authors [12,16] and indicated by the results in Table 3, for simpler loading conditions, the implicit method takes a shorter solution time. In the case of loading conditions involving contact between the muscles and the eyeball or even incorporating the orbital fat, the explicit method will be the preferable choice [14,15]. Furthermore, the problem solved with this method can be easily parallelized in separated computer processors since the inverse of the lumped mass matrix can be split in decoupled set of equations.

5. Conclusions

In this paper, a comparison between implicit and explicit dynamic algorithms is presented and applied to model the 3D motion of the eyeball subjected to the action of EOMs. Our high-speed simulations showed that the dynamic implicit algorithm offers a substantial reduction in the required computational time in a model with no contact interactions between the surfaces. Although mass-scaling can provide a reduction in the computational time with the explicit algorithm, it is not recommended for high-speed movements taking into consideration the activation of the muscle tissue, due to the system increment of mass inertia.

Author Contributions: Conceptualization, methodology, and writing—review and editing, J.G. and B.C.; software, J.G.; and project administration, B.C. Both authors have read and agreed to the published version of the manuscript.

Funding: This research was funded by Spanish Ministerio de Ciencia, Innovación y Universidades grant number DPI2017-84047-R and the Department of Industry and Innovation (Government of Aragon) through the research group Grant T24-20R (co-financed by Feder 2014-2020: Construyendo Europa desde Aragon). Part of the work was performed by the ICTS "NANBIOSIS" specifically by the High Performance Computing Unit (U27), of the CIBER in Bioengineering, Biomaterials & Nanomedicne (CIBER-BBN at the University of Zaragoza).

Conflicts of Interest: The authors declare no conflict of interest.

References

1. Shumway, C.L.; Motlagh, M.; Wade, M. *Anatomy, Head and Neck, Eye Superior Rectus Muscle*; StatPearls Internet: Treasure Island, FL, USA, 2020.
2. Clark, R.A. The Role of Extraocular Muscle Pulleys in Incomitant Non-Paralytic Strabismus. *Middle East Afr. J. Ophthalmol.* **2015**, *22*, 279–285. [CrossRef] [PubMed]
3. Gao, Z.; Guo, H.; Chen, W. Initial tension of the human extraocular muscles in the primary eye position. *J. Theor. Biol.* **2014**, *353*, 78–83. [CrossRef] [PubMed]
4. Iskander, J.; Hossny, M.; Nahavandi, S.; Del Porto, L. An ocular biomechanic model for dynamic simulation of different eye movements. *J. Biomech.* **2018**, *71*, 208–216. [CrossRef] [PubMed]
5. Calvo, B.; Ramírez, A.; Alonso, A.; Grasa, J.; Soteras, F.; Osta, R.; Mu noz, M.J. Passive nonlinear elastic behaviour of skeletal muscle: Experimental results and model formulation. *J. Biomech.* **2010**, *43*, 318–325. [CrossRef]
6. Grasa, J.; Ramírez, A.; Osta, R.; Mu noz, M.J.; Soteras, F.; Calvo, B. A 3D active-passive numerical skeletal muscle model incorporating initial tissue strains. Validation with experimental results on rat tibialis anterior muscle. *Biomech. Model. Mechanobiol.* **2011**, *10*, 779–787. [CrossRef]
7. Hernández-Gascón, B.; Pe na, E.; Grasa, J.; Pascual, G.; Bellón, J.M.; Calvo, B. Mechanical response of the herniated human abdomen to the placement of different prostheses. *J. Biomech. Eng.* **2013**, *135*, 51004. [CrossRef]
8. Hernández-Gascón, B.; Grasa, J.; Calvo, B.; Rodríguez, J.F. A 3D electro-mechanical continuum model for simulating skeletal muscle contraction. *J. Theor. Biol.* **2013**, *335*, 108–118. [CrossRef]
9. Grasa, J.; Sierra, M.; Lauzeral, N.; Mu noz, M.J.; Miana-Mena, F.J.; Calvo, B. Active behavior of abdominal wall muscles: Experimental results and numerical model formulation. *J. Mech. Behav. Biomed. Mater.* **2016**, *61*, 444–454. [CrossRef]
10. Karami, A.; Eghtesad, M.; Haghpanah, S.A. Prediction of muscle activation for an eye movement with finite element modeling. *Comput. Biol. Med.* **2017**, *89*, 368–378. [CrossRef]
11. Wei, Q.; Sueda, S.; Pai, D.K. Physically-based modeling and simulation of extraocular muscles. *Prog. Biophys. Mol. Biol.* **2010**, *103*, 273–283. [CrossRef]
12. Naghibi Beidokhti, H.; Janssen, D.; Khoshgoftar, M.; Sprengers, A.; Perdahcioglu, E.S.; Van den Boogaard, T.; Verdonschot, N. A comparison between dynamic implicit and explicit finite element simulations of the native knee joint. *Med. Eng. Phys.* **2016**, *38*, 1123–1130. [CrossRef]
13. Borst, R.; Crisfield, M.; Remmers, J.; Verhoosel, C. Non-Linear Finite Element Analysis of Solids and Structures: Second Edition. In *Non-Linear Finite Element Analysis of Solids and Structures: Second Edition*; John Wiley & Sons Inc.: Hoboken, NJ, USA, 2012; doi:10.1002/9781118375938. [CrossRef]
14. Choi, H.H.; Hwang, S.M.; Kang, Y.H.; Kim, J.; Kang, B.S. Comparison of Implicit and Explicit Finite-Element Methods for the Hydroforming Process of an Automobile Lower Arm. *Int. J. Adv. Manuf. Technol.* **2002**, *20*, 407–413. [CrossRef]
15. Sun, J.; Lee, K.; Lee, H. Comparison of implicit and explicit finite element methods for dynamic problems. *J. Mater. Process. Technol.* **2000**, *105*, 110–118. [CrossRef]
16. Harewood, F.; McHugh, P. Comparison of the implicit and explicit finite element methods using crystal plasticity. *Comput. Mater. Sci.* **2007**, *39*, 481–494. [CrossRef]

17. *ABAQUS User's Manual, Version 6.14*; Dassault Systèmes Simulia Corp.: Providence, RI, USA, 2014.
18. Stålhand, J.; Klarbring, A.; Holzapfel, G.A. A mechanochemical 3D continuum model for smooth muscle contraction under finite strains. *J. Theor. Biol.* **2011**, *268*, 120–130. [CrossRef]
19. Grasa, J.; Sierra, M.; Mu noz, M.J.; Soteras, F.; Osta, R.; Calvo, B.; Miana-Mena, F.J. On simulating sustained isometric muscle fatigue: A phenomenological model considering different fiber metabolisms. *Biomech. Model. Mechanobiol.* **2014**, *13*, 1373–1385. [CrossRef]
20. Arruda, E.M.; Mundy, K.; Calve, S.; Baar, K. Denervation does not change the ratio of collagen I and collagen III mRNA in the extracellular matrix of muscle. *Am. J. Physiol. Regul. Integr. Comp. Physiol.* **2007**, *292*, R983–R987. [CrossRef]
21. Ramírez, A.; Grasa, J.; Alonso, A.; Soteras, F.; Osta, R.; Mu noz, M.J.; Calvo, B. Active response of skeletal muscle: In vivo experimental results and model formulation. *J. Theor. Biol.* **2010**, *267*, 546–553. [CrossRef]
22. Weiss, J.A.; Maker, B.N.; Govindjee, S. Finite element implementation of incompressible, transversely isotropic hyperelasticity. *Comput. Methods Appl. Mech. Eng.* **1996**, *135*, 107–128. [CrossRef]
23. Hilber, H.M.; Hughes, T.J.R.; Taylor, R.L. Improved numerical dissipation for time integration algorithms in structural dynamics. *Earthq. Eng. Struct. Dyn.* **1977**, *5*, 283–292. [CrossRef]
24. Mitsuhashi, N.; Fujieda, K.; Tamura, T.; Kawamoto, S.; Takagi, T.; Okubo, K. BodyParts3D: 3D structure database for anatomical concepts. *Nucleic Acids Res.* **2009**, *37*, D782–D785. [CrossRef]
25. Demer, J.L.; Miller, J.M.; Poukens, V.; Vinters, H.V.; Glasgow, B.J. Evidence for fibromuscular pulleys of the recti extraocular muscles. *Investig. Ophthalmol. Vis. Sci.* **1995**, *36*, 1125–1136.
26. Heymsfield, S.B.; Gonzalez, M.C.; Thomas, D.; Murray, K.; Jia, G.; Cattrysse, E. Adult Human Ocular Volume: Scaling to Body Size and Composition. *Anat. Physiol.* **2016**, *6*. [CrossRef]
27. Ward, S.R.; Lieber, R.L. Density and hydration of fresh and fixed human skeletal muscle. *J. Biomech.* **2005**, *38*, 2317–2320. [CrossRef]

mathematics

Article

A General Mechano-Pharmaco-Biological Model for Bone Remodeling Including Cortisol Variation

Rabeb Ben Kahla [1,2], Abdelwahed Barkaoui [3], Moez Chafra [1] and João Manuel R. S. Tavares [4,*

[1]   Laboratoire de Systèmes et de Mécanique Appliquée (LASMAP), Ecole Polytechnique de Tunis Université De Carthage, La Marsa 2078, Tunisia; rabeb.benkahla@enit.utm.tn (R.B.K.); moez.chafra@utm.tn (M.C.)
[2]   Laboratoire de Mécanique Appliquée et Ingénierie (LR-MAI), Ecole Nationale d'Ingénieurs de Tunis Université Tunis El Manar, BP 37, Tunis 1002, Tunisia
[3]   Laboratoire des Energies Renouvelables et Matériaux Avancés (LERMA), Université Internationale de Rabat, Rocade Rabat Salé 11100, Morocco; abdelwahed.barkaoui@uir.ac.ma
[4]   Instituto de Ciência e Inovação em Engenharia Mecânica e Engenharia Industrial, Departamento de Engenharia Mecânica, Faculdade de Engenharia, Universidade do Porto, Rua Dr. Roberto Frias, s/n, 4200-465 Porto, Portugal
*   Correspondence: tavares@fe.up.pt; Tel.: +351-22-041-3472

Abstract: The process of bone remodeling requires a strict coordination of bone resorption and formation in time and space in order to maintain consistent bone quality and quantity. Bone-resorbing osteoclasts and bone-forming osteoblasts are the two major players in the remodeling process. Their coordination is achieved by generating the appropriate number of osteoblasts since osteoblastic-lineage cells govern the bone mass variation and regulate a corresponding number of osteoclasts. Furthermore, diverse hormones, cytokines and growth factors that strongly link osteoblasts to osteoclasts coordinated these two cell populations. The understanding of this complex remodeling process and predicting its evolution is crucial to manage bone strength under physiologic and pathologic conditions. Several mathematical models have been suggested to clarify this remodeling process, from the earliest purely phenomenological to the latest biomechanical and mechanobiological models. In this current article, a general mathematical model is proposed to fill the gaps identified in former bone remodeling models. The proposed model is the result of combining existing bone remodeling models to present an updated model, which also incorporates several important parameters affecting bone remodeling under various physiologic and pathologic conditions. Furthermore, the proposed model can be extended to include additional parameters in the future. These parameters are divided into four groups according to their origin, whether endogenous or exogenous, and the cell population they affect, whether osteoclasts or osteoblasts. The model also enables easy coupling of biological models to pharmacological and/or mechanical models in the future.

Keywords: biomechanics; mathematical model; cell dynamics; bone physiology; bone disorders

Citation: Kahla, R.B.; Barkaoui, A.; Chafra, M.; Tavares, J.M.R.S. A General Mechano-Pharmaco-Biological Model for Bone Remodeling Including Cortisol Variation. *Mathematics* **2021**, *9*, 1401. https://doi.org/10.3390/math9121401

Academic Editor: Mauro Malvè

Received: 9 April 2021
Accepted: 25 May 2021
Published: 17 June 2021

Publisher's Note: MDPI stays neutral with regard to jurisdictional claims in published maps and institutional affiliations.

1. Introduction

Fragility fracture rates are growing exponentially, mainly due to population aging. The World Health Organization has recorded a substantial increase in population growth and aging, with a life expectancy rising from about 65 years old in 2005 up to 73 years old in 2019; while in Africa this latter age is around 64 years old, and it is around 78 years old in Europe and the Western Pacific. Recently, various countries worldwide have experienced a fragility fracture crisis with this increase in life expectancy. In 2017, the International Osteoporosis Foundation reported an estimated 2.7 million fragility fractures in six European countries, mainly Germany, Italy, France, Spain, Sweden and the United Kingdom, which resulted in an associated annual cost of 37.5 billion euros for their health care systems. On an individual level, these fragility fractures affected the independence and quality of life of thousands of people in each of these countries. By 2030, the number of annual fragility fractures is expected to have increased by 23%, reaching 3.3 million, with projected costs of approximately 47.4 billion euros [1]. This makes bone fractures, mainly fragility fractures, a major public health problem.

The evolution of bone mineral density is related to bone metabolism and the different factors affecting it. In fact, throughout life, bone is constantly being renewed through the resorption of old and damaged bone and the formation of new bone. This renewal is what is known as bone remodeling, a process requiring a strict coordination in time and space to maintain bone quantity and quality. This coordination mainly involves bone-resorbing osteoclasts and bone-forming osteoblasts, which are the two major players in the remodeling process. The delicate balance between the amount of resorbed bone and the subsequent deposited new bone requires a close coordination of the resorption and formation activities. This coordination controls the generation of the appropriate number of osteoblasts for remodeling, which is referred to as the coupling mechanism. Moreover, several hormones, cytokines and growth factors are involved in linking the osteoblast- and osteoclast-lineage cells (via a complex network) throughout the remodeling cycle.

The remodeling phenomenon has been a research and discussion subject for over a century. In 1892, Julius Wolff stated that bones will adapt to the degree of mechanical loading. This statement is currently known as Wolff's Law and represents the first explicit statement that directly links bone microstructure to mechanical loading. This law establishes that the tendencies of bone trabeculae are aligned with the principal directions of stress. Since then, the mathematical description of bone behavior has experienced a remarkable development. At first, most models were phenomenological or purely mechanical, lacking any biological foundation. Afterwards, models started gaining insight into the biological processes of bone remodeling and the relationship between them and the mechanical processes. Concepts such as bone multicellular unit activity, metabolic activity, mineralization processes as well as damage accumulation have now been introduced and are included in the new mathematical models. These new bone remodeling models have been classified into mechanological, biological and mechanobiological, according to the main mechanism of the process.

The mechanological models show the mechanical stimulus and the final bone response (net resorption or formation) in a black-box manner [2–8]. These models are mainly based on the mechanostat theory formulated by Frost [9], which represents a sophisticated version of the statement of Wolff's law [10]. Additionally, they only focus on bone biomechanical properties, without taking the biochemical aspects and the cell dynamics into account. This does not provide a clear understanding of these internal processes, and there is no inclusion of bone tissue responses to drugs or hormonal dysfunctions. Later models were then developed as semi-mechanistic versions of this bone remodeling theory to provide more realistic outcomes [11–13].

Two research groups were mainly responsible for initiating the biological models. The first research group was Kamarova et al. [14], who combined the autocrine and paracrine factors into a single exponential parameter. Thus, bone remodeling could be studied using power-law approximations, which was an almost perfect fit, since osteoblast and osteoclast activities could be temporally describe throughout a single remodeling cycle. This Kamarova et al. model [14] has been further developed by several other authors, such as Ryser et al. [15], who predicted the spatiotemporal evolution of a single Bone Multicellular Unit (BMU), and Ayati et al. [16] who reformulated the approach to develop a diffusion bone remodeling model, which enabled the study of myeloma bone disease. The second research group was Lemaire et al. [17], who developed a model including the RANK-RANKL-OPG signaling pathway, combined with the influence of the transforming growth factor-β (TGFβ) and the parathyroid hormone (PTH) on bone cell activities, while taking into account the differentiation stages of the osteoblasts and osteoclasts. Owing to its success, this second approach has been developed further by other groups of authors, such as Maldonado et al. [18], who added the osteocyte function as a mechanotransducer, and Pivonka et al. [19], who used a set of activator/repressor functions for a more appropriate description of the receptor–ligand interactions that occur throughout the remodeling process. The first family of models is simplistic, neglecting the cell differentiation stages that has an important role in the osteoblast–osteoclast interplay, but has the advantage of requiring a relatively low number of parameters, facilitating the numerical and computational calculations. On the other hand, the second family of models reduces the number of simplifications, but requires a significantly high number of parameters, which may affect the results [20].

The mechanobiological models follow the models of either Lemaire et al. [17] or Komarova et al. [14]. Owing to the complexity of simultaneously modeling mechanical and biological components, progress in the implementation of this third category of mathematical models has been successful only in recent years. In this third category of mathematical bone remodeling models, the modeling process starts by calculating the mechanical parameters such as stress, strain, pressure and fluid velocity, which then regulate the biological process that includes cell activities and tissue formation [21–23]. However, a primary limitation of mechanobiological models is the need to solve a series of Partial Differential Equations (PDEs) and to develop a Finite Element (FE) formulation to implement within an iterative procedure [24–26].

Few models tackle bone remodeling as a mechano-chemo-biological process, going from the mechanical stimulus applied to bone up to the generated chemical reactions and followed by bone cell responses [27–33]. All these works were developed using a number of simple assumptions to model, the so-called mechanotransduction mechanism/process. In addition, coupling mechanical and chemical phenomenon together with mechanosensing [34], which is a lesser-known component of the remodeling process [35–37], requires various simplified assumptions.

Following these steps, with the aim to create a novel mechano-pharmaco-biological model, the current article provides a first step towards this goal. This work presents a pharmaco-biological approach that is coupled with a previously developed mechanical approach [38] and the hormonal effects of another previous study [39,40]. Both of these previous works were focused on modelling the effects of cyclic loading and sex hormones on the remodeling process throughout the lifecycle of a person. However, the current work focuses on the effects of sex hormones combined with that of cortisol, TGFβ and PTH on bone remodeling in the case of healthy persons, and in the case of persons with an endogenous hypercortisolism known as Cushing disease. Then, the effects of bisphosphonates and denosumab on the enhancement of the pathologic remodeling process were compared. Since osteoblastic-lineage cells have been found to govern the bone mass variation, the mechanical stimulus was included in the proposed model as an exogenous paracrine model acting on osteoblast concentrations, which, in turn, act on the osteoclast concentrations.

2. Materials and Methods

2.1. Development of the Bone Remodeling Mathematical Model

As aforementioned, the proposed approach is based on combining both of the biological models of Komarova et al. [14] and Pivonka et al. [19]. Figure 1 summarizes the cell dynamics according to the model of Komarova et al. [14], and Figure 2 summarizes that of Pivonka et al. [19].

The proposed approach retains the structure of the Komarova et al. model [14] and includes the model parameters according to the Pivonka et al. approach [19]. Then, the effects of sex hormones (estradiol in women and testosterone in men), cortisol and antiresorptive drugs (bisphosphonates and denosumab) are taken into consideration. The evolution of each of these parameters is described according to the concept of Hill functions.

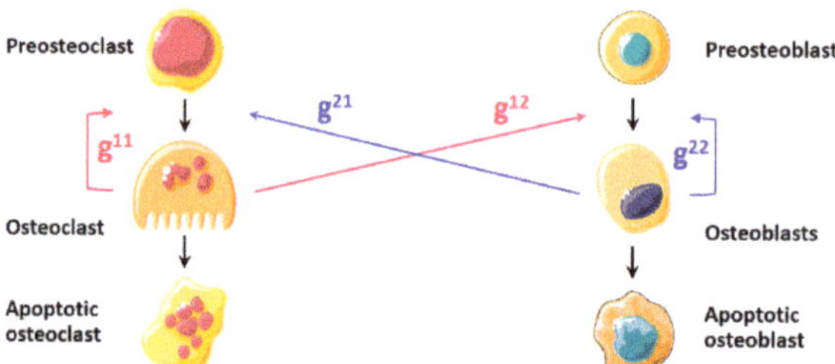

Figure 1. Diagram of the cell dynamics according to the model of Komarova et al. [14]. Here, g11 reflects the action of TGFβ and g22 the action of IGF, both of them representing autocrine factors secreted by one cell to influence the other cells of its own lineage. g12 reflects the actions of TGFβ and IGF, acting as paracrine factors secreted by osteoclasts to modulate osteoblasts. g21 reflects the actions of RANKL and OPG, acting as paracrine factors secreted by osteoblasts to modulate osteoclasts.

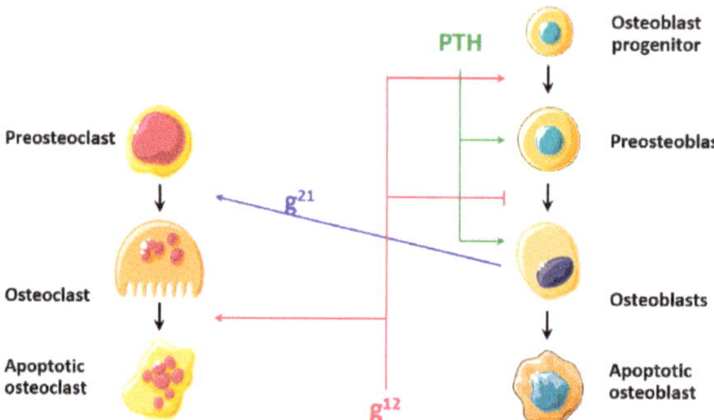

Figure 2. Diagram of the cell dynamics according to the model of Pivonka et al. [19], which considers that osteoblasts modulate osteoclast differentiation through g21. In this diagram, g12 only reflects the action of TGFβ stored in the bone matrix and released by osteoclasts during the bone resorption phase, and g21 reflects the actions of RANKL and OPG expressed by osteoblasts to modulate osteoclast differentiation. The model also involves the effect of PTH, as an external factor to the BMU.

2.1.1. Development of the Pharmaco-Biological Model

Several hormones, cytokines and growth factors influence cell behavior within an active BMU. The development of the proposed biological model was based on including each factor according to the concept of Hill functions [41].

Transforming growth factor beta (TGFβ)

TGFβ plays a critical role in bone remodeling. It stimulates the synthesis of matrix protein, dramatically affects osteoblasts and osteoclasts, and is abundant in bone. During bone formation, osteoblasts accumulate TGFβ in a latent form in the bone matrix. During bone resorption, osteoclasts release and activate TGFβ that, in turn, induces the activation of preosteoblasts recruited for the following formation phase and their migration to prior resorptive sites [42–44]. Thus, the preosteoblast concentration increases due to the differentiation of osteoblast progenitors into preosteoblasts, promoted by TGFβ [45,46], and decreases due to the differentiation of preosteoblasts into active osteoblasts, suppressed by TGFβ [45–49]. Additionally, the active osteoclast concentration decreases owing to active osteoclast apoptosis (Table 1). Mathematically, these effects can be expressed through functions: $Ta(OC)$ expressing the stimulation of osteoclast apoptosis by TGFβ; $Ta(OB)$ expressing the stimulation of osteoblast progenitor differentiation to preosteoblasts by

TGFβ; and $Tr(OB)$ conveying the repression of preosteoblast differentiation into active osteoblasts by $TGF\beta$:

$$Ta(OC) = \frac{TGF\beta}{K_{Ta1} + TGF\beta} \tag{1}$$

$$Ta(OB) = \frac{TGF\beta}{K_{Ta2} + TGF\beta} \tag{2}$$

$$Tr(OB) = \frac{K_{Tr2}}{K_{Tr2} + TGF\beta} \tag{3}$$

where $TGF\beta$ denotes $TGF\beta$ concentration, K_{Ta1} the activation coefficient of active osteoclast apoptosis mediated by $TGF\beta$, K_{Ta2} the activation coefficient of osteoblast progenitor differentiation mediated by $TGF\beta$ and K_{Tr2} the repression coefficient of preosteoblast differentiation mediated by $TGF\beta$.

Table 1. TGFβ actions and their mathematical formulation.

TGFβ	
Actions	Stimulation of osteoclast apoptosis ($Ta(OC)$). Stimulation of osteoblast progenitor differentiation into preosteoblasts ($Ta(OB)$). Inhibition of preosteoblast differentiation into active osteoblasts ($Tr(OB)$).
Diagram	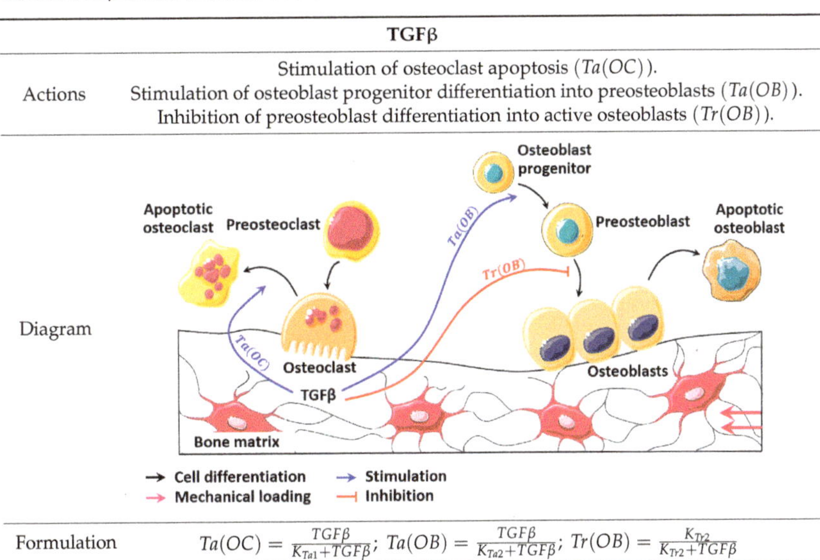
Formulation	$Ta(OC) = \frac{TGF\beta}{K_{Ta1}+TGF\beta}$; $Ta(OB) = \frac{TGF\beta}{K_{Ta2}+TGF\beta}$; $Tr(OB) = \frac{K_{Tr2}}{K_{Tr2}+TGF\beta}$

Parathyroid hormone (PTH)

PTH is one of the main endocrine regulators of extracellular phosphate and calcium levels. It up-regulates the RANKL expression by osteoblasts and down-regulates the OPG expression by the same cells, which leads to osteoclast maturation and activity. Moreover, PTH stimulates the production of osteocytic soluble RANK (sRANK) that also promotes osteoclast activity. Meanwhile, PTH suppresses the production of osteocytic sclerostin (SOST) that suppresses osteoblast activity. Although PTH is likely to promote bone resorption and formation, its precise mechanism remains unclear, because of the limited in vivo data [50,51].

In the current model, two actions of PTH are considered: its action on stimulating RANKL production (Equation (4)) and its action on inhibiting OPG production (Equation (5)):

$$Pa(OC) = \frac{PTH}{K_{Pa1} + PTH} \tag{4}$$

$$Pr(OC) = \frac{K_{Pr1}}{K_{Pr1} + PTH} \tag{5}$$

where PTH denotes the PTH concentration, K_{Pa1} the activation coefficient of PTH action on RANKL production rate, increasing the preosteoclast differentiation rate, and K_{Pr1} the repression coefficient of PTH action on OPG production rate, which also increases the preosteoclast differentiation rate, Table 2.

Table 2. PTH actions and their mathematical formulation.

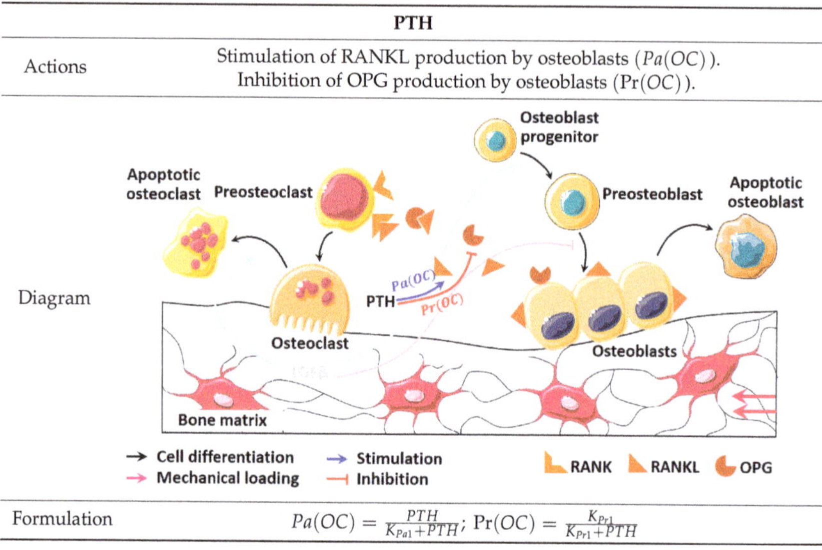

	PTH
Actions	Stimulation of RANKL production by osteoblasts ($Pa(OC)$). Inhibition of OPG production by osteoblasts ($Pr(OC)$).
Diagram	
Formulation	$Pa(OC) = \frac{PTH}{K_{Pa1}+PTH}$; $Pr(OC) = \frac{K_{Pr1}}{K_{Pr1}+PTH}$

Estradiol (Es)

Although the mechanism of estrogen transcriptional activity is not fully understood, it has been suggested that estrogen regulates bone matrix formation by acting on key factors involved in osteoblast differentiation and maturation [52]. The main effect of estrogens is the suppression of bone turnover, probably via osteocytes. Yet, they also inhibit bone resorption, through direct effects on osteoclasts, although the estrogen effects on osteoblasts/osteocytes are also likely to take part. A gap between the resorption and the formation activities has been linked to estrogen deficiency, probably because of the loss of estrogen effects on reducing the osteoblast apoptosis rate, decreasing NF-κB osteoblastic activity, lowering oxidative stress and perhaps other as yet undefined effects [53].

Estrogens may be classified into three main classes: estriol, estrone and estradiol. The latter represents the most potent human endogenous estrogen, with a high affinity for estrogen receptors. Thus, the proposed model is based on the effect of estradiol concentration (Table 3) on cells and bone remodeling through the following two repressive functions:

$$Es(OC) = \frac{K_{Es1}}{K_{Es1} + Es(t)} \tag{6}$$

$$Es(OB) = \frac{K_{Es2}}{K_{Es2} + Es(t)} \tag{7}$$

where $Es(t)$ denotes the estradiol concentration function, K_{Es1} the repression coefficient of estradiol action on osteoclast differentiation and K_{Es2} the repression coefficient of estradiol action on osteoblast apoptosis.

Table 3. Estradiol actions and their mathematical formulation.

Estradiol	
Actions	Inhibition of osteoclast formation and activity ($Es(OC)$). Inhibition of osteoblast apoptosis ($Es(OB)$).
Diagram	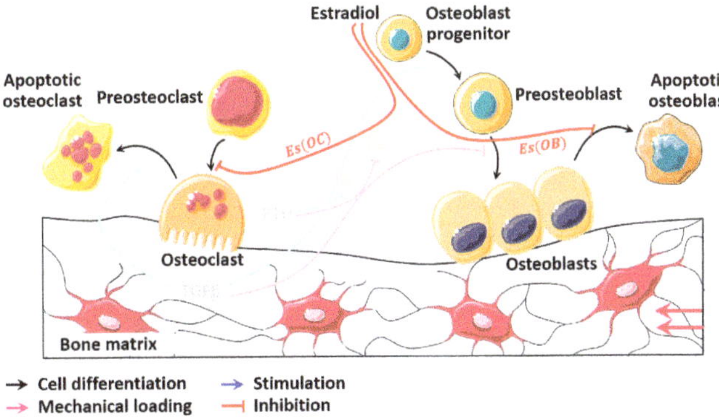
Concentration evolution	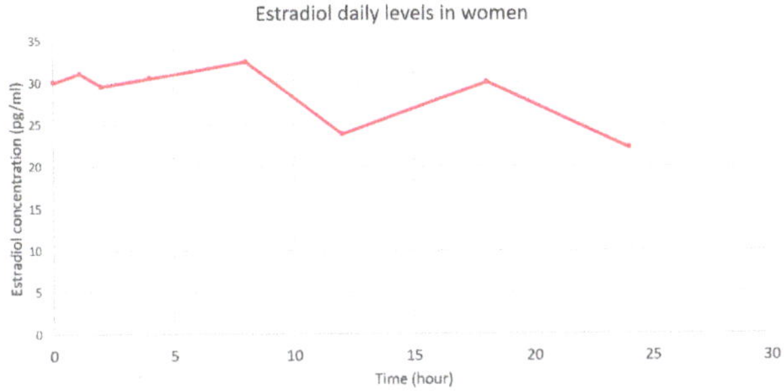
Formulation	$Es(OC) = \dfrac{K_{Es1}}{K_{Es1}+Es(t)}$; $Es(OB) = \dfrac{K_{Es2}}{K_{Es2}+Es(t)}$

Testosterone (Ts)

Androgen receptors have been identified in cultured human fetal osteoblasts using a nuclear-binding assay. Subsequently, mRNA and the androgen receptor protein have been identified in osteoblasts. Almost all studies have shown that androgens enhance the expression of androgen receptors in osteoblasts. Testosterone and 5α-dihydrotestosterone have been found to stimulate the proliferation of cultured preosteoblasts in distinctive species, and the collected evidence generally suggests that androgens stimulate osteoblast differentiation [54].

Moreover, androgen deficiency is most likely to be associated with osteoclast proliferation after orchiectomy. Androgens exert their bone defensive effects indirectly via osteoblasts, and orchidectomy generates preosteoblast proliferation, which increases RANKL secretion, thereby stimulating osteoclast proliferation and activation and resulting in bone loss. In vitro studies have shown that dihydrotestosterone interacts with androgen receptors on osteoclasts and inhibits bone resorption in human osteoclasts [54].

Since testosterone is the main androgen produced by Leydig cells and represents the most well-known male sex hormone, the effect of its concentrations (Table 4) on bone cells and remodeling is included in the model, through the repression function and the activation function, as follows:

$$Ts(OC) = \frac{K_{Ts1}}{K_{Ts1} + Ts(t)} \tag{8}$$

$$Ts(OB) = \frac{Ts(t)}{K_{Ts2} + Ts(t)} \tag{9}$$

where $Ts(t)$ denotes the testosterone concentration function, K_{Ts1} the repression coefficient of testosterone action on osteoclast activity and K_{Ts2} the activation coefficient of testosterone action on preosteoblasts.

Table 4. Testosterone actions and their mathematical formulation.

	Testosterone
Actions	Inhibition of osteoclast activity ($Ts(OC)$). Stimulation of preosteoblast differentiation into active osteoblasts ($Ts(OB)$).
Diagram	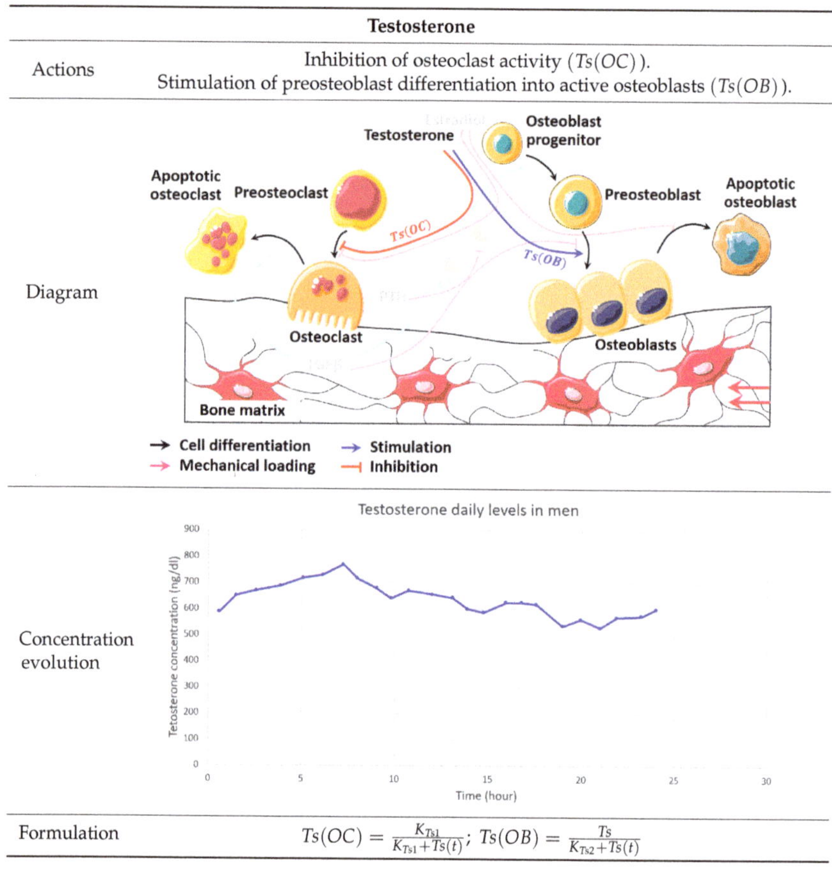
Concentration evolution	
Formulation	$Ts(OC) = \dfrac{K_{Ts1}}{K_{Ts1} + Ts(t)}; \; Ts(OB) = \dfrac{Ts}{K_{Ts2} + Ts(t)}$

Cortisol (Co)

Cortisol has well-established implications for diverse body systems. Specifically, it exerts direct negative effects on bone mineral density (BMD) by affecting bone cell growth, disrupting the bone remodeling process, impairing calcium intestinal absorption and renal reabsorption, as well as by inhibiting activity of sex steroids [55]. Indeed, an imbalance in the serum calcium homeostasis increases osteoclast-resorptive activity and eventually reduces BMD. Even a short bout of elevated cortisol levels may result in a reduced bone formation rate and a lower BMD. Prolonged cortisol oversecretion is consistently associated with a high prevalence of osteoporosis and may be linked to a decrease in BMD with age and to an increase in fragility fracture risk in elderly people [56]. The negative association between cortisol and BMD and the positive association between cortisol and fracture risk have been reported in several studies conducted on healthy older adults [57,58]. Moreover, signs of glucocorticoid-induced osteoporosis (GIOP), including a BMD reduction in the spine, altered ultrasound bone characteristics, as well as a higher number of morphometric fractures than healthy individuals, were found in patients with adrenal incidentaloma, and diagnosed as having subclinical hypercortisolism [59].

Indeed, it is unclear to date whether physiological cortisol levels also contribute to bone diseases or not. When performing a four-year analysis of single-point serum cortisol levels, no correlation was found between BMD, bone markers and bone loss. However, when analyzing integrated serum cortisol levels over 24 h, a correlation was found with BMD at the femur and the spine. These findings point out that physiological cortisol concentrations affect bone density. However, analyzing its effects may be difficult owing to diurnal variations of serum cortisol [60].

Therefore, the proposed model considers that the cortisol concentration (Table 5) increases osteoclast activity through the activation function; $Co(OC)$ inhibits osteoblast formation through the repression function; and $Co(OB)$ inhibits estradiol actions and testosterone actions according to:

$$Co(OC) = \frac{Co(t)}{K_{Co1} + Co(t)} \tag{10}$$

$$Co(OB) = \frac{K_{Co2}}{K_{Co2} + Co(t)} \tag{11}$$

$$CoE(OC) = -Es(OC) \tag{12}$$

$$CoE(OB) = -Es(OB) \tag{13}$$

$$CoT(OC) = -Ts(OC) \tag{14}$$

$$CoT(OB) = -Ts(OB) \tag{15}$$

where $Co(t)$ denotes the cortisol concentration function, K_{Co1} the activation coefficient of cortisol action on preosteoclasts and K_{Co2} the repression coefficient of cortisol action on preosteoblasts.

Bisphosphonates (Bp)

Bisphosphonates affect osteoclasts, but not their precursors. In fact, bisphosphonates are internalized in osteoclasts, probably by endocytosis, and inhibit the synthesis of a key enzyme in the mevalonate (MVA) pathway. This alters the intracellular proteins, accumulates cytotoxic intermediates, and alters the function of osteoclasts, which presumably increase their apoptosis rate [61]. Thus, bone resorption is inhibited (Table 6). The inhibition of osteoclast activity is given by the following repressive function:

$$Bp(OC) = k_{Bp} \times Bp \tag{16}$$

where Bp denotes the administered bisphosphonate concentration and k_{Bp} the Hill function parameter for drug regulation.

Denosumab (De)

Denosumab acts in the same way as OPG does, which is the natural antagonist receptor for RANKL. Denosumab binds to RANKL, thereby deterring the binding of RANKL to its receptor, RANK, expressed on both osteoclast and preosteoclast membranes. Subsequently, the RANK-RANKL signaling pathway is not activated, which impairs preosteoclast differentiation and osteoclast function, and possibly increases the osteoclast apoptosis rate. All of these effects inhibit bone resorption [61]. Thus, the following repression function describes the negative effect of denosumab on osteoclast differentiation and activity (Table 7):

$$De(OC) = \frac{C_D^{ser}}{0.6667 \times d_D \times 3 \times 10^{-16}} \tag{17}$$

where C_D^{ser} denotes the serum concentration of denosumab and d_D its administered dose.

Table 5. Cortisol actions and their mathematical formulation.

	Cortisol
Actions	Stimulation of preosteoclast differentiation into active osteoclasts ($Co(OC)$). Inhibition of osteoblast activity ($Co(OB)$). Inhibition of estradiol actions on osteoblasts and osteoclasts (CoE). Inhibition of testosterone actions on osteoblasts and osteoclasts (CoT).
Diagram	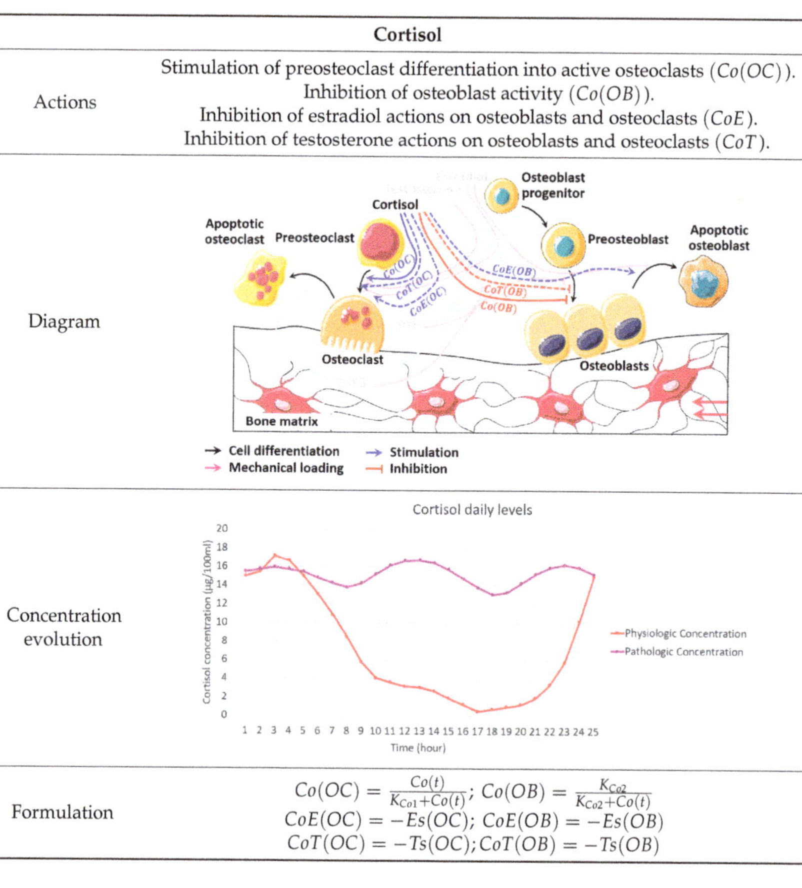
Concentration evolution	
Formulation	$Co(OC) = \dfrac{Co(t)}{K_{Co1}+Co(t)}; \; Co(OB) = \dfrac{K_{Co2}}{K_{Co2}+Co(t)}$ $CoE(OC) = -Es(OC); \; CoE(OB) = -Es(OB)$ $CoT(OC) = -Ts(OC); CoT(OB) = -Ts(OB)$

Table 6. Bisphosphonate actions and their mathematical formulation.

	Bisphosphonates
Actions	Inhibition of osteoclast activity ($Bp(OC)$).
Diagram	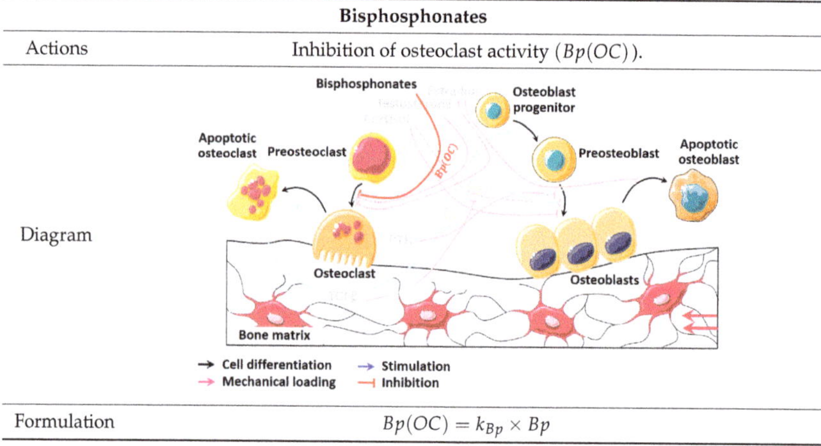
Formulation	$Bp(OC) = k_{Bp} \times Bp$

Table 7. Denosumab actions and their mathematical formulation.

Denosumab		
Actions	Inhibition of preosteoclast differentiation into active osteoclasts, and their activity $(De(OC))$.	

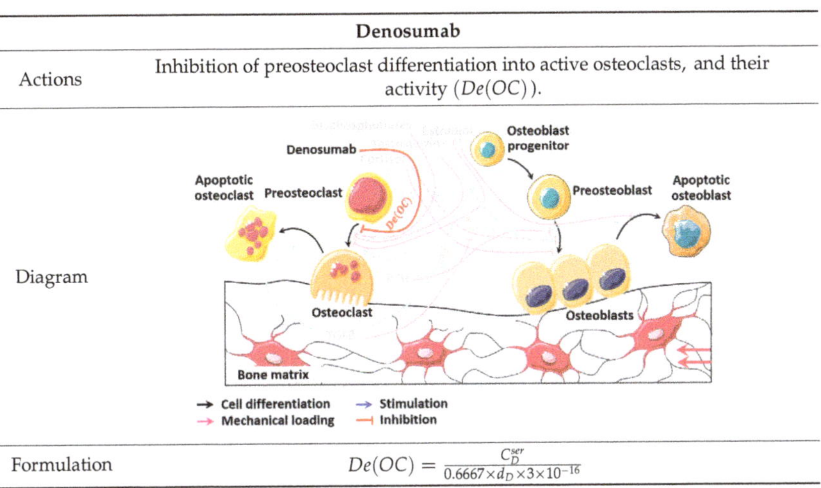

Formulation	$De(OC) = \dfrac{C_D^{ser}}{0.6667 \times d_D \times 3 \times 10^{-16}}$	

2.1.2. Mechanical Model

The mechanical approach developed by Bonfoh et al. [62] was adopted on a macroscopic scale. The local dimension sites of the BMU, the total number of osteocytes and their sensitivity as well as the interactions between bone cells were not considered when formulating the expression of the mechanical stimulus. However, when assuming an elastic isotropic behavior for bone, the mechanical stimulus can be expressed by:

$$\Delta\psi(\vec{x}) = w\frac{\left(\vec{x}^{(i)}\right)}{\rho} - W^* \tag{18}$$

where W^* denotes the balance stimulus [63], $w\left(\vec{x}^{(i)}\right)$ the strain energy density and ρ the apparent density of bone. The mechanical signal w detected by an osteocyte i at its location $\vec{x}^i$ is described by:

$$w\left(\vec{x}^{(i)}\right) = \frac{1}{2}\sigma\left(\vec{x}^{(i)}\right) : \varepsilon\left(\vec{x}^{(i)}\right) \tag{19}$$

where σ and ε denote the stress and strain tensors, respectively.

Bone is naturally damaged under the effect of the daily stresses it is subjected to, which leads to its fatigue and thereby to its aging and the occurrence and propagation of microcracks. To describe the evolution of bone mechanical properties over time, a fatigue damage formulation can be used [64–66], and the damage resulting from cyclic loading can be modeled using the life cycle approach suggested by Chaboche [67]. The damage can reach a maximum value of 1 (one), referring to material failure, and can be expressed in terms of the number of failure cycles, N_f, given by [68–70]:

$$N_f = C\Delta\varepsilon^{-\vartheta} \tag{20}$$

where $\Delta\varepsilon$ denotes the amplitude of the applied microstrains, which refer to the equivalent strain defined by $\varepsilon_{eq} = \sqrt{\frac{2}{3}\varepsilon_{ij}\varepsilon_{ij}}$, and C and ϑ are constants obtained from experimental data in the literature. These two variables, C and ϑ, depend on the nature of the solicitation [71]:

$$N_{f,c} = 1.479 \cdot 10^{-21} \Delta\varepsilon^{-10.3} \text{ for compressive loads} \tag{21}$$

$$N_{f,t} = 3.630 \cdot 10^{-32} \Delta\varepsilon^{-14.1} \text{ for tensile loads} \tag{22}$$

The damage at failure ($d = 1$) represents the accumulated damage at a given cycle expressed as:

$$d^{n+1} = d^n + \delta d \tag{23}$$

where δd denotes the incremental damage to the cycle ($n + 1$).

Nonlinear cumulative damage is generally characterized using the expression $\delta d = \left(\frac{1}{N_f}\right)^{\beta}$. However, here the following nonlinear simplistic evolution is used:

$$\delta d = \frac{1}{N_f} \tag{24}$$

Afterwards, cumulative damage is expressed using accumulated stress cycles:

$$d^{n+1} = d^n + \delta d = \frac{N+1}{N_f} \tag{25}$$

where N denotes the number of loading cycles.

When isotropic properties are assigned to the adopted material, the incremental damage can be expressed using the compressive and/or the tensile fatigue cycles. Hence, total damage δd_i, which is induced by an osteocyte i at its location $\vec{x}^{(i)}$, refers to the sum of both compressive $\delta d_{i,\,c}$ and tensile $\delta d_{i,\,t}$ damages. The latter depends on the microstrain amplitude $\Delta\varepsilon_i$ detected by each osteocyte i [69]:

$$\delta d_{i,\,c} = \frac{1}{N_{f,\,c}} = \frac{1}{1.479 \cdot 10^{-21} \Delta\varepsilon^{-10.3}} \tag{26}$$

$$\delta d_{i,\,t} = \frac{1}{N_{f,\,t}} = \frac{1}{3.630 \cdot 10^{-32} \Delta\varepsilon^{-14.1}} \tag{27}$$

$$\delta d_i = \delta d_{i,\,c} + \delta d_{i,\,t} \tag{28}$$

Using the example given by Baste et al. [72], an isotropic formulation for the damage affecting bone properties was developed by multiplying blank modules by $(1 - d_i)$. Therefore, the elastic moduli are expressed as:

$$E = E_i^0 (1 - d_i)^2 \tag{29}$$

Since the damage is only considered in the longitudinal directions, the values of the shear moduli remain constant.

2.2. Overview of the Whole Model

In order to visualize the effects of all the included parameters more clearly, Figure 3 summarizes the developed model, and it depicts the level at which each parameter acts and whether this action is a stimulation or an inhibition.

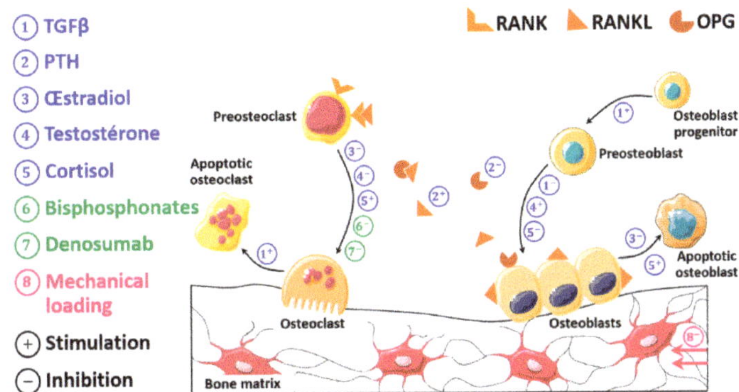

Figure 3. Summary of the proposed bone remodeling model. The shown numbers indicate the involved parameter; the positive and negative signs indicate stimulation or inhibition, respectively, exerted by each factor; the blue color indicates the biological parameters of the model, the green the pharmacological parameters and the purple the mechanical loading.

The pharmaco-biological parameters mentioned above are grouped into autocrine parameters, A^i, and paracrine parameters that are subdivided into endogenous factors, $P_{E_N}^i$, produced by the human body, and exogenous ones, $P_{E_X}^i$, introduced into the human body. Hence, the differential equations of cell dynamics can be written as follows:

$$\Rightarrow \begin{cases} \frac{dOC}{dt} = \alpha_{OC} OC^{A^{OC}} OB^{P_{E_N}^{OC}+P_{E_X}^{OC}} - \beta_{OC} OC \\ \frac{dOB}{dt} = \alpha_{OB} OB^{A^{OB}} OC^{P_{E_N}^{OB}} - \beta_{OB} OB \end{cases} \tag{30}$$

where OC denotes the osteoclast concentration, OB the osteoblast concentration, α_i the cell production activities and β_i the cell elimination activities, with:

$$P_{E_N}^{OC} \leftarrow Ta(OC) + Pr(OC) + Pa(OC) + Es(OC) + Ts(OC) + Co(OC) + CoE(OC) + CoT(OC)$$

$$P_{E_N}^{OB} \leftarrow Ta(OB) + Tr(OB) + Es(OB) + Ts(OB) + Co(OB) + CoE(OB) + CoT(OB)$$

$$P_{E_X}^{OC} \leftarrow Bp(OC) + De(OC)$$

2.3. Mechanobiological Coupling

Nowadays, osteoclast activity is acknowledged as being modulated by the osteoblastic cell lineage. By adapting the model of Pastrama et al. [31] to that of Bonfoh et al. [62], the proposed approach considers that the mechanical stimulus, ψ, that the bone is subjected to, acts as an exogenous paracrine factor on the variation of osteoblast concentration, according to:

$$\Rightarrow \begin{cases} \frac{dOC}{dt} = \alpha_{OC} OC^{A^{OC}} OB^{P_{E_N}^{OC}+P_{E_X}^{OC}} - \beta_{OC} OC \\ \frac{dOB}{dt} = \alpha_{OB} OB^{A^{OB}+P_{E_X}^{OB}} OC^{P_{E_N}^{OB}} - \beta_{OB} OB \end{cases} \tag{31}$$

with:

$$P_{E_X}^{OB} \leftarrow a + be^{-\gamma\Delta\psi} \tag{32}$$

where $a = 1.6$, $b = -0.49$ and $\gamma = 16.67 \ g/J$.

As mentioned above, the mechanical stimulus was included as an exogenous paracrine model acting on the osteoblast concentration since osteoblastic-lineage cells govern the bone mass variation. The osteoblast ability to influence osteoclast formation in a paracrine manner has been clearly demonstrated over the years. In fact, osteoblasts modulate osteoclast formation and activity by synthesizing a number of cytokines and growth factors.

This is achieved through direct contact between the two cells, established by exchanging small water-soluble molecules through gap junctions [73].

In summary, the mechanical loading that the bone is subjected to activates osteoblasts and increases their concentration according to its intensity. The increase in osteoblast concentration generates higher RANKL and OPG production rates, which modulates the appropriate osteoclast concentration for the subsequent bone resorption activity. Bone resorption releases the matrix-embedded factors, TGFβs, that act on both cell populations to regulate their differentiation, activity and apoptosis. Hormones, including PTH, cortisol, estradiol and testosterone, are also important factors that are external to the BMU, but which regulate the cell dynamics during the remodeling event (Figure 4). The estradiol parameter is considered when the study is focused on the remodeling process in women, and the testosterone parameter is considered when the study is focused on the male remodeling process. Furthermore, when the cortisol concentration is higher than under physiologic conditions, RANKL production by osteoblastic cells increases, inducing an increase in the osteoclast differentiation rate, and thereby leading to a higher resorption rate. This affects the bone response to the mechanical loading by decreasing the bone mass density and altering its mechanical response.

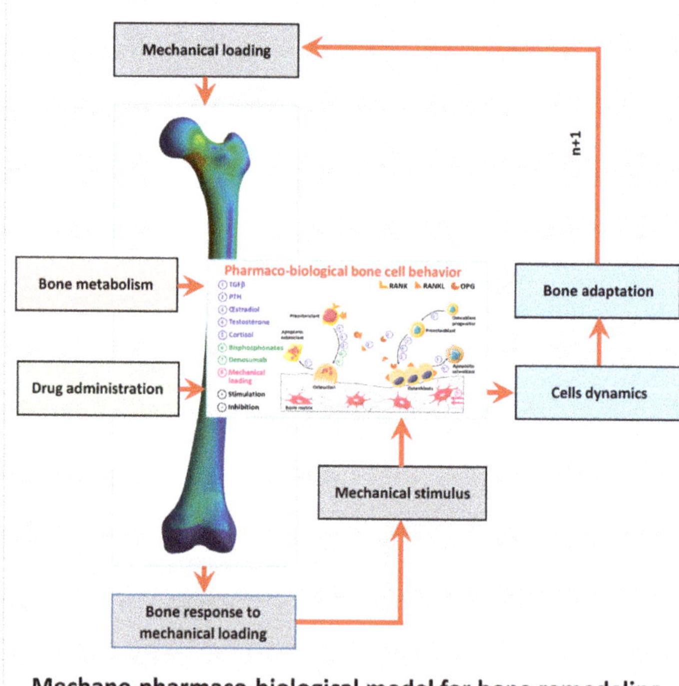

Figure 4. Diagram summarizing the proposed mechano-pharmaco-biological model.

2.4. Bone Mass Evolution

Bone mass evolution throughout time is giving by:

$$\frac{dm}{dt} = (k_{OB} \times OB) - (k_{OC} \times OC) \tag{33}$$

where k_{OB} and k_{OC} denote the normalized formation and resorption activities, respectively, and OB and OC denote the osteoblast and osteoclast concentrations, respectively.

3. Discussion and Conclusions

The focus of this article was to provide a pharmaco-biological bone remodeling model that could be easily coupled with mechanical models and was extendable to be able to include various parameters, and consequently allowing the simulation of bone physiologic metabolism for pathologic disorders. At present, the model is designed to simulate the influence of an endogenous hypercortisolism, which is caused by an excessive secretion of cortisol, on bone response to mechanical stimulus and fatigue damage to which it is subjected. The current model combines two of the most current biological models used to predict the evolution of bone mechanical properties.

The most important parameters considered were the autocrine, endogenous paracrine or exogenous paracrine parameters; in short, these are the different ways any parameter may act on bone cell dynamics during the remodeling event. Previous models either neglected various parameters [14,16,74], or the number of parameters were fixed [17,19,75,76]. However, the current model is easily extendable and can include various other parameters that can provide support in the remodeling process, since these parameters primarily act on osteoblasts or on osteoclasts, whether they are endogenous or exogenous. The proposed approach is able to track the changes in bone remodeling that are specific to each parameter. Consequently, it gives a better overview of the remodeling process by regrouping several parameters at once, instead of simulating one or a limited number of parameters each time. Furthermore, it is able to choose to neglect any unneeded parameter, according to the goal of the study in question.

The current model has some limitations since it assumes an isotropic homogenous material, which idealizes the bone behavior and can affect the outcome. Additionally, the model does not consider any differentiation stages of the bone cells in the mathematical formulations of the cell dynamics. Still, the aim was to provide a mathematical model that applied to any metabolic bone disorder, any drug administered to treat such disorder, and at the same time to track the evolution of hormones and growth factors incorporated in the model, in order to be able to adjust the drug dosage specifically for each patient.

In summary, the current model was developed based on the role of osteoblasts and osteoclasts in renewing bone, as the main players in the remodeling process. All the other factors involved were considered according to their effects on each of the cell lineages involved. Yet, it is well known that a number of these factors, such as the hormonal factors, are provided to osteoblasts and osteoclasts through blood microvessels across bone tissue. This draws the attention to the general problem of the pressure that the microvessel networks exert on bone, and their role in the bone remodeling response to the mechanical stimuli applied to it. Therefore, more accurate results may be found when taking the effect of microvessel networks into account in the mathematical and numerical bone modeling equations.

In forthcoming studies, the results of the current model will be analyzed, and the work will be extended to implement the finite element method, and to visualize the effects of hypercortisolism on a virtual 3D bone model. The proposed model can be confirmed and validated by conducting an experimental study. This will reveal its accuracy in simulating and predicting bone strength under cyclic loading, considering the physiologic conditions and the disorder related to hypercortisolism as described.

Author Contributions: All the authors contributed to this work: R.B.K. and A.B. for the mathematical model formulation, search for information about the model, and writing; M.C. and J.M.R.S.T. in the analysis, correction, discussion and writing. All authors have read and agreed to the published version of the manuscript.

Funding: This research received no external funding.

Institutional Review Board Statement: Not applicable.

Informed Consent Statement: Not applicable.

Conflicts of Interest: The authors declare no conflict of interest.

References

1. Borgström, F.; Karlsson, L.; Ortsäter, G.; Norton, N.; Halbout, P.; Cooper, C.; Lorentzon, M.; McCloskey, E.V.; Harvey, N.C.; Javaid, M.K.; et al. Fragility fractures in Europe: Burden, management and opportunities. *Arch. Osteoporos.* **2020**, *15*, 1–21. [CrossRef]
2. Beaupre, G.S.; Orr, T.E.; Carter, D.R. An approach for time-dependent bone modeling and remodeling-theoretical development. *J. Orthop. Res.* **1990**, *8*, 651–661. [CrossRef]
3. Carter, D.R.; Fyhrie, D.P.; Whalen, R. Trabecular bone density and loading history: Regulation of connective tissue biology by mechanical energy. *J. Biomech.* **1987**, *20*, 785–794. [CrossRef]
4. Carter, D.; Orr, T.; Fyhrie, D. Relationships between loading history and femoral cancellous bone architecture. *J. Biomech.* **1989**, *22*, 231–244. [CrossRef]
5. Doblaré, M.; García, J. Anisotropic bone remodelling model based on a continuum damage-repair theory. *J. Biomech.* **2002**, *35*, 1–17. [CrossRef]
6. Jacobs, C.R.; Levenston, M.E.; Beaupré, G.S.; Simo, J.C.; Carter, D.R. Numerical instabilities in bone remodeling simulations: The advantages of a node-based finite element approach. *J. Biomech.* **1995**, *28*, 449–459. [CrossRef]
7. Mullender, M.; Huiskes, R.; Weinans, H. A physiological approach to the simulation of bone remodeling as a self-organizational control process. *J. Biomech.* **1994**, *27*, 1389–1394. [CrossRef]
8. Mullender, M.; Huiskes, R. Osteocytes and bone lining cells: Which are the best candidates for mechano-sensors in cancellous bone? *Bone* **1997**, *20*, 527–532. [CrossRef]
9. Frost, H.M. The mechanostat: A proposed pathogenic mechanism of osteoporoses and the bone mass effects of mechanical and non mechanical agents. *Bone Miner.* **1987**, *2*, 73–85. [PubMed]
10. Wolff, J. Das Gesetz der Transformation der Knochen. *DMW Dtsch. Med. Wochenschr.* **1893**, *19*, 1222–1224. [CrossRef]
11. Huiskes, H.R.; Ruimerman, R.R.; Van Lenthe, G.H.; Janssen, J.D. Effects of mechanical forces on maintenance and adaptation of form in trabecular bone. *Nature* **2000**, *405*, 704–706. [CrossRef]
12. Rouhi, G.; Vahdati, A.; Li, X.; Sudak, L. A three-dimensional computer model to simulate spongy bone remodeling under overload using a semi-mechanistic bone remodeling theory. *J. Mech. Med. Biol.* **2015**, *15*, 1550061. [CrossRef]
13. Van Schaick, E.; Zheng, J.; Ruixo, J.J.P.; Gieschke, R.; Jacqmin, P. A semi-mechanistic model of bone mineral density and bone turnover based on a circular model of bone remodeling. *J. Pharmacokinet. Pharmacodyn.* **2015**, *42*, 315–332. [CrossRef]
14. Komarova, S.V.; Smith, R.J.; Dixon, S.; Sims, S.M.; Wahl, L.M. Mathematical model predicts a critical role for osteoclast autocrine regulation in the control of bone remodeling. *Bone* **2003**, *33*, 206–215. [CrossRef]
15. Ryser, M.D.; Nigam, N.; Komarova, S.V. Mathematical Modeling of Spatio-Temporal Dynamics of a Single Bone Multicellular Unit. *J. Bone Miner. Res.* **2009**, *24*, 860–870. [CrossRef]
16. Ayati, B.P.; Edwards, C.M.; Webb, G.F.; Wikswo, J.P. A mathematical model of bone remodeling dynamics for normal bone cell populations and myeloma bone disease. *Biol. Direct* **2010**, *5*, 28. [CrossRef] [PubMed]
17. Lemaire, V.; Tobin, F.L.; Greller, L.D.; Cho, C.R.; Suva, L.J. Modeling the interactions between osteoblast and osteoclast activities in bone remodeling. *J. Theor. Biol.* **2004**, *229*, 293–309. [CrossRef]
18. Maldonado, S.; Borchers, S.; Findeisen, R.; Allgower, F. Mathematical modeling and analysis of force induced bone growth. In Proceedings of the 2006 International Conference of the IEEE Engineering in Medicine and Biology Society, New York, NY, USA, 30 August–3 September 2006; pp. 3154–3157.
19. Pivonka, P.; Zimak, J.; Smith, D.W.; Gardiner, B.S.; Dunstan, C.; Sims, N.A.; Martin, T.J.; Mundy, G.R. Model structure and control of bone remodeling: A theoretical study. *Bone* **2008**, *43*, 249–263. [CrossRef] [PubMed]
20. Peyroteo, M.M.A.; Belinha, J.; Dinis, L.M.J.S.; Jorge, R.M.N. Bone remodeling: An improved spatiotemporal mathematical model. *Arch. Appl. Mech.* **2019**, *90*, 635–649. [CrossRef]
21. Ghiasi, M.S.; Chen, J.; Vaziri, A.; Rodriguez, E.K.; Nazarian, A. Bone fracture healing in mechanobiological modeling: A review of principles and methods. *Bone Rep.* **2017**, *6*, 87–100. [CrossRef] [PubMed]
22. Aitoumghar, I.; Barkaoui, A.; Chabrand, P. Mechanobiological Behavior of a Pathological Bone. In *BioMechanics and Functional Tissue Engineering*; IntechOpen: London, UK, 2021.
23. Kahla, R.B.; Barkaoui, A.; Salah, F.Z.B.; Chafra, M. Cell Interaction and Mechanobiological Modeling of Bone Remodeling Process. In *BioMechanics and Functional Tissue Engineering*; IntechOpen: London, UK, 2020.
24. Anderson, D.D.; Thomas, T.P.; Marin, A.C.; Elkins, J.M.; Lack, W.D.; Lacroix, D. Computational techniques for the assessment of fracture repair. *Injury* **2014**, *45*, S23–S31. [CrossRef]
25. Betts, D.C.; Müller, R. Mechanical regulation of bone regeneration: Theories, models, and experiments. *Front. Endocrinol.* **2014**, *5*, 211. [CrossRef]
26. Isaksson, H. Recent advances in mechanobiological modeling of bone regeneration. *Mech. Res. Commun.* **2012**, *42*, 22–31. [CrossRef]
27. Avval, P.T.; Klika, V.; Bougherara, H. Predicting bone remodeling in response to total hip arthroplasty: Computational study using mechanobiochemical model. *J. Biomech. Eng.* **2014**, *136*, 051002. [CrossRef] [PubMed]
28. Klika, V.; Pérez, M.A.; García-Aznar, J.M.; Maršík, F.; Doblaré, M. A coupled mechano-biochemical model for bone adaptation. *J. Math. Biol.* **2014**, *69*, 1383–1429. [CrossRef]

29. Lerebours, C.; Buenzli, P.R.; Scheiner, S.; Pivonka, P. A multiscale mechanobiological model of bone remodelling predicts site-specific bone loss in the femur during osteoporosis and mechanical disuse. *Biomech. Model. Mechanobiol.* **2016**, *15*, 43–67. [CrossRef] [PubMed]
30. Martin, M.; Sansalone, V.; Cooper, D.M.L.; Forwood, M.R.; Pivonka, P. Mechanobiological osteocyte feedback drives mechanostat regulation of bone in a multiscale computational model. *Biomech. Model. Mechanobiol.* **2019**, *18*, 1475–1496. [CrossRef]
31. Pastrama, M.-I.; Scheiner, S.; Pivonka, P.; Hellmich, C. A mathematical multiscale model of bone remodeling, accounting for pore space-specific mechanosensation. *Bone* **2018**, *107*, 208–221. [CrossRef]
32. Rouhi, G.; Epstein, M.; Sudak, L.; Herzog, W. Modeling bone resorption using Mixture Theory with chemical reactions. *J. Mech. Mater. Struct.* **2007**, *2*, 1141–1155. [CrossRef]
33. Scheiner, S.; Pivonka, P.; Hellmich, C. Coupling systems biology with multiscale mechanics, for computer simulations of bone remodeling. *Comput. Methods Appl. Mech. Eng.* **2013**, *254*, 181–196. [CrossRef]
34. Rouhi, G.; Vahdati, A.; Li, X.; Sudak, L.J. An investigation into the effects of osteocytes density and mechanosensitivity on trabecular bone loss in aging and osteoporotic individuals. *Biomed. Eng. Lett.* **2015**, *5*, 302–310. [CrossRef]
35. Hinton, P.V.; Rackard, S.M.; Kennedy, O.D. In Vivo Osteocyte Mechanotransduction: Recent Developments and Future Directions. *Curr. Osteoporos. Rep.* **2018**, *16*, 746–753. [CrossRef]
36. Klein-Nulend, J.; Bakker, A.D.; Bacabac, R.G.; Vatsa, A.; Weinbaum, S. Mechanosensation and transduction in osteocytes. *Bone* **2013**, *54*, 182–190. [CrossRef]
37. Ashrafi, M.; Gubaua, J.E.; Pereira, J.T.; Gahlichi, F.; Doblaré, M. A mechano-chemo-biological model for bone remodeling with a new mechano-chemo-transduction approach. *Biomech. Model. Mechanobiol.* **2020**, *19*, 2499–2523. [CrossRef]
38. Barkaoui, A.; Ben Kahla, R.; Merzouki, T.; Hambli, R. Age and gender effects on bone mass density variation: Finite elements simulation. *Biomech. Model. Mechanobiol.* **2016**, *16*, 521–535. [CrossRef]
39. Ben Kahla, R.; Barkaoui, A.; Merzouki, T. Age-related mechanical strength evolution of trabecular bone under fatigue damage for both genders: Fracture risk evaluation. *J. Mech. Behav. Biomed. Mater.* **2018**, *84*, 64–73. [CrossRef] [PubMed]
40. Barkaoui, A.; Ben Kahla, R.; Merzouki, T.; Hambli, R. Numerical Simulation of Apparent Density Evolution of Trabecular Bone under Fatigue Loading: Effect of Bone Initial Properties. *J. Mech. Med. Biol.* **2019**, *19*, 1950041. [CrossRef]
41. Hill, A.V. The possible effects of the aggregation of the molecules of haemoglobin on its dissociation curves. *J. Physiol.* **1910**, *40*, 4–7.
42. Tamma, R.; Zallone, A. Osteoblast and osteoclast crosstalks: From OAF to Ephrin. *Inflamm. Allergy Drug Targets* **2012**, *11*, 196–200. [CrossRef]
43. Tang, Y.; Wu, X.; Lei, W.; Pang, L.; Wan, C.; Shi, Z.; Zhao, L.; Nagy, T.; Peng, X.; Hu, J.; et al. TGF-β1–induced migration of bone mesenchymal stem cells couples bone resorption with formation. *Nat. Med.* **2009**, *15*, 757–765. [CrossRef]
44. Oumghar, I.A.; Barkaoui, A.; Chabrand, P. Toward a Mathematical Modeling of Diseases' Impact on Bone Remodeling: Technical Review. *Front. Bioeng. Biotechnol.* **2020**, *8*, 1236. [CrossRef]
45. Erlebacher, A.; Filvaroff, E.H.; Ye, J.-Q.; Derynck, R. Osteoblastic Responses to TGF-β during Bone Remodeling. *Mol. Biol. Cell* **1998**, *9*, 1903–1918. [CrossRef]
46. Janssens, K.; Dijke, P.T.; Janssens, S.; Van Hul, W. Transforming Growth Factor-β1 to the Bone. *Endocr. Rev.* **2005**, *26*, 743–774. [CrossRef]
47. Alliston, T.; Choy, L.; Ducy, P.; Karsenty, G.; Derynck, R. TGF-β-induced repression of CBFA1 by Smad3 decreases cbfa1 and osteocalcin expression and inhibits osteoblast differentiation. *EMBO J.* **2001**, *20*, 2254–2272. [CrossRef] [PubMed]
48. Bonewald, L.F.; Dallas, S.L. Role of active and latent transforming growth factor β in bone formation. *J. Cell. Biochem.* **1994**, *55*, 350–357. [CrossRef]
49. Mundy, G.R.; Boyce, B.F.; Yoneda, T.; Bonewald, L.F.; Roodman, G.D. *Osteoporosis*; Marcus, R., Feldman, D., Kelsey, J., Eds.; Academic Press: New York, NY, USA, 1996.
50. Siddiqui, J.A.; Johnson, J.; Le Henaff, C.; Bitel, C.L.; Tamasi, J.A.; Partridge, N.C. Catabolic Effects of Human PTH (1–34) on Bone: Requirement of Monocyte Chemoattractant Protein-1 in Murine Model of Hyperparathyroidism. *Sci. Rep.* **2017**, *7*, 15300. [CrossRef]
51. Wein, M.N.; Kronenberg, H.M. Regulation of Bone Remodeling by Parathyroid Hormone. *Cold Spring Harb. Perspect. Med.* **2018**, *8*, a031237. [CrossRef]
52. Khosla, S.; Monroe, D.G. Regulation of Bone Metabolism by Sex Steroids. *Cold Spring Harb. Perspect. Med.* **2018**, *8*, a031211. [CrossRef] [PubMed]
53. Okman-Kilic, T. Estrogen deficiency and osteoporosis. In *Advances in Osteoporosis*; IntechOpen: London, UK, 2015.
54. Mohamad, N.V.; Soelaiman, I.-N.; Chin, K.-Y. A concise review of testosterone and bone health. *Clin. Interv. Aging* **2016**, *11*, 1317–1324. [CrossRef]
55. Chiodini, I.; Torlontano, M.; Carnevale, V.; Trischitta, V.; Scillitani, A. Skeletal involvement in adult patients with endogenous hypercortisolism. *J. Endocrinol. Investig.* **2008**, *31*, 267–276. [CrossRef] [PubMed]
56. Mathis, S.L.; Farley, R.S.; Fuller, D.K.; Jetton, A.E.; Caputo, J.L. The Relationship between Cortisol and Bone Mineral Density in Competitive Male Cyclists. *J. Sports Med.* **2013**, *2013*, 1–7. [CrossRef]
57. Çetin, A.; Gökçe-Kutsal, Y.; Çeliker, R. Predictors of bone mineral density in healthy males. *Rheumatol. Int.* **2001**, *21*, 85–88. [CrossRef]

58. Reynolds, R.M.; Dennison, E.M.; Walker, B.R.; Syddall, H.E.; Wood, P.J.; Andrew, R.; Phillips, D.I.; Cooper, C. Cortisol Secretion and Rate of Bone Loss in a Population-Based Cohort of Elderly Men and Women. *Calcif. Tissue Int.* **2005**, *77*, 134–138. [CrossRef]

59. Osella, G.; Ventura, M.; Ardito, A.; Allasino, B.; Termine, A.; Saba, L.; Vitetta, R.; Terzolo, M.; Angeli, A. Cortisol secretion, bone health, and bone loss: A cross-sectional and prospective study in normal nonosteoporotic women in the early postmenopausal period. *Eur. J. Endocrinol.* **2012**, *166*, 855–860. [CrossRef]

60. Siggelkow, H.; Etmanski, M.; Bozkurt, S.; Groß, P.; Koepp, R.; Brockmöller, J.; Tzvetkov, M.V. Genetic polymorphisms in 11β-hydroxysteroid dehydrogenase type 1 correlate with the postdexamethasone cortisol levels and bone mineral density in patients evaluated for osteoporosis. *J. Clin. Endocrinol. Metab.* **2014**, *99*, E293–E302. [CrossRef]

61. Anastasilakis, A.D.; Polyzos, S.A.; Makras, P. Therapy of endocrine disease: Denosumab vs bisphosphonates for the treatment of postmenopausal osteoporosis. *Eur. J. Endocrinol.* **2018**, *179*, R31–R45. [CrossRef]

62. Bonfoh, N.; Novinyo, E.; Lipinski, P. Modeling of bone adaptative behavior based on cells activities. *Biomech. Model. Mechanobiol.* **2010**, *10*, 789–798. [CrossRef] [PubMed]

63. García-Aznar, J.M.; Rueberg, T.; Doblare, M. A bone remodelling model coupling microdamage growth and repair by 3D BMU-activity. *Biomech. Model. Mechanobiol.* **2005**, *4*, 147–167. [CrossRef] [PubMed]

64. Martin, B. A theory of fatigue damage accumulation and repair in cortical bone. *J. Orthop. Res.* **1992**, *10*, 818–825. [CrossRef]

65. Nagaraja, S.; Couse, T.L.; Guldberg, R.E. Trabecular bone microdamage and microstructural stresses under uniaxial compression. *J. Biomech.* **2005**, *38*, 707–716. [CrossRef] [PubMed]

66. Zioupos, P.; Casinos, A. Cumulative damage and the response of human bone in two-step loading fatigue. *J. Biomech.* **1998**, *31*, 825–833. [CrossRef]

67. Chaboche, J.-L. Continuous damage mechanics—A tool to describe phenomena before crack initiation. *Nucl. Eng. Des.* **1981**, *64*, 233–247. [CrossRef]

68. Pattin, C.A.; Caler, W.E.; Carter, D.R. Cyclic mechanical property degradation during fatigue loading of cortical bone. *J. Biomech.* **1996**, *29*, 69–79. [CrossRef]

69. Martin, R.B.; Burr, D.B.; Sharkey, N.A. *Skeletal Tissue Mechanics*; Springer: Berlin/Heidelberg, Germany, 1998; Volume 190.

70. Martin, R.B. Fatigue damage, remodeling, and the minimization of skeletal weight. *J. Theor. Biol.* **2003**, *220*, 271–276. [CrossRef]

71. Hambli, R. Connecting mechanics and bone cell activities in the bone remodeling process: An integrated finite element modeling. *Front. Bioeng. Biotechnol.* **2014**, *2*, 6. [CrossRef]

72. Baste, S.; el Guerjouma, R.; Gérard, A. Mesure de l'endommagement anisotrope d'un composite céramique-céramique par une méthode ultrasonore. *Rev. Phys. Appliquée* **1989**, *24*, 721–731. [CrossRef]

73. Chen, X.; Wang, Z.; Duan, N.; Zhu, G.; Schwarz, E.M.; Xie, C. Osteoblast–osteoclast interactions. *Connect. Tissue Res.* **2018**, *59*, 99–107. [CrossRef] [PubMed]

74. Martin, M.; Buckland-Wright, J. Sensitivity analysis of a novel mathematical model identifies factors determining bone resorption rates. *Bone* **2004**, *35*, 918–928. [CrossRef]

75. Bahia, M.T.; Hecke, M.B.; Mercuri, E.G.F. Image-based anatomical reconstruction and pharmaco-mediated bone remodeling model applied to a femur with subtrochanteric fracture: A subject-specific finite element study. *Med. Eng. Phys.* **2019**, *69*, 58–71. [CrossRef] [PubMed]

76. Scheiner, S.; Pivonka, P.; Smith, D.W.; Dunstan, C.; Hellmich, C. Mathematical modeling of postmenopausal osteoporosis and its treatment by the anti-catabolic drug denosumab. *Int. J. Numer. Methods Biomed. Eng.* **2014**, *30*, 1–27. [CrossRef]

Article

Methodology to Calibrate the Dissection Properties of Aorta Layers from Two Sets of Experimental Measurements

Itziar Ríos-Ruiz [1,*], Myriam Cilla [1,2,3], Miguel A. Martínez [1,3] and Estefanía Peña [1,3,*]

1   Applied Mechanics and Bioengineering, Aragón Institute of Engineering Research (I3A), University of Zaragoza, 50018 Zaragoza , Spain; mcilla@unizar.es (M.C.); miguelam@unizar.es (M.A.M.)
2   Centro Universitario de la Defensa, Academia General Militar, 50090 Zaragoza, Spain
3   CIBER de Bioingeniería, Biomateriales y Nanomedicina (CIBER-BBN), 50018 Zaragoza, Spain
*   Correspondence: itziar@unizar.es (I.R.-R.); fany@unizar.es (E.P.)

Abstract: Aortic dissection is a prevalent cardiovascular pathology that can have a fatal outcome. However, the mechanisms that trigger this disease and the mechanics of its progression are not fully understood. Computational models can help understand these issues, but they need a proper characterisation of the tissues. Therefore, we propose a methodology to obtain the dissection parameters of all layers in aortic tissue via the computational modelling of two different delamination tests: the peel and mixed tests. Both experimental tests have been performed in specimens of porcine aorta, where the intima-media and media-adventitia interfaces, as well as the medial layer, were dissected. These two tests have been modelled using a cohesive zone formulation for the separating interface and a hyperelastic anisotropic material model via an implicit static analysis. The dissection properties of each interface have been calibrated by reproducing the force-displacement curves obtained in the experimental tests. The values of peak and mean force of the experiments were fitted with an error below 10%. With this methodology, we intend to contribute to the development of reliable numerical tools for simulating aortic dissection and aortic aneurysm rupture.

Keywords: aortic dissection; delamination tests; cohesive zone model; porcine aorta; vascular mechanics

Citation: Ríos-Ruiz, I.; Cilla, C.; Martínez, M.A.; Peña, E. Methodology to Calibrate the Dissection Properties of Aorta Layers from Two Sets of Experimental Measurements. *Mathematics* **2021**, *9*, 1593. https://doi.org/10.3390/math9141593

Academic Editor: Rafael Sebastian

Received: 31 May 2021
Accepted: 2 July 2021
Published: 7 July 2021

Publisher's Note: MDPI stays neutral with regard to jurisdictional claims in published maps and institutional affiliations.

1. Introduction

According to the World Health Organization, cardiovascular disease, the treatment of which results in a major economic burden, remains the principal cause of mortality and morbidity worldwide [1]. Among all, aortic dissection and aortic aneurysm rupture are acute life threatening events. Aortic dissection usually begins with an intimal tear in the wall, followed by a fissure in a radial direction. The crack then advances into the medial layer, or between the media and the adventitia, causing the separation of the wall layers and creating a false lumen through which blood can flow [2]. Aortic aneurysms lead to Stanford type A dissections—affecting the ascending aorta—or type B dissections—affecting the descending thoracic aorta [3]. The fissure of the intima that leads to dissection of the ascending aorta is usually located a few centimetres above the coronary arteries, while those leading to dissection of the descending aorta are located a few centimetres below the left subclavian artery [4]. Therefore, it is a location specific disease and its study should consider the particular biomechanical environment and properties of each site. In addition to the above pathologies, a trauma of the thoracic aorta during traffic accidents can initiate the dissection process or cause an instantaneous rupture. Mortality estimates suggest that 20% of cases of Type A acute aortic dissection die before reaching the hospital [5]. There is about 1% mortality per hour within the first 48 h upon arrival [5] and postoperative survival at 1 year postdischarge after surgical repair is evaluated at 96.1% [6].

Existing studies in the literature have investigated the dissection of aortic media, but mechanical investigations of arterial wall delamination of the layer interfaces and the

177

numerical simulations have been limited. Sommer et al. [7] worked on understanding the mechanisms of fissure propagation during aortic dissection. To do so, they performed uniaxial tension tests in the radial direction to study the dissection strength throughout the lamellae of the vessels and medial layer peeling tests to obtain the fracture energy required for fracture propagation, both from healthy human abdominal aortas. The same group also investigated the mechanical properties of the aneurysmal media and dissected human thoracic aortas, including the less studied dissection behaviour in shear mode or mode II [8]. Pasta et al. [9] investigated the dissection properties of non-aneurysmal and aneurysmal human ascending thoracic aortas and Angouras et al. [10] analysed the dissection properties of aneurysmal ascending thoracic aortas. Both groups also performed microstructure-based models of ascending thoracic aortic aneurysms to further characterise and understand the pathological process [11,12]. Manopoulos et al. [13] characterised the mechanical properties of specimens from ascending aortas after type A dissection. Amabili et al. [14] investigated the mechanical properties of the dissected layers of one single human aneurysmal aorta after chronic Type A (Stanford) dissection. Leng et al. [15] quantified the energy release rate of the medial layer of a porcine abdominal aorta via two delamination experiments: the mixed-mode delamination experiment and the "T"-shaped delamination experiment. All these studies analysed the dissection properties of the medial layer for porcine or humans, but not the dissection properties of the intima-media or media-adventitia interfaces, which are necessary to understand crack propagation along the arterial wall. Recently, FitzGibbon and McGarry [16] presented an experimental technique to generate and characterise mode II crack initiation and propagation on excised ascending bovine aorta. Regarding the numerical simulations, several studies have analysed the behaviour of arterial tissue under delamination mode I [15,17,18] or mixed mode along the medial layer [15,16], but few investigations have focused on comparing the contributions of these two failure modes to the process of delamination of the layer interfaces. In each case, a cohesive zone (CZ) formulation has been used to model the propagation of tissue crack. Gasser and Holzapfel [17] used a cohesion law within the extended finite element (FE) method to simulate the controlled peeling (dissection) experiments by Sommer et al. [7]. Subsequently, Ferrara and Pandolfi [18] applied an anisotropic cohesion law to reproduce the anisotropic behavior observed in the peeling tests. Noble et al. [19] computationally investigated arterial perforation or dissection by an external body. Leng et al. [15] used a CZ model to simulate the arterial wall delamination under shear mode-dominated failure and the opening "T"-shaped delamination modes. Recently, FitzGibbon and McGarry [16] calibrated the mode II fracture energy based on measurement of crack propagation rates by a CZ model. However, a methodology to combine "T"-shaped and mixed delamination experiments with CZ models in order to fit the normal delamination properties for media and interface layers has not been presented yet in the literature.

Therefore, the aim of this paper is to provide a computational framework to analyse the normal delamination properties in order to gain a more in-depth understanding of the possible mechanisms leading to these fatal events. Using data taken from "T"-shaped (also known as peel) and mixed delamination experiments, together with a FE model that includes a CZ formulation to model interface properties, we estimate the normal failure properties of the medial and interface layers of the descending thoracic aorta. We are motivated by the need for reliable numerical tools for simulating aortic dissection and aortic aneurysm rupture.

2. Materials and Methods

All the specimens used in the experimental testing were obtained from one healthy porcine aorta harvested post-mortem. The 45 kg, 3.5 months old female pig was sacrificed for a different study that does not interfere with the aorta or the circulatory system. The elastic properties of the arteries were fitted from uniaxial tensile tests. The dissection properties among different layers in the aorta were calibrated via two dissection tests: the peel and the mixed tests.

2.1. Experiments

From a proximal porcine descending thoracic aorta, a total of seven 5×20 mm strips were cut in each longitudinal and circumferential directions, all of them located in a close position. One sample from each direction was used to characterise the elastic properties of the aorta by means of uniaxial tension tests. The other 12 samples were divided into two sets of 6 samples, one for the peeling test and the other for the mixed test. The specimens were dissected throughout each different arterial layer, i.e., intima-media (IM), media-adventitia (MA) and within the media (M).

Simple uniaxial tension tests were performed in a high precision drive Instron Microtester 5548 system using a 10 N load cell. A non-contact Instron 2663-281 video-extensometer was used to measure the strain during the tests. Three loading and unloading stress levels were performed (60, 120 and 240 kPa uniaxial stress) at 30%/min of strain rate. Five preconditioning cycles at all load levels were applied. The engineering stress (first Piola Kirchhoff stress tensor **P**) was computed as $P_i = \frac{F_i}{t_i\,w_i}$, where F_i is the load registered by the Instron machine and t_i and w_i are the initial thickness and width of each strip in circumferential and longitudinal directions. Only the elastic properties of the tissue were considered, therefore only the experimental data at the second loading level (120 kPa) after preconditioning was considered. For further details about the uniaxial tests, see Peña et al. [20].

For the dissection tests, an initial incision of around 5 mm was performed in each strip, facilitating the separation of the layers of interest. This selective incision was carefully performed with the aid of magnifying eyeglasses, which facilitated the perception of the different layers. In the peel test, each separated part of the specimen was pulled away by clamps. These clamps moved in opposite directions at a speed of 1 mm/min each, separating the layers of the specimen in the direction normal to the interface plane. These tests were carried out in an Instron BioPuls^{TM} low-force planar-biaxial Testing System. In the mixed test, the intimal side of the strip was glued to a clamp plate and fixed during the test and therefore only the other flap was gripped in a moving clamp. This clamp moved at a speed of 1 mm/min almost parallel to the fixed and not-yet delaminated interface, via the high precision drive Instron Microtester 5548 system.

The experiments were approved by the Ethical Committee for Animal Research of the University of Zaragoza and all procedures were carried out in accordance with the "Principles of Laboratory Animal Care" (86/609/EEC Norm).

2.2. Elastic Properties of Aortic Tissue

The Gasser-Ogden-Holzapfel (GOH) model presented by Gasser et al. [21] is used to reproduce the elastic response of the aorta tissue. This model proposed by the application of a generalised structure tensor $\mathbf{H} = \kappa\mathbf{1} + (1 - 3\kappa)\mathbf{M}_0$ (where $\mathbf{1}$ is the identity tensor and $\mathbf{M}_0 = \mathbf{m}_0 \otimes \mathbf{m}_0$ is a structure tensor defined using unit vector $\mathbf{m}_0$ to specify the mean orientation of fibres) is considered. The strain energy function (SEF) of the GOH model is as follows:

$$\Psi = \mu(I_1 - 3) + \sum_{i=4,6}\left[\frac{k_1}{2k_2}\left(exp\{k_2\hat{E}_i\}\right) - 1)\right], \tag{1}$$

where $I_1 = tr\bar{\mathbf{C}}$ represents the first invariant of the Cauchy-Green tensor ($\mathbf{C} = \mathbf{F}^T\mathbf{F}$), $\mathbf{F}$ is the deformation gradient [22] and

$$\hat{E}_i = \kappa I_1 + (1 - 3\kappa)I_i - 1 \quad i = 4,6 \tag{2}$$

where

$$I_4 = \lambda_\theta^2\cos^2(\theta) + \lambda_z^2\sin^2(\theta), \qquad I_6 = \lambda_\theta^2\cos^2(-\theta) + \lambda_z^2\sin^2(-\theta). \tag{3}$$

In this equation, I_1 represents the first invariant of the Cauchy-Green tensor [22], $\mu > 0$ and $k_1 > 0$ are stress-like parameters and $k_2 > 0$ and κ are dimensionless. Here, θ is the orientation angle relative to the circumferential direction. $\kappa \in [0, 1/3]$ is a dispersion parameter (the same for each collagen fibre family).

2.3. Fracture Properties of Aortic Tissue

To model the fracture behaviour of the interface between the arterial layers, we propose a Traction Separation Law (TSL) that relates the interfacial traction **t** (normal and shear) with interfacial displacement δ [23]. The components of the traction stress vector (τ_n, τ_s, and τ_t) represent the normal and the two shear tractions along the interface and the related displacements are δ_n, δ_s, and δ_t.

The elastic behavior of the interface is defined by

$$\boldsymbol{\tau} = \begin{pmatrix} K_{nn} & 0 & 0 \\ 0 & K_{ss} & 0 \\ 0 & 0 & K_{tt} \end{pmatrix} \boldsymbol{\delta} = \mathbf{K}\boldsymbol{\delta}. \tag{4}$$

A triangular TSL was considered to model cohesive properties of the tissue, see Figure 1. The initial interface displacement $\delta_{0_{n,s,t}}$, the tissue maximum strength $\tau_{n,s,t_{max}}$ and the energy release rate (the energy dissipated by the cohesive zone) $G_{0_{n,s,t}}$ define the mechanics of the cohesive zone following

$$K_{ii} = \frac{\tau_{i_{max}}}{\delta_{0_i}} \qquad G_{0_i} = \frac{\delta_{r_i} \cdot \tau_{i_{max}}}{2}, \tag{5}$$

where $i = n, s, t$.

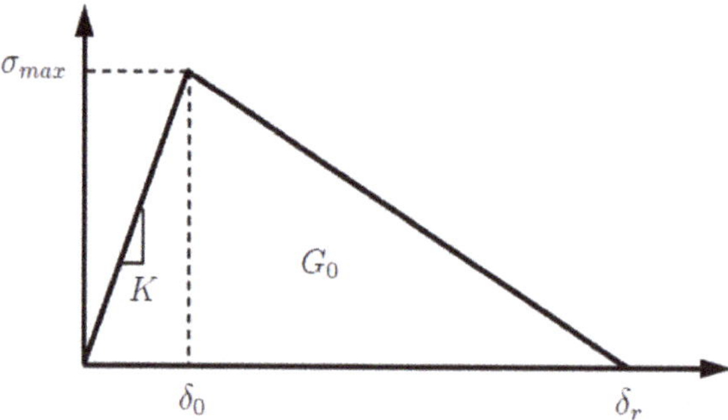

Figure 1. Traction Separation Law considered. The cohesive strength σ_{max}, the initial (reversible) interface displacement δ_0 and the maximum cohesive displacement, δ_r, are parameters to be defined.

The evolution of damage can be defined by specifying either the effective displacement at complete failure, $\delta_{r_{n,s,t}}$, related to the effective displacement at the initiation of damage, $\delta_{0_{n,s,t}}$, or the energy dissipated due to failure $G_{0_{n,s,t}}$. Damage law is defined in the context of Continuum Damage Mechanics Theory [24]. Damage is assumed to initiate when the maximum nominal stress ratio reaches a value of one

$$max\left\{ \frac{\tau_n}{\tau_{n_{max}}}, \frac{\tau_s}{\tau_{s_{max}}}, \frac{\tau_t}{\tau_{t_{max}}} \right\} = 1. \tag{6}$$

$D \in [0, 1]$ is a scalar variable that represents the damage of the material and combines the effects of all the mechanisms. D monotonically progresses from 0 to 1 upon further loading after the initiation of damage. In the context of linear softening, the evolution of the damage variable, D, is computed as

$$D = \frac{\delta_m^f(\delta_m^{max} - \delta_m^0)}{\delta_m^{max}(\delta_m^f - \delta_m^0)}, \tag{7}$$

where δ_m is computed as

$$\delta_m = \sqrt{\langle \delta_n \rangle^2 + \delta_s^2 + \delta_t^2},\tag{8}$$

and δ_m^f is the displacement at complete failure relative to the displacement at damage initiation, δ_m^0, and δ_m^{max} refers to the maximum value of the displacement reached during the loading history. The Macaulay brackets ($\langle x \rangle = 0$ if $x < 0$ or $\langle x \rangle = x$ if $x > 0$) are used to enforce that there is no damage initiation at pure compressive deformation or stress.

Finally, the stress of the traction-separation model is computed according to

$$\tau_n = \begin{cases} (1-D)\tau_n' & \text{if } \tau_n' > 0 \\ \tau_n' & \text{otherwise} \end{cases}\tag{9}$$

$$\tau_s = (1-D)\tau_s'\tag{10}$$

$$\tau_t = (1-D)\tau_t'\tag{11}$$

where τ_n', τ_s' and τ_t' are the effective stress components of the undamaged material computed by the elastic traction-separation law.

2.4. Methodology to Calibrate the Failure Properties from Experimental Measurements

The elastic properties of the aorta were fitted with the uniaxial tension tests data by using a Nelder and Mead type minimisation algorithm [25] defining the objective function $\chi^2 = \Sigma_{i=1}^{n}\left[\left(P_{\theta\theta} - P_{\theta\theta}^{\Psi}\right)_i^2 + \left(P_{zz} - P_{zz}^{\Psi}\right)_i^2\right]$ using HyperFit software. (www.hyperfit.wz.cz, accessed on 31 March 2021) . The tissue was assumed incompressible [26], i.e., $\det(\mathbf{F}) = \lambda_1\lambda_2\lambda_3 = 1$, where $\mathbf{F}$ represents the deformation gradient tensor and λ_i, $i = 1,2,3$, the stretches in the principal directions. $P_{\theta\theta}$ and P_{zz} are the First Piola-Kirchhoff (engineering) stress data obtained from the tests, and $P_{\theta\theta}^{\Psi} = \frac{\partial \Psi_{iso}}{\partial \lambda_\theta}$ and $P_{zz}^{\Psi} = \frac{\partial \Psi_{iso}}{\partial \lambda_z}$ are the First Piola-Kirchhoff stresses for the ith point for a homogeneous pure uniaxial state Ψ. The normalised root mean square error, $\varepsilon \in [0,1]$, was computed for the fitting of the material model, following

$$\varepsilon = \frac{\sqrt{\frac{\chi^2}{n-q}}}{\omega},\tag{12}$$

where $\omega = \Sigma_{i=1}^{n}\frac{P_i}{n}$ is the mean value of the measured engineering stresses, n is the number of data points, q is the number of parameters of the SEF and, therefore, $n - q$ represents the number of degrees of freedom.

The normal values of cohesive properties δ_{0_n}, δ_{r_n}, $\tau_{n_{max}}$, K_{nn} and G_{0_n} were calibrated by an iterative fitting of the experimental measurements of the peeling and mixed tests. In each iteration, the values of peak and mean force of the computational modelling were compared to the experimental data until their difference was below 10%.

2.5. Numerical Implementation

The peel and mixed tests were used to identify the normal cohesive material parameters that model purely normal failure at the interface of IM, MA and M (δ_{0_n}, δ_{r_n}, $\tau_{n_{max}}$, K_{nn} and G_{0_n}). A cohesive zone was introduced in the interface to analyse, where tissue delamination was expected. A refined mesh in this contact area was needed in order to obtain appropriate results. A TSL was postulated [27], see Figure 1. The FE geometry was specific for each experimental strip. The total thickness of the specimens was measured and the ratio of thickness per layer was obtained from Peña et al. [20]. The corresponding interfaces were defined with the cohesive contact model previously presented and the models were meshed with hybrid eight-node linear bricks (C3D8H), see Figure 2. A mesh sensitivity analysis was performed in both models to achieve the compromise between accuracy and computational time. The dimensions and number of elements of each specimen are included in Table 1. The symmetry of the problem was taken into account and only half of the width of the specimens was modelled. The total length of the strips in all models was of 20 mm.

Table 1. Dimensions and number of elements of each model.

Specimen		Width [mm]	Thickness [mm]	Number of Elements
Peel test	IM	4.0	1.70	50,340
	MA	4.0	2.00	54,700
	M	4.0	2.10	57,040
Mixed test	IM	5.0	1.57	25,265
	MA	4.8	2.30	20,370
	M	5.0	2.00	27,962

Regarding the boundary conditions, in the peel test, the non-separated end of the strip was fixed to avoid its movement as solid rigid, and in both flaps at the other end the same displacement was imposed in opposite directions, causing delamination at the interface of the layer. As for the mixed test, the inner surface of the intima of the specimen was fixed and a displacement loading parallel to the strip length was applied to the free end of the other layer, causing the desired delamination at the interface.

The FE model was computed with Abaqus/Standard v6.14. An iterative trial-error procedure was performed to fit the delamination properties of aorta layers. The mechanical data in terms of the load vs. displacement curve from peel and mixed tests was compared in each iteration. The displacements applied in the free surfaces were prescribed and the cohesive properties updated in order to fit the mean force recorded during the tests. The hyperelastic material model was implemented via the in-built material model in Abaqus. The preferred fibre directions were included manually in the input files. A static implicit analysis was carried out for all models, as the strips are pulled apart slow enough to exclude inertial effects.

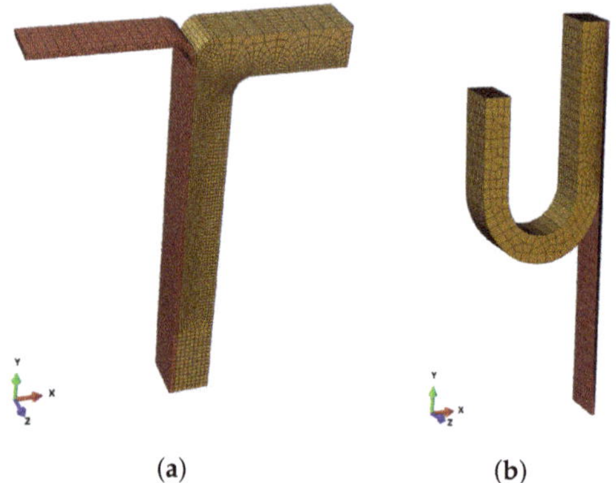

(a)   (b)

Figure 2. FE models of delamination experimental tests (a) Peel test and (b) Mixed test. Both models represent the IM separation. The thinner copper-coloured parts represent the intimal layer and the thicker brass-coloured parts, the media and adventitia.

3. Results

3.1. Elastic Properties of Aortic Tissue

The elastic mechanical data obtained by uniaxial tests experiments in each direction—longitudinal and circumferential—were fitted using the SEF represented in (1). The material constants resulting from the fitting to the SEF are shown in Table 2. The low value obtained of $\varepsilon = 0.0652$ demonstrates the goodness of the fitting.

Table 2. Material parameters obtained from the uniaxial stress-stretch curves. Constants μ and k_1 in kPa, θ in degrees, k_2, κ and ε dimensionless.

μ [kPa]	k_1 [kPa]	k_2 $[-]$	κ $[-]$	θ [°]	R^2	ε
18.0606	504.9060	44.8462	0.24299	35	0.9893080	0.0652

Plots of the fitted stress-stretch behaviour for the longitudinal and circumferential directions, together with the underlying experimental data are depicted in Figure 3 for the constitutive law in (1).

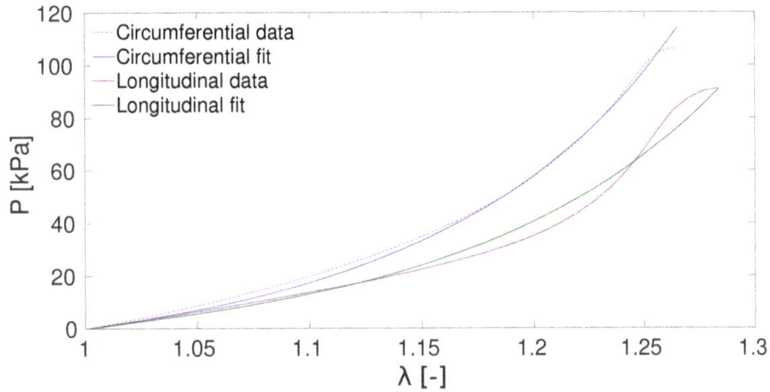

Figure 3. Experimental data of uniaxial tension tests and computational fitting obtained with the proposed constitutive law.

3.2. Fracture Properties of Aortic Tissue

Several iterations were needed to identify the sets of constants ($\tau_{n_{max}}$, K_{nn} and G_{0_n}) that minimise the differences between the value of load vs. displacement curve obtained by the experimental peel and mixed tests and that obtained by the FE model. These parameters are presented in Tables 3 and 4. The ranges of force achieved in the simulation could mostly be fitted by modifying the damage parameters $\tau_{n_{max}}$ and G_{0_n}. The parameter related to the cohesive behaviour K_{nn} had a reduced impact in the level of force and was found to account for the convergence of the models.

Table 3. Normal cohesive material parameters obtained by the fitting of the peel test and used to model normal failure at the interface.

	Interface	δ_{0_n} [mm]	δ_{r_n} [mm]	$\tau_{n_{max}}$ [kPa]	K_{nn} [mN/mm^3]	G_{0_n} [mN/mm]
IM	Longitudinal	0.023	0.070	230	10,000	8
	Circumferential	0.014	0.100	200	14,000	10
MA	Longitudinal	0.019	0.086	185	10,000	8
	Circumferential	0.020	0.063	160	8000	5
M	Longitudinal	0.013	0.092	130	10,000	6
	Circumferential	0.010	0.100	80	8000	4

In comparison with the experimental peel data of the interfaces intima-media, media-adventitia and media, the simulation using the fitted parameters is in good agreement with results with an error below 10%, see Figure 4. The initial elastic part of the curves is well reproduced in all cases except for the MA separation in the circumferential direction. This part of the tests is mainly affected by the modelling of the material and the preferred fibre directions. Due to convergence issues, this test could not be computed for a clamp displacement of more than 2 mm per side. Damage properties shown in Table 3 are consistently lower in the circumferential direction than in the longitudinal direction in each interface. This is in accordance with the dissection in the circumferential direction reportedly being easier, as it can propagate separating lamellar layers and not tearing them [7]. Furthermore, damage properties are notably smaller in the dissection within the medial layer compared to the dissection of both interfaces. The cohesive behaviour K_{nn} is similar in all cases.

In comparison with the experimental mixed data of the interfaces intima-media, media-adventitia and media, the simulation using the fitted parameters is in good agreement with results with an error below 10%, see Figure 5. In this case, the initial elastic part of the curves is well reproduced in all cases. The modelling of the mixed test allowed for higher convergence, up to 10 mm of clamp displacement. All fitted parameters in this test shown in Table 4 are consistently higher than those obtained for the simulation of the peel test. The dissection of the IM provided the lowest damage parameters, in accordance with this separation presenting the lowest dissection forces. For the IM and M dissections, the longitudinal direction presented higher values in its properties than the circumferential direction. The effective displacement at complete failure, δ_{r_n}, is of around 0.1 mm in all cases. The cohesive behaviour K_{nn} is the same in all cases, except for the IM separation in the longitudinal direction.

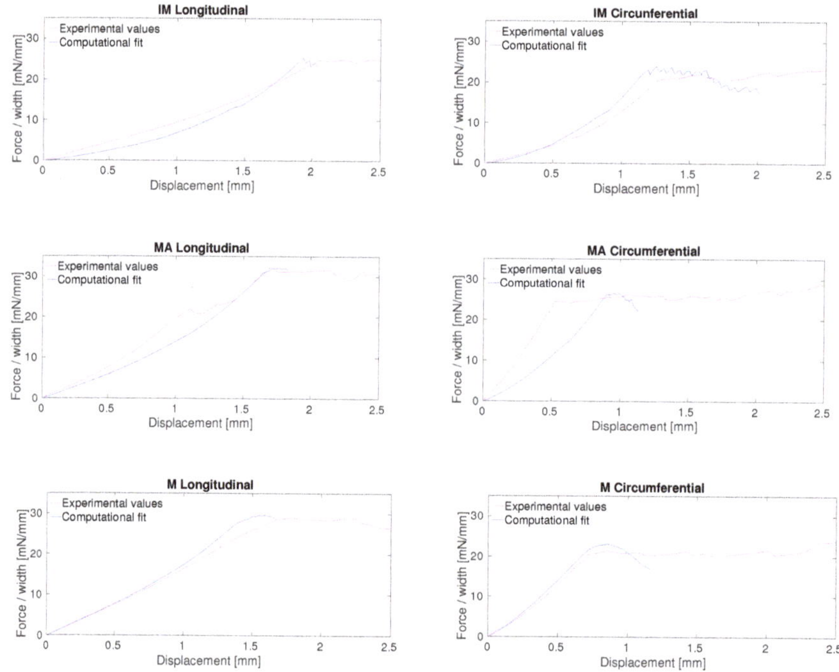

Figure 4. Correlation between force/width vs. displacement curves and computation of the peel test of the interfaces intima-media, media-adventitia and media for longitudinal and circumferential directions.

Table 4. Normal cohesive material parameters obtained by the fitting of the mixed test and used to model normal failure at the interface.

	Interface	δ_{0_n} [mm]	δ_{r_n} [mm]	$\tau_{n_{max}}$ [kPa]	K_{nn} [mN/mm^3]	G_{0_n} [mN/mm]
IM	Longitudinal	0.040	0.100	800	20,000	40
	Circumferential	0.034	0.073	550	16,000	20
MA	Longitudinal	0.078	0.104	1250	16,000	65
	Circumferential	0.088	0.107	1400	16,000	75
M	Longitudinal	0.081	0.100	1300	16,000	65
	Circumferential	0.075	0.100	1200	16,000	60

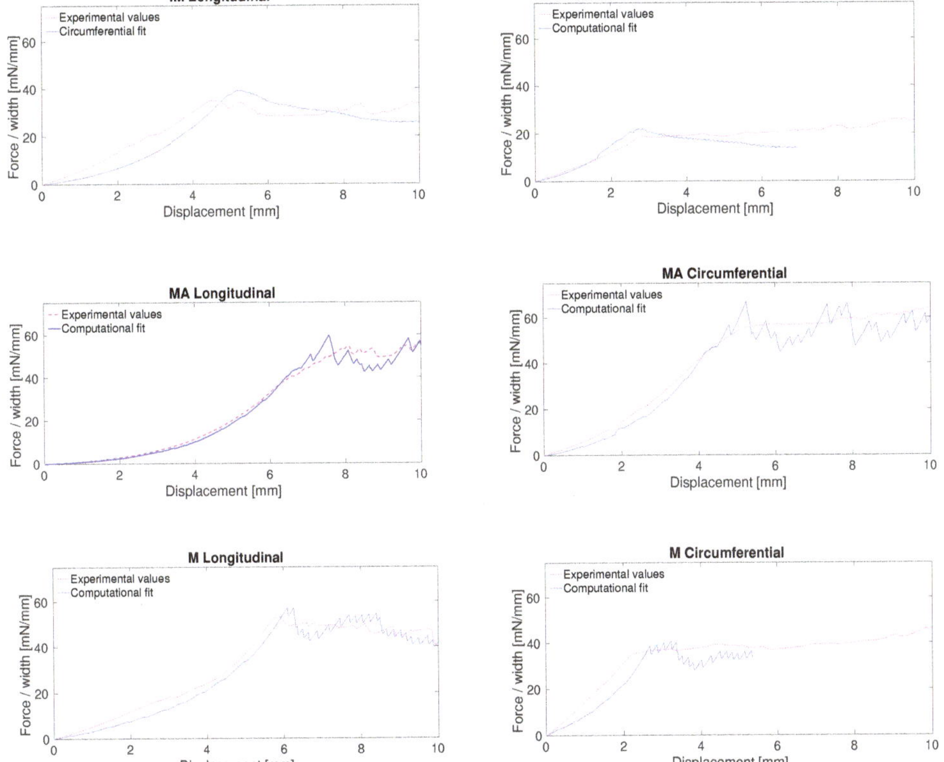

Figure 5. Correlation between force/width vs. displacement curves and computation of the mixed test of the interfaces intima-media, media-adventitia and media for longitudinal and circumferential directions.

4. Discussion

Aortic dissection is an important cardiovascular pathology and its triggering mechanism and development mechanics are not fully comprehended. In particular, the delamination properties of aortic tissue, which could provide insight into the development of this disease, have been sparsely studied. To contribute to this field, in this study, we have numerically reproduced two tissue dissection tests—the peel and mixed tests—of porcine aorta specimens. These numerical studies allow obtaining different dissection parameters that characterise the behaviour of the tissue.

The experimental forces to be fitted are predominantly higher in the mixed test than in the peel test, with the exception of the separation of the intima-media in the circumferential direction, in which the mixed test obtained a lower dissection force. The dissection parameters obtained in all simulations of the mixed tests are notably higher—sometimes one order of magnitude—than those obtained in the peel tests. The main parameter that affects the reaction force of the simulations was $\tau_{n_{max}}$ and therefore is the one that varies most throughout the simulations. Ferrara and Pandolfi [18] had checked the relevant influence of this parameter in the numerical results of a dissection process. The marked difference between the parameters obtained in both types of simulations is not convenient. The small impact of the tangential components of the cohesive model in the mixed test simulations lead to the assumption that not all the damage phenomena are being captured in the simulations, which could lead to these differences. Moreover, the rupture stress obtained in the uniaxial tension tests was of 1090 and 660 kPa for the circumferential and longitudinal directions, respectively. The values of $\tau_{n_{max}}$ obtained in the mixed test models are more similar to these fracture stresses. This could also imply that the low values of these parameters in the peel test could be due to the specimens experiencing damage in the tissue and not the specific separation of layers.

This study reproduces the results of two different tests which were carried out once in each condition and therefore the numerical results depend completely in the data of only one repetition and do not account for the deviation present in the mechanical testing of biological tissues. However, the objective was not to determine the cohesive properties of the porcine descending aorta, but to develop two computational models that could perform such determination.

When compared to the literature, Leng et al. [15] reproduced computationally these same two tests for the separation of the medial layer of a porcine abdominal aorta. They established a $\tau_{n_{max}}$ of 440 kPa for both tests, which lies in the same order of magnitude of our results for the peel test. The energy rate they obtain however is higher—220 and 186 mN/mm for the mixed and peel test, respectively.

The complexity of the numerical models here presented entails some convergence issues. The simulation of contact and damage has always been a tricky challenge, even more when combined with hyperelastic anisotropic material models. Carrying out these computations with a static implicit analysis further hinders a full convergence of the models. In order to solve this matter, in future studies, these models are to be defined via a dynamic explicit analysis. As yet another future development, the cohesive model could be modified to include a more real contribution of shear stresses in the dissection properties of the tissue. An anisotropic damage model with a different dependence on the separation direction could probably provide a more uniform fitting for the peel and mixed tests.

This study has provided a calibrated methodology to obtain delamination properties of arteries. The characterisation of these properties is relevant to achieve a better understanding of the mechanical behaviour of vessels in general and of the process of aortic dissection in particular. Furthermore, numerical studies can benefit from this type of data to reproduce with more accuracy the physiology and pathology of the cardiovascular system.

Author Contributions: Conceived and designed the study: M.A.M. and E.P. Development of experiments: I.R.-R. and E.P. Computational implementation of model: I.R.-R. and M.C. Writing, review and editing: I.R.-R., M.C., M.A.M. and E.P. Funding acquisition, M.A.M. and E.P. All authors have read and agreed to the published version of the manuscript.

Funding: This research was funded by the Spanish Ministry of Science and Technology through research projects DPI2016-76630-C2-1-R and PID2019-107517RB-I00 and by the regional Government of Aragón through research project T24-20R and grant IIU/1408/2018.

Data Availability Statement: Not applicable.

Acknowledgments: The authors gratefully acknowledge research support from the University of Zaragoza for the use of the Servicio General de Apoyo a la Investigación-SAI. Part of the work was performed by the ICTS "NANBIOSIS" specifically by the Tissue & Scaffold Characterization Unit

(U13), of the CIBER in Bioengineering, Biomaterials & Nanomedicne (CIBER-BBN at the University of Zaragoza). CIBER Actions are financed by the Instituto de Salud Carlos III with assistance from the European Regional Development Fund.

Conflicts of Interest: The authors declare no conflict of interest.

References

1. World Health Organization. *Global Health Estimates 2019: Global Health Estimates: Life Expectancy and Leading Causes of Death and Disability*; Technical Report; World Health Organization (WHO): Geneva, Switzerland, 2020.
2. Mikich, M. Dissection of the aorta: A new approach. *Heart* **2003**, *89*, 6–8. [CrossRef] [PubMed]
3. Sherifova, S.; Holzapfel, G.A. Biochemomechanics of the thoracic aorta in health and disease. *Prog. Biomed. Eng.* **2020**, *99*, 1–17.
4. Elefteriades, J.A. Thoracic aortic aneurysm: Reading enemy's playbook. *J. Biol. Med.* **2008**, *81*, 175–186. [CrossRef] [PubMed]
5. Erbel, R.; Alfonso, F.; Boileau, C.; Dirsch, O.; Eber, B.; Haverich, A.; Rakowski, H.; Struyven, J.; Radegran, K.; Sechtem, U.; et al. Diagnosis and management of aortic dissection: Recommendations of the Task Force in Aortic Dissection. *Eur. Heart J.* **2001**, *22*, 1642–1681. [CrossRef] [PubMed]
6. Evangelista, A.; Isselbacher, E.M.; Bossone, E.; Gleason, T.G.; Eusanio, M.D.; Sechtem, U.; Ehrlich, M.P.; Trimarchi, S.; Braverman, A.C.; Myrmel, T.; et al. Insights From the International Registry of Acute Aortic Dissection. *Circulation* **2018**, *137*, 1846–1860. [CrossRef] [PubMed]
7. Sommer, G.; Gasser, T.C.; Regitnig, P.; Auer, M.; Holzapfel, G.A. Dissection Properties of the Human Aortic Media: An Experimental Study. *J. Biomech. Eng.* **2008**, *130*, 021007. [CrossRef]
8. Sommer, G.; Sherifova, S.; Oberwalder, P.; Dapunt, O.; Ursomanno, P.; De Anda, A.; Griffith, B.; Holzapfel, G.A. Mechanical strength of aneurysmatic and dissected human thoracic aortas at different shear loading modes. *J. Biomech.* **2016**, *49*, 2374–2382. [CrossRef]
9. Pasta, S.; Phillippi, J.A.; Gleason, T.G.; Vorp, D.A. Effect of aneurysm on the mechanical dissection properties of the human ascending thoracic aorta. *Thoracic Cardiovasc. Surg.* **2012**, *143*, 460–467. [CrossRef]
10. Angouras, D.C.; Kritharis, E.; Sokolis, D. Regional distribution of delamination strength in ascending thoracic aortic aneurysms. *J. Mech. Behav. Biomed. Mater.* **2019**, *98*, 58–70. [CrossRef]
11. Pasta, S.; Phillippi, J.A.; Tsamis, A.; D'Amore, A.; Raffa, G.M.; Pilato, M.; Scardulla, C.; Watkins, S.C.; Wagner, W.R.; Gleason, T.G.; Vorp, D.A. Constitutive modeling of ascending thoracic aortic aneurysms using microstructural parameter. *Med. Eng. Phys.* **2016**, *38*, 121–130. [CrossRef]
12. Sassani, S.G.; Tsangaris, S.; Sokolis, D.P. Layer- and region-specific material characterization of ascending thoracic aortic aneurysms by microstructure-based models. *J. Biomech.* **2015**, *48*, 3757–3765. [CrossRef] [PubMed]
13. Manopoulos, C.; Karathanasis, I.; Kouerinis, I.; Angouras, D.; Lazaris, A.; Tsangaris, S.; Sokolis, D. Identification of regional/layer differences in failure properties and thickness as important biomechanical factors responsible for the initiation of aortic dissections. *J. Biomech.* **2018**, *80*, 102–110. [CrossRef]
14. Amabili, M.; Arena, G.O.; Balasubramanian, P.; Breslavsky, I.D.; Cartier, R.; Ferrari, G.; Holzapfel, G.A.; Kassab, A.; Mongrain, R. Biomechanical characterization of a chronic type a dissected human aorta. *J. Biomech.* **2020**, *110*, 109978. [CrossRef]
15. Leng, X.; Zhou, B.; Deng, X.; Davis, L.; Lessner, S.M.; Sutton, M.A.; Shazly, T. Experimental and numerical studies of two arterial wall delamination modes. *J. Mech. Behav. Biomed. Mater.* **2018**, *77*, 321–330. [CrossRef] [PubMed]
16. FitzGibbon, B.; McGarry, P. Development of a test method to investigate mode II fracture and dissection of arteries. *Acta Biomat.* **2021**, *121*, 444–460. [CrossRef] [PubMed]
17. Gasser, T.C.; Holzapfel, G.A. Modeling the propagation of arterial dissection. *Eur. J. Mech. Solids* **2006**, *25*, 617–633. [CrossRef]
18. Ferrara, A.; Pandolfi, A. A numerical study of arterial media dissection processes. *Int. J. Fract.* **2010**, *166*, 21–33. [CrossRef]
19. Noble, C.; van der Sluis, O.; Voncken, R.M.; Burke, O.; Franklin, S.E.; Lewis, R.; Taylor, Z.A. Simulation of arterial dissection by a penetrating external body using cohesive zone modelling. *J. Mech. Behav. Biomed.* **2017**, *71*, 95–105. [CrossRef]
20. Peña, J.A.; Martínez, M.A.; Peña, E. Layer-specific residual deformations and uniaxial and biaxial mechanical properties of thoracic porcine aorta. *J. Mech. Behav. Biomed.* **2015**, *50*, 55–69. [CrossRef]
21. Gasser, T.C.; Ogden, R.W.; Holzapfel, G.A. Hyperelastic modelling of arterial layers with distributed collagen fibre orientations. *J. R. Soc. Interface* **2006**, *3*, 15–35. [CrossRef]
22. Spencer, A.J.M. Theory of Invariants. In *Continuum Physics*; Academic Press: New York, NY, USA, 1971; pp. 239–253.
23. Forsell, C.; Gasser, T.C. Numerical simulation of the failure of ventricular tissue due to deeper penetration, the impact of constitutive properties. *J. Biomech.* **2011**, *44*, 45–51. [CrossRef] [PubMed]
24. Lemaitre, J. A continuous damage mechanics model for ductile fracture. *J. Eng. Mater. Tchnol.* **1985**, *107*, 83–89. [CrossRef]
25. Nelder, J.A.; Mead, R. A simplex method for function minimization. *Comput. J.* **1965**, *7*, 308–313. [CrossRef]
26. Carew, T.E.; Vaishnav, R.N.; Patel, D.J. Compressibility of the arterial wall. *Circ. Res.* **1968**, *23*, 61–86. [CrossRef]
27. Hernández, Q.; Peña, E. Failure properties of vena cava tissue due to deep penetration during filter insertion. *Biomech. Model. Mechanobiol.* **2016**, *15*, 845–856. [CrossRef] [PubMed]

Article

Numerical Assessment of the Structural Effects of Relative Sliding between Tissues in a Finite Element Model of the Foot

Marco A. Martínez Bocanegra [1,2], Javier Bayod López [1,3], Agustín Vidal-Lesso [2,*], Andrés Mena Tobar [1] and Ricardo Becerro de Bengoa Vallejo [4]

1   Group of Applied Mechanics and Bioengineering (AMB), Aragon Institute of Engineering Research (i3A), Universidad de Zaragoza, 50018 Zaragoza, Spain; 772636@unizar.es or ma.martinezbocanegra@ugto.mx (M.A.M.B.); jbayod@unizar.es (J.B.L.); mena@unizar.es (A.M.T.)
2   Mechanical Engineering Department, Engineering Division of the Irapuato-Salamanca Campus (DICIS), Universidad de Guanajuato, Salamanca 36885, Guanajuato, Mexico
3   Biomedical Research Networking Center in Bioengineering, Biomaterials and Nanomedicine (CIBER-BBN), 50018 Zaragoza, Spain
4   Nursing Department, Faculty of Nursing, Physiotherapy and Podiatry, Universidad Complutense de Madrid, 28040 Madrid, Spain; ribebeva@ucm.es
*   Correspondence: agustin.vidal@ugto.mx

Abstract: Penetration and shared nodes between muscles, tendons and the plantar aponeurosis mesh elements in finite element models of the foot may cause inappropriate structural behavior of the tissues. Penetration between tissues caused using separate mesh without motion constraints or contacts can change the loading direction because of an inadequate mesh displacement. Shared nodes between mesh elements create bonded areas in the model, causing progressive or complete loss of load transmitted by tissue. This paper compares by the finite element method the structural behavior of the foot model in cases where a shared mesh has been used versus a separated mesh with sliding contacts between some important tissues. A very detailed finite element model of the foot and ankle that simulates the muscles, tendons and plantar aponeurosis with real geometry has been used for the research. The analysis showed that the use of a separate mesh with sliding contacts and a better characterization of the mechanical behavior of the soft tissues increased the mean of the absolute values of stress by 83.3% and displacement by 17.4% compared with a shared mesh. These increases mean an improvement of muscle and tendon behavior in the foot model. Additionally, a better quantitative and qualitative distribution of plantar pressure was also observed.

Keywords: foot finite element method; foot and ankle model; shared nodes; separated mesh; plantar pressure

Citation: Bocanegra, M.A.M.; López, J.B.; Vidal-Lesso, A.; Tobar, A.M.; Vallejo, R.B.d.B. Numerical Assessment of the Structural Effects of Relative Sliding between Tissues in a Finite Element Model of the Foot. *Mathematics* **2021**, *9*, 1719. https://doi.org/10.3390/math9151719

Academic Editor: Mauro Malvè

Received: 29 May 2021
Accepted: 25 June 2021
Published: 22 July 2021

Publisher's Note: MDPI stays neutral with regard to jurisdictional claims in published maps and institutional affiliations.

1. Introduction

The biomechanical behavior and load transfer of muscles, tendons, ligaments, and plantar aponeurosis in a finite element foot model (FEFM) are relevant factors that influence the model's functional response [1–5]. Inappropriate structural behavior of these biological tissues may provide inaccurate or limited information about the case study.

Nowadays, several FEFMs have been developed, some of them very complete and with a degree of detail and complexity superior to others. The creation of a FEFM significantly depends on the study to be carried out with it. However, it is important to clarify that this is not the only factor to consider. The case study or investigation will define aspects of our model, such as the CAD modeling of the tissues, the meshing of the model, the mechanical characterization of the tissue´s behavior, the application of boundary conditions, and the selection of the loads.

All scientific research is subject to limitations of all kinds, and of course, these limitations are reflected in the creation of the finite element model. The creation of simplified models is a good option in these cases. Some of these simplified foot models simulate ligaments, tendons, the aponeurosis plantar and muscles with one-dimensional elements [6–9], while other models simulate these soft tissues with three-dimensional solid elements with scanned human geometry [10–12]. We also see that some of the simplified finite element foot models mentioned above lack some muscles, tendons, or skin (fat pad). Furthermore, it is possible to obtain acceptable results. Therefore, what are the difficulties that can arise when trying to simulate the biomechanical behavior of muscles and tendons in the foot model? In addition, why is it important to simulate the proper behavior of these tissues?

There are commonly encountered problems when simulating the structural behavior of muscles and tendons with both three-dimensional solid elements and one-dimensional elements in FEFM. The first is the penetration between the mesh elements that make up these biological tissues. Penetration between mesh elements is due to the non-application of contact and motion constraints between groups of nodes of the mesh elements of the tissues. In foot models where there is mesh penetration between the tissues, the direction of the tissue loading can change considerably so that the muscles do not properly transmit the mechanical stimuli to the areas where they insert with the bone. The second refers to the shared nodes between the mesh elements of the tissues that do not allow the complexity of sliding contact between them. These groups of shared nodes between the mesh of tissues generate bonded areas throughout the model. These involuntary bonded areas between muscles and tendons cause a progressive loss of load as they travel through the tissue to the insertion zone with the bone.

The present work compares the structural behavior of muscles, tendons, aponeurosis plantar, and fat-pad when the finite element model includes a shared mesh versus separated mesh with a relative displacement between the soft tissues. For this purpose, this paper evaluates three case studies in which the plantar pressure and the maximum principal stress in some tissues are used as comparison parameters.

The cases evaluated are the following:

- Case 1: Model with shared mesh and a linearly elastic, homogeneous, and isotropic mechanical behavior of most of its tissues.
- Case 2: Model with shared mesh and a linearly elastic, homogeneous, and hyperelastic mechanical behavior of most of its tissues.
- Case 3: Model with separate mesh and a linearly elastic, homogeneous, and hyperelastic mechanical behavior of most of its tissues.

The work done in this paper aims to generate knowledge on how the use of a shared mesh with isotropic properties as opposed to the use of a separate mesh with contacts and hyperelastic properties could cause changes in the structural response of the tissues and the model in general. The knowledge generated in this research will allow those who work in simulation using numerical models based on finite element method to know the advantages and disadvantages of using a shared mesh versus a separated mesh both during the model construction process and the simulation of the cases.

2. Materials and Methods

2.1. Model Conformation

The tissue geometries of the foot and ankle model used in this research were generated through three-dimensional scanning techniques by Morales Orcajo Enrique [13] of the Applied Mechanics and Bioengineering group (AMB) at the University of Zaragoza, Spain. The foot and ankle model includes cartilage, skin (fat pad), and the cortical and trabecular tissue for bones which comprise 15 phalanges (proximal, medial, and distal), 5 metatarsals, 2 sesamoids (medial and lateral), 3 cuneiforms (medial, intermediate, and lateral), navicular, cuboid, talus, calcaneus, and part of the tibia and fibula (see Figure 1a). Most muscles and tendons are included with a more realistic three-dimensional morphology. These are the extensor hallucis longus, extensor hallucis brevis, extensor digitorum longus, extensor digitorum brevis, dorsal interossei, plantar interossei, adductor hallucis (transverse and oblique head), abductor hallucis, lumbricals, flexor hallucis brevis, flexor hallucis longus, flexor digitorum longus, flexor digitorum brevis, flexor digiti minimi brevis, peroneus longus, peroneus brevis, tibialis anterior, quadratus plantae, tibialis posterior, aponeurosis plantar (lateral and central), Achilles, and part of the soleus (see Figure 1b,c). The peroneus tertius and opponens digiti minimi are the only tissues missing from the foot model.

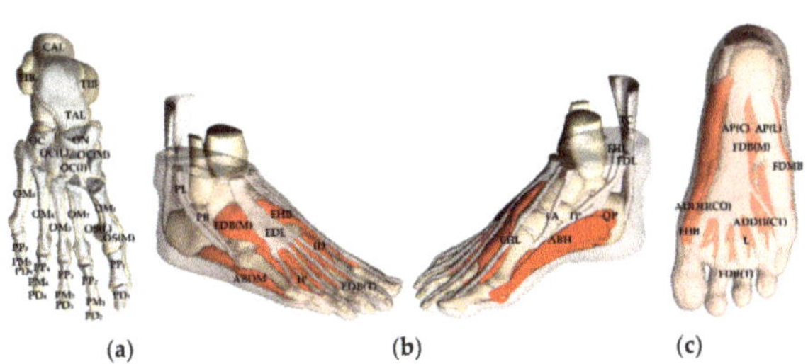

Figure 1. Foot model composition. (**a**) Skeletal system; (**b**) musculoskeletal system; (**c**) skin.

Table 1 shows the abbreviations of the names of each of the tissues that make up the model in Figure 1.

Table 1. Abbreviations of the names of the tissues of the foot model.

Tissue	Abbreviation	Tissue	Abbreviation
Phalanx distalis	PD_n	Abductor digiti minimi	ABDM
Phalanx media	PM_n	Interossei dorsales	ID
Phalanx proximalis	PP_n	Interossei plantares	IP
Ossa sesamoidea (laterale)	OS(L)	Extensor hallucis longus	EHL
Ossa sesamoidea (mediale)	OS(M)	Extensor hallucis brevis	EHB
Ossa metatarsalia	OM_n	Aponeurosis plantar (central)	AP(C)
Os cuneiforme (mediale)	OC(M)	Aponeurosis plantar (laterale)	AP(L)
Os cuneiforme (intermedium)	OC(I)	Flexor hallucis longus	FHL
Os cuneiforme (laterale)	OC(L)	Flexor digitorum longus	FDL
Os cuboideum	OC	Tendo calcaneus (soleus)	TC
Os naviculare	ON	Abductor hallucis	ABH
Talus	TAL	Quadratus plantae	QP
Calcaneus	CAL	Tibialis anterior	TA
Tibia	TIB	Tibialis posterior	TP
Fibula	FIB	Peroneus longus	PL
Flexor digitorum brevis (musculi)	FDB(M)	Peroneus brevis	PB
Flexor digitorum brevis (tendo)	FDB(T)	Flexor digiti minimi brevis	FDMB
Extensor digitorum brevis (musculi)	EDB(M)	Adductor hallucis (caput obliquum)	ADDH(CO)
Extensor digitorum brevis (tendo)	EDB(T)	Adductor hallucis (capt transversum)	ADDH(CT)
Extensor digitorum longus	EDL	Lumbricales	L

2.2. Mesh

All the tissues in the foot model were meshed using ICEM® CFD (ANSYS® Inc., Canonsburg, PA, USA). An auto mesh (unstructured) with the octree method was used. First-order tetrahedral elements of variable sizes were used to avoid the loss of morphology and the appearance of singularities in the tissues. An element size of 1 mm was used for small or thin tissues, while for large tissues element sizes of 2 to 4 mm were used.

During the meshing stage of the foot model tissues, the software reads the imported files (stereolithography extension) of each of the tissues as a set of points. When the tissues are being meshed, the meshing algorithm cannot identify the boundaries between each tissue. Therefore, the software creates a continuous mesh. Because of that, two problems arose that would have adverse biomechanical effects on the final FEFM response. The first was the generation of undesirable tetrahedral elements at the periphery of the areas in contact between the tissues and the small gaps between them, causing a loss of the model morphology (see Figure 2a). The second was the shared nodes between mesh elements of the tissues in contact, causing bonded areas that should not be in the model.

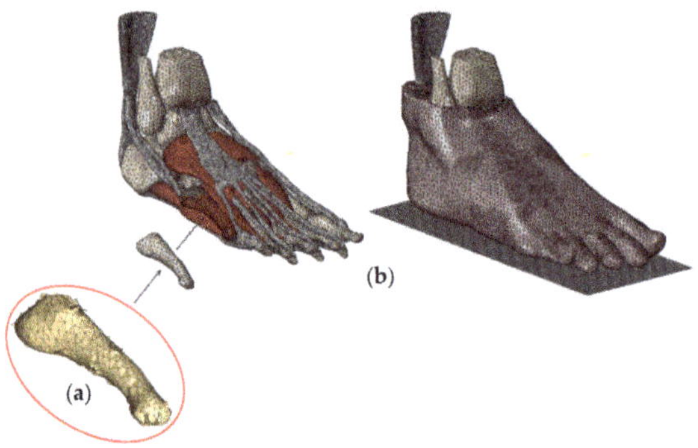

Figure 2. Tissue mesh. (**a**) Mesh without fixing; (**b**) fixed mesh.

Because of these two problems, a reworking of the mesh had to be done for most tissues. In the first stage, which we will call tissue mesh cleaning, those tetrahedral elements of the meshes that were incorrectly assigned were returned to the mesh of the correct tissues. In addition, all those tetrahedral elements of the mesh which modified the morphology of the original tissue were eliminated and, in some cases, new elements were created manually. This mesh reworking of the foot model was done with the same meshing software. The final model was made up of 1,128,602 elements and 208,721 nodes (see Figure 2b).

In a second stage, which we will call mesh separation between tissues, a separation process of the shared nodes between mesh elements of the tissues was done. During the meshing stage, the software generated a continuous mesh. Continuous mesh generates a sharing of nodes between the mesh elements of the tissues in contact, causing bonded areas. Bonded areas between the FEFM tissues are not physiologically possible because they increase the stiffness of the general response of the foot model, removing the ability of the soft tissues to transfer the load appropriately to the insertion areas on the bone besides erroneously generating the activation of secondary tissues.

The mesh shared between the tissues of the foot model was separated by an algorithm programmed in C++ language in an open-source integrated development environment called Code:Blocks.

The program created within the integrated development environment is provided with an input file, where there is information on the mesh of the bonded tissues as node groups, node coordinates, and a group of tetrahedral elements with four nodes (C3D4). The algorithm identifies the nodes of the shared faces between tetrahedral elements by the direction of their normal. Once the group of shared nodes has been identified, the program duplicates all the nodes of that group, assigning them the same position as the shared nodes but with a different identification number. The old group of nodes is assigned to one tissue, while the new group of nodes is assigned to the opposite tissue (see Figure 3). Additionally, the program creates a surface shell with the new nodes between the tissues in contact where the contact between the shell surface and the group of old nodes is subsequently declared. Finally, the program generates an output file with the information of the new nodes, their coordinates, and the three-node triangular shell elements (S3) created. The information contained in the output file is updated in the main file, and in this way, the separation between two meshes is performed.

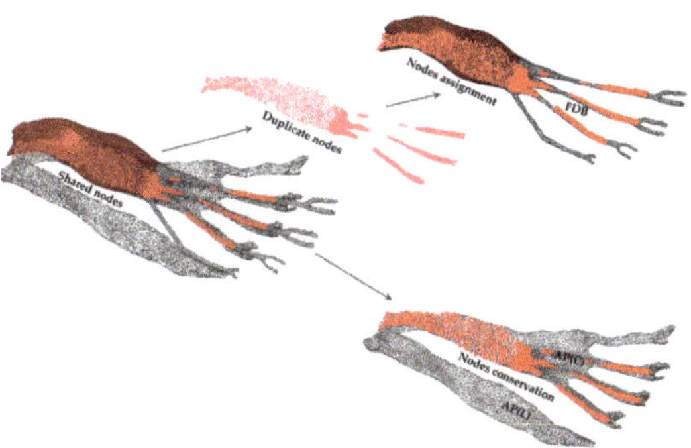

Figure 3. Separation of shared nodes between mesh elements.

It is important to mention that no other meshing software, method, or algorithm was used at this stage. Some bonded areas between tissues (such as the bone-fat pad) favored the transmission of loads of the model and avoiding the application of contacts. We only focused on separating those tissues with greater relevance in the analysis.

2.3. Mechanical Properties of Tissues

The mechanical properties used to model the behavior of the foot model tissues in this research were taken from the literature. Cortical bone, trabecular bone, hard cartilage, and soft cartilage were modeled with an isotropic, homogeneous, and linearly elastic mechanical behavior [14–16]. For the rest of the tissues, a nonlinear isotropic and homogeneous behavior was used with several hyperelastic deformation energy density functions: fourth-order Ogden function for the lateral plantar aponeurosis, a fifth-order reduced polynomial function for the central plantar aponeurosis [4], a second-order polynomial function for the skin [17], and a first-order Ogden function for muscles [18] and tendons [19].

2.4. Boundary Conditions and Loads

Several boundary conditions were considered to simulate the environment during standing. First, in the three directions, the displacement and rotation were constrained in the tibia and fibula through a set of nodes taken from the upper cross-section (see Figure 4a). To simulate the support of the foot model against the ground, a flat surface was generated with three-dimensional rigid elements of four nodes (rigid body), allowing only vertical displacement (Y). A force (350 N in the "Y" direction) equivalent to half the subject's body weight was applied on the flat surface to simulate the reaction force against the ground.

Finally, a series of loads were applied through eight of the main stabilizing tendons activated during the analysis (see Figure 4a). The magnitudes of these loads were taken from a research paper published by Morales-Orcajo Enrique et al. (2017).

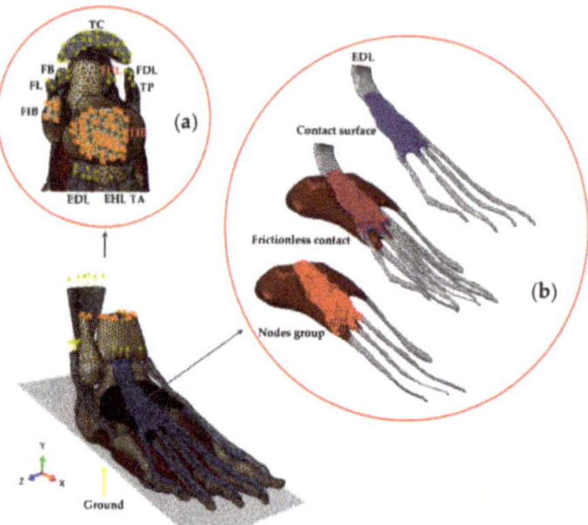

Figure 4. Load and boundary conditions. (**a**) Loads application and motion constraints; (**b**) contact application.

2.5. Contact between Tissues and Ground

Shell surfaces with the shape and extent of the zone of interaction between tissues were created. Later, contacts between the elements whose mesh was separated were applied to avoid penetration and ensure load transfer. Only those sliding contacts between the most relevant tissues of the foot model were generated (see Figure 4b), that is, in those tissues with greater participation in the transfer of load in the simulation (short extensor-long extensor, short flexor-long flexor, and short flexor-plantar fascia).

Finally, the foot (sole) interaction with the ground was simulated with frictional contact. In this case, a friction coefficient of 0.6 was used [20].

2.6. Case Studies

Three case studies have been evaluated with the finite element foot model, where we have calculated the contact pressure, displacements, and the maximum principal stress in some tissues to determine the effect that the use of a shared mesh versus a separated mesh can have on these parameters. In addition, we evaluated the biomechanical, convergence, and computational effects of using a shared versus separated mesh and isotropic versus hyperelastic properties in some tissues.

The cases evaluated in this work are detailed below:

- Case 1: Model with shared mesh and a linearly elastic, homogeneous, and isotropic mechanical behavior of most of its tissues. For this model, a shared mesh was used (a product of the software), i.e., a mesh with joints (undesired in some cases) between groups of nodes of the tetrahedral elements that make up the mesh of the tissues. The joints presented in this model are in all areas of contact between a tissue and its adjacent tissues. Therefore, a general stiffened FEFM response is expected. Finally, in this simulation, most of the tissues of the model are considered to have an isotropic behavior, except the tendons and skin (hyperelastic behavior).
- Case 2: Model with shared mesh and a linearly elastic, homogeneous, and hyperelastic mechanical behavior of most of its tissues. For this model, the same shared mesh is used as for Case 1 is still used. The difference is that in this model, hyperelastic properties are assigned to some tissues (skin, tendons, muscles and the plantar aponeurosis). The purpose of this model is to evaluate the effect on contact pressure when we used hyperplastic instead of isotropic properties.
- Case 3: Model with separate mesh and a linearly elastic, homogeneous, and hyperelastic mechanical behavior of most of its tissues. Here the mesh was reworked to undo some joints between muscles, tendons and the plantar aponeurosis.

It is important to understand that some bonded areas between the mesh of the tissues of the foot model are necessary for the transmission of mechanical stimuli between them, as in the case of the musculoskeletal model with the skin (fat) or tendons with some bones. In some cases, applying a contact or a motion constraint instead of leaving the joint generated by the software increase the computation time for convergence. However, some other joints as muscle-to-muscle or muscle-to-bone cause a deficient or abnormal load transmission, thus generating an inaccurate response from the model. Finally, some sliding contacts that were considered important (as short extensor-long extensor, short flexor-long flexor, and short flexor-plantar fascia) were applied in the model to avoid penetration between the tissues and to guarantee an appropriate interaction between them.

2.7. Foot Model Validation

ABAQUS software (ABAQUS Inc., Pawtucket, RI, USA) was used to solve our FEFM. The distribution and magnitude of the contact pressure (plantar pressure) on the sole, specifically under the calcaneus (heel) and the metatarsals heads, reported in previous experimental and numerical analysis were used to compare and validate the numerical foot model developed in this work [8,9,17,21–24]. In addition, a study was used of the plantar pressure (footprint) of the patient from which the model was obtained [19].

Table 2 shows a comparison of the plantar pressure (specifically on the talus and metatarsal heads) of our FEFM with FEFMs, and experimental foot data (EFD) reported by other authors. In this validation stage, the plantar pressure results were taken from those FEFMs whose analyses were performed during standing and under boundary and loading conditions similar to our analysis. Additionally, the table shows in brackets below of our FEFM results (Martínez et al.) the percentage difference between our plantar pressure values and the mean value of the experimental data (ED) and by the finite element method (FEM) reported in the literature by the authors mentioned in the same Table 2.

Table 2. Foot model validation (maximum contact pressure, CPRESS).

	Area	Martínez et al. (Present Work) MPa	Morales et al. (2018) MPa	Cheung et al. (2004) MPa	Cheung et al. (2005) MPa	Chen et al. (2010) MPa	Zhang et al. (2014) MPa	Mao et al. (2017) MPa	Li et al. (2017) MPa	Wang et al. (2018) MPa
OH	FEFM	0.192 (−13.41% FEM) (7.6% ED)	0.16	0.26	0.23	0.168	0.23	0.233	0.325	0.168
	EFD	NEFD	0.17	0.15	0.17	0.13	0.15	0.204	0.3	0.157
UMH	FEFM	0.112 (−22.75% FEM) (−10.70% ED)	0.17	0.19	0.097	0.077	0.18	0.07	0.325	0.051
	EFD	NEFD	0.16	0.07	0.09	0.08	0.12	NEFD	0.3	0.058

FEFM = finite element foot model; EFD = experimental foot data; NEFD = no experimental foot data; ED = experimental data; OH = on the heel; UMH = under metatarsal heads; FEM = finite element method.

3. Results

Plantar pressure, principal stress, and displacements are the comparison parameters that will enable us to quantify the importance of the work done in optimizing the behavior of the tendons and muscles in the foot model.

Table 3 shows the maximum principal stress and the displacement in the three directions for the most relevant tissues in the model without a shared mesh (Case 1) and the model with a separate mesh and sliding contacts (Case 3). In addition, in each column of the sliding foot model, the percentage difference with the values obtained from the model without sliding contacts is shown in brackets.

Table 3. Maximum principal stress (MPS) and displacement (U_n) obtained in some relevant tissues.

Tissue	Non-Sliding Foot Model (Case 1)				Sliding Foot Model (Case 3)			
	MPS (MPa)	U_x (mm)	U_y (mm)	U_z (mm)	MPS (MPa)	U_x (mm)	U_y (mm)	U_z (mm)
Sole(skin)	−0.1169 OH −0.0892 U_3MH	−2.821	−6.579	2.823	−0.1753 OH (49.95%) −0.0635 U_1MH (−28.81%)	−2.690 (−4.64%)	−6.092 (−7.40%)	2.896 (2.58%)
TC	3.411	2.516	−1.676	−3.186	3.441 (0.87%)	2.638 (4.84%)	−3.483 (107.81%)	−3.038 (−4.64%)
EHL	1.953	−0.3354	−3.967	−1.230	2.127 (8.90%)	−1.243 (270.60%)	−5.450 (37.38%)	−2.032 (65.20%)
EDL	0.6850	−0.3927	−4.267	−1.352	0.7598 (10.91%)	−2.181 (455.38%)	−4.720 (10.61%)	−1.734 (28.25%)
EDB/EHB	3.367	−0.3055	−3.354	−1.122	0.4190 (−87.55%)	−1.620 (430.27%)	−4.861 (44.93%)	−1.910 (70.23%)
FHL	2.529	0.6404	−4.011	−1.722	2.668 (5.49%)	−1.059 (−265.36%)	−5.461 (36.15%)	−2.235 (29.79%)
FHB	0.8868	−0.1773	−3.208	−1.646	0.09633 (−89.13%)	−0.6678 (276.64%)	−4.830 (50.56%)	−2.322 (41.06%)
FDL	2.821	0.7357	−4.251	−3.317	3.016 (6.91%)	−1.761 (−339.36%)	−4.897 (15.19%)	−3.683 (11.03%)
FDB	0.7442	−0.3423	−4.005	−1.480	2.133 (186.61%)	−1.728 (404.82%)	−5.021 (25.36%)	−1.728 (−16.75%)
TA	0.1521	−0.4180	−2.020	−0.7874	0.5087 (234.45%)	−0.5349 (27.96%)	−2.339 (15.79%)	−1.100 (39.70%)
TP	4.551	1.288	−3.773	−3.613	4.779 (5%)	−3.957 (−407.22%)	−4.226 (12.03%)	−2.957 (−18.15%)
PL	1.458	0.2389	−1.805	−0.8958	2.970 (103.70%)	−0.3972 (−266.26%)	−2.240 (24.09%)	−1.204 (34.40%)
PB	0.8265	−0.1556	−1.211	−0.7911	0.7964 (−3.64%)	−0.3812 (144.98%)	−2.564 (111.72%)	−1.253 (58.38%)
AP(C)	1.191	0.3878	−3.60	−1.613	2.862 (140.30%)	−1.389 (−458.17%)	−5.255 (45.97%)	−2.273 (40.91%)
AP(L)	0.6715	0.3841	−2.729	−1.233	5 (644.6%)	−0.966 (−243.85%)	−1.318 (−51.70%)	0.8430 (−168.36%)

OH = on the heel; U_nMH = under the metatarsal head.

Table 3 shows that the maximum principal stress obtained on the heel for the non-sliding foot model was −0.1169 MPa, while for the sliding foot model it was −0.1753 MPa. This means an increase of 49.95% for the principal stress in the model with the separated mesh compared to the model with the shared mesh. On the other hand, in the support zone of the metatarsal bones, the maximum principal stress obtained for the non-sliding foot model was −0.0892 MPa, located under the head of the third metatarsal. The maximum principal stress for the sliding foot model was −0.0635 MPa, under the head of the first metatarsal. In this case, the change of the area under greater stress under the metatarsals, as well as the variation of the maximum principal stress of the model with the separate mesh compared to the model with the shared mesh, is attributed to the greater freedom of movement of the tissues resulting from the separation of their bonded areas with the mesh of the tissues in contact. In addition, this greater degree of freedom of movement of the model with the separate mesh can be reflected in the displacements of some tissues (see Table 3), where there is a tendency towards a greater displacement of up to 458.1% in the model with the separate mesh in comparison with the model with the shared mesh. Moreover, the soft tissue with the greatest variation in the principal stress when considering the condition of mesh separation and slippage, such as aponeurosis plantar (lateral), shows variations of up to 644.6%.

Overall, the soft tissue results showed a mean absolute variation and standard deviation (SD) in stress values of 83.3% (SD 186.9) for the model with the separate mesh (Case 3) compared with the model with the shared mesh (Case 1), while the mean absolute value and standard deviation for the displacements was 17.4% (SD 190.9).

As can be seen in Figure 5, the distribution of the maximum contact pressure in the three models was on the heel, with a magnitude of 0.162 MPa for the model with the shared mesh and an isotropic behavior of most of its tissues (Figure 5a) and 0.187 MPa and 0.192 MPa for the models with a hyperelastic behavior of most of its tissues but with a shared mesh (Figure 5b) and a separated mesh (Figure 5c), respectively. For the two latter models, we have a percentage variation of the maximum plantar pressure of 2.6% in the model with the separated mesh (Case 3) compared to the model with the shared mesh (Case 2).

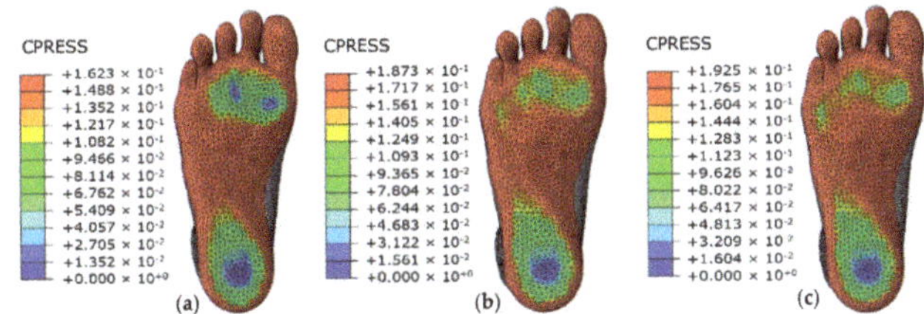

Figure 5. Contact pressure (CPRESS). (**a**) Isotropic non-sliding foot model; (**b**) hyperelastic non-sliding foot model; (**c**) hyperelastic sliding foot model.

Comparing the maximum principal stress in the metatarsal bones, the maximum stress in the model with the shared mesh and isotropic properties was in the fourth metatarsal bone with a value of 5.095 MPa (see Figure 6a), while for the model with the separate mesh it was in the second metatarsal bone with a value of 5.510 MPa (see Figure 6b).

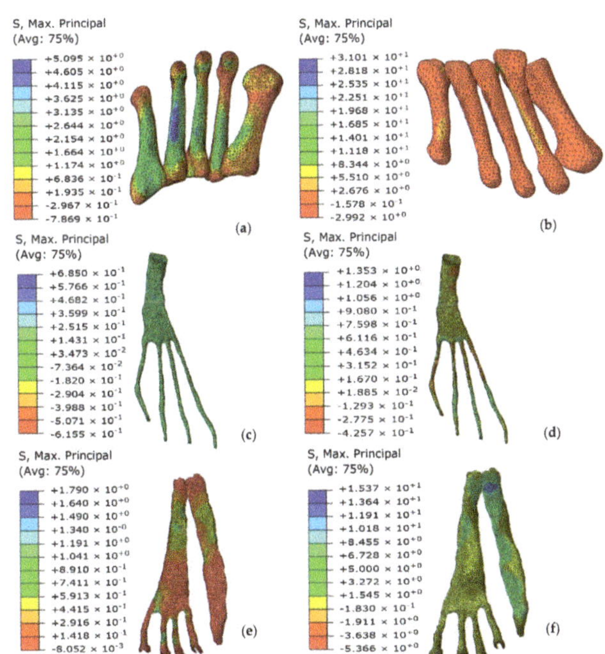

Figure 6. Maximum principal stress on soft and hard tissues. (**a**) Metatarsal bones of the non-sliding foot model; (**b**) metatarsal bones of the sliding foot model; (**c**) extensor hallucis longus of the non-sliding foot model; (**d**) extensor hallucis longus of the sliding foot model; (**e**) flexor hallucis brevis of the non-sliding foot model; (**f**) flexor hallucis brevis of the sliding foot model.

It is important to emphasize the non-homogenous distribution of the maximum principal stress in the extensor digitorum longus and aponeurosis plantar of the model with the shared mesh and isotropic properties when compared with the model with the separated mesh and hyperelastic properties. A better distribution of stress was obtained for the tissues shown in Figure 6d,f (separate mesh) compared with those shown in Figure 6c,e (shared mesh) where it was observed a higher stress concentration and singularities. A better distribution of stress is indicative of more appropriate behavior of the tissues by eliminating anchorage areas due to the shared mesh.

4. Discussion

Simulating proper muscle and tendon behavior in a finite element model of a foot is a challenge today. The load transmitted by these tissues to the bone and surrounding tissues can be compromised as early as the meshing stage. During the mesh stage, the software used may or may not generate joints by sharing nodes between mesh elements in all areas in contact between tissues. In both cases, if this fact is overlooked, the behavior of the model may be considerably affected. Before starting the creation of the finite element model, the researcher must define which tissues are necessary, which are relevant, and which others might not be considered in the analysis. The larger the number of elements, the more complicated it becomes to work with the model. In addition, the greater the number of contacts and hyperelastic characterizations of the tissues, the longer the computation time when trying to achieve convergence of the model. In short, the more complete and complex the model, the greater the complexity to obtain a result.

This research shows the structural effect that a shared mesh can have compared to a separate mesh with sliding contacts on the structural behavior of the fabrics of a foot model. For this purpose, one of the most complete finite element foot models available today is used. This model includes almost all muscles and tendons in the foot, with the exception of the peroneus tertius and opponens digiti minimi, which are not necessary for the present analysis. In addition, the muscles, tendons and plantar fascia are modeled with real geometry and simulated with deformable solid elements within the analysis.

According to the values shown in Table 3, there is a mean absolute increase of 10.6% in the maximum principal stress on the sole for the model with the separate mesh compared with the model with the shared mesh. This increase in stress translates into a better load distribution through the soft tissues. A better stress distribution in the foot model can be observed in a lower number of stress concentrations in different areas of the tissues, as well as in greater freedom of displacement of the tissues. The increased magnitude in displacements may be due to the freedom of movement of muscles, tendons and the aponeurosis plantar resulting from eliminating bonded areas with other tissues. In the tissues of the non-sliding foot model (shared mesh foot model), a greater number of tissues with stress concentrations (singularities) were observed. This translates into a progressive loss of load in these tissues due to the mesh bonding zones of other tissues. This progressive loss of load generates a global stiffness response of the foot model. Strong evidence that soft tissues such as the aponeurosis plantar, peroneus brevis, tibialis posterior and extensor digitorum longus are working better is provided by the values obtained for the mean absolute variation of displacement (17.4%) and stress (83.3%) in the tissues belonging to the model in Case 3 compared to Case 1.

The contact pressure analysis of the models showed that the hyperelastic behavior assigned to muscles of the models corresponding to Figure 5b,c, compared to the model shown in Figure 5a with an isotropic behavior of most of its tissues, improved the qualitative distribution of the plantar pressure on the sole, resulting in a more homogeneous pressure. According to the mean of the experimental data taken from the literature [8,9,17,19,21–24], mesh separation in the model in Figure 5c improved plantar pressure under the heel and metatarsal head by 1.6% and 18.6%, respectively, compared to the model in Figure 5a. However, the mesh separation between the tissues in the model in Figure 5c, compared to the shared mesh in the model in Figure 5b, did not make a considerable difference either quantitatively or qualitatively in the variation of the maximum plantar pressure on the sole. However, there was certainly less stress concentration under the metatarsal head of the third and fourth toes in the model in Figure 5c compared to the model in Figure 5b. This is attributed to separating the mesh of the long extensor tendons with adjacent tissues, allowing a large part of the load on these tissues to reach the insertion in the distal phalanges, generating a slight elevation of the forefoot.

As can be seen in the tissues in Figure 6d,f (where their mesh was separated) compared to the tissues in Figure 6c,e (with the shared continuous mesh), there is a better distribution of the maximum principal stresses (fewer singularities), which is indicative of a more adequate behavior of these muscles in the foot model.

The results obtained in the model presented in this work, where the mesh among bones, muscles, tendons and plantar fascia was separated, showed a significant improvement in the general behavior of the model. It is, therefore, imperative that in this type of model used for the analysis of biomechanical phenomena, the soft tissues should be separate from their surrounding tissues, especially in those analyses focused on the structural behavior of muscles, tendons and the plantar fascia.

When creating a model made up of 101 elements (tissues), it is to be expected to find a high degree of interaction between the tissues. In total, 138 bonded areas were counted in the foot model. Some of the tissues had more than four bonds. Analyzing the behavior of the case studies, specifically where it was observed that the separation of the mesh does not have a considerable effect on the variation of plantar pressure. Furthermore, in some cases, when separating the mesh between two bonded tissues, it was necessary to apply a contact, and in turn, the contact considerably increased the calculation time. The decision was made to separate only those bonded areas that were relevant to the analysis. Given that skin is a tissue that covers the entire musculoskeletal model, its mesh generates joints with bone, tendons, muscles and the plantar aponeurosis. Despite considering the skin as a hyperelastic tissue, the junction with the tendons causes a loss of load. However, it is difficult to generate an internal contact of the skin with the musculoskeletal model so that the latter transmits the mechanical stimuli more adequately to the adipose tissue. For this reason, one of the important limitations in our analysis is the loss of load that the skin causes on the active tendons in the stance phase due to the shared mesh between them.

Finally, as a result of the analysis, we can suggest for future work that for foot models which include muscles, tendons, the plantar fascia and ligaments simulated with solid deformable elements, it is advisable to use a separated mesh in tissues such as bone (cortical bone) and skin (fat pad). Furthermore, and subsequently, we use specialized meshing software to generate the joints or separations between the mesh elements where later a contact condition or motion constraints will be applied that will help to better transmit the effects from one mesh to another. In this way, it will be possible to obtain a better behavior of the tissues in the model.

The use of a shared or separate mesh of the skin with the musculoskeletal tissue makes it very difficult to adequately simulate the behavior of the rest of the soft tissues. Therefore, depending on the case study, it could be advisable to generate simplified models where a separated mesh and only some tissues or a portion of them could be used, as shown in Figure 7.

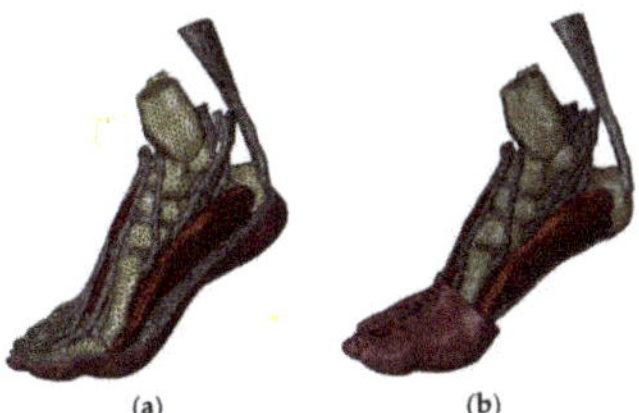

Figure 7. Simplified models. (**a**) Model with skin (fat pad) on the sole of the foot. (**b**) Model with skin (fat pad) on the forefoot.

5. Conclusions

The results obtained in this work showed that a more appropriate mechanical characterization of tissues such as muscles, tendons, plantar aponeurosis and skin (fat-pad) improved the quantitative and qualitative distribution of plantar pressure. What is referenced in the text above, can be seen by comparing the results between Figure 5a,c. Contrary to expectations, the mesh separation work performed did not have a significant impact on plantar pressure. The above can be corroborated by comparing the model results in Figure 5b with Figure 5c.

However, the work of mesh separation between the tissues of most relevance for analysis (plantar aponeurosis and extrinsic musculature) showed an improvement in the structural behavior of the individual tissues. Quantitatively in the data shown in Table 3, displacement differences of up to 458.1% can be found. Qualitatively we can see in Figure 6b,d,f (Case 3) a low concentration of the maximum principal stress on the tissues compared with the tissues in Figure 6a,c,e (Case 1). The result is a better structural behavior of the tissues and better load transmission to the areas where they are inserted into the bone. The improvement made in the model can lead to a better simulation of specific tissues when a specific pathology is analyzed.

The work carried out in this paper generated knowledge on how the use of a separate mesh with contacts and hyperelastic properties improves the structural behavior of the tissues and the model in general, compared to the use of a shared mesh with isotropic properties. In addition, it could be proven that the use of a very detailed finite element model generates more computational time in the solution, requires equipment with greater capacity to solve the system of equations, increases the level of difficulty in finding convergence, and increases the complexity of simulating each of the tissues appropriately.

Therefore, it is recommended to take into consideration, whenever possible, the creation of simplified models to avoid or reduce some of the problems mentioned in the research. A simplified model could have a good approximation of the case study, but without having to face the problems that working with a more detailed model could bring.

Author Contributions: Conceptualization, M.A.M.B.; methodology, J.B.L.; software, A.M.T.; validation, M.A.M.B. and R.B.d.B.V.; formal analysis, M.A.M.B.; investigation, M.A.M.B., J.B.L. and A.V.-L.; resources, J.B.L. and A.V.-L.; data curation, M.A.M.B.; writing—original draft preparation, M.A.M.B.; writing—review and editing, M.A.M.B., J.B.L. and A.V.-L.; visualization, M.A.M.B.; supervision, J.B.L. and A.V.-L.; project administration, M.A.M.B., J.B.L. and A.V.-L.; funding acquisition, J.B.L. and A.V.-L. All authors have read and agreed to the published version of the manuscript.

Funding: This research received no external funding.

Institutional Review Board Statement: Not applicable.

Informed Consent Statement: Not applicable.

Data Availability Statement: Data and codes available under request to the authors.

Acknowledgments: The authors gratefully acknowledge the support of the Ministry of Economy and competitiveness of the Government of Spain through the project PID2019-108009RB-I00 and the Universidad de Guanajuato for all facilities to this work and CONACYT, Mexico (CVU No. 591520) for the PhD scholarship granted.

Conflicts of Interest: The authors declare no conflict of interest.

References

1. Morales-Orcajo, E.; Becerro de Bengoa, R.; Losa, M.; Bayod, J.; Barbosa, E. Foot internal stress distribution during impact in barefoot running as function of the strike pattern. *Comput. Methods Biomech. Biomed. Eng.* **2018**, *21*, 471–478. [CrossRef]
2. Salathe, E.P.; Arangio, G.A. A Biomechanical Model of the Foot: The Role of Muscles, Tendons, and Ligaments. *J. Biomech. Eng.* **2002**, *124*, 281–287. [CrossRef]
3. Chen, W.-M.; Park, J.; Park, S.-B.; Shim, V.P.-W.; Lee, T. Role of gastrocnemius–soleus muscle in forefoot force transmis-sion at heel rise—A 3D finite element analysis. *J. Biomech.* **2012**, *45*, 1783–1789. [CrossRef]

4. Guo, J.; Liu, X.; Ding, X.; Wang, L.; Fan, Y. Biomechanical and mechanical behavior of the plantar fascia in macro and microstructures. *J. Biomech.* **2018**, *76*, 160–166. [CrossRef]

5. Akrami, M.; Qian, Z.; Zou, Z.; Howard, D.; Nester, C.J.; Ren, L. Subject-specific finite element modelling of the human foot complex during walking: Sensitivity analysis of material properties, boundary and loading conditions. *Biomech. Model. Mechanobiol.* **2018**, *17*, 559–576. [CrossRef]

6. Martínez, M.A.; Bayod, J.; Vidal-Lesso, A.; Becerro de Bengoa, R.; Lesso-Arroyo, R. Structural interaction between bone and implants due to arthroplasty of the first metatarsophalangeal joint. *Foot Ankle Surg.* **2019**, *25*, 150–157. [CrossRef] [PubMed]

7. Isvilanonda, V.; Dengler, E.; Iaquinto, J.M.; Sangeorzan, B.J.; Ledoux, W.R. Finite element analysis of the foot: Model valida-tion and comparison between two common treatments of the clawed hallux deformity. *Clin. Biomech.* **2012**, *27*, 837–844. [CrossRef] [PubMed]

8. Mao, R.; Guo, J.; Luo, C.; Fan, Y.; Wen, J.; Wang, L. Biomechanical study on surgical fixation methods for minimally invasive treatment of hallux valgus. *Med. Eng. Phys.* **2017**, *46*, 21–26. [CrossRef] [PubMed]

9. Wang, Y.; Li, Z.; Wong, D.W.-C.; Cheng, C.-K.; Zhang, M. Finite element analysis of biomechanical effects of total ankle ar-throplasty on the foot. *J. Orthop. Translat.* **2018**, *12*, 55–65. [CrossRef] [PubMed]

10. Chen, W.-M.; Lee, S.-J.; Lee, P.V.S. Plantar pressure relief under the metatarsal heads–Therapeutic insole design using three-dimensional finite element model of the foot. *J. Biomech.* **2015**, *48*, 659–665. [CrossRef] [PubMed]

11. Cifuentes-De La Portilla, C.; Larrainzar-Garijo, R.; Bayod, J. Analysis of biomechanical stresses caused by hindfoot joint ar-throdesis in the treatment of adult acquired flatfoot deformity: A finite element study. *Foot Ankle Surg.* **2019**, *S1268–S7731*, 30073–30076.

12. La Portilla, C.C.-D.; Larrainzar-Garijo, R.; Bayod, J. Biomechanical stress analysis of the main soft tissues associated with the development of adult acquired flatfoot deformity. *Clin. Biomech.* **2019**, *61*, 163–171. [CrossRef]

13. Morales, E.; Barbosa, E.; Bayod, J. Computational Foot Modeling for Clinical Assessment. Ph.D. Dissertation, Universidad de Zaragoza, Zaragoza, Spain, 2015.

14. Athanasiou, K.A.; Liu, G.T.; Lavery, L.A.; Lanctot, D.R.; Schenk, R.C., Jr. Biomechanical topography of human articular carti-lage in the first metatarsophalangeal joint. *Clin. Orthop. Relat. Res.* **1998**, *348*, 269–281. [CrossRef]

15. Gefen, A. Stress analysis of the standing foot following surgical plantar fascia release. *J. Biomech.* **2002**, *35*, 629–637. [CrossRef]

16. Garcia-Aznar, J.M.; Bayod, J.; Rosas, A.; Larrainzar, R.; García-Bógalo, R.; Doblaré, M.; Llanos, L.F. Load Transfer Mechanism for Different Metatarsal Geometries: A Finite Element Study. *J. Biomech. Eng.* **2008**, *131*, 021011. [CrossRef] [PubMed]

17. Chen, W.-M.; Lee, T.; Lee, P.V.-S.; Lee, J.W.; Lee, S.-J. Effects of internal stress concentrations in plantar soft-tissue-a prelimi-nary three-dimensional finite element analysis. *Med. Eng. Phys.* **2010**, *32*, 324–331. [CrossRef] [PubMed]

18. Petre, M.; Erdemir, A.; Panoskaltsis, V.P.; Spirka, T.A.; Cavanagh, P.R. Optimization of Nonlinear Hyperelastic Coefficients for Foot Tissues Using a Magnetic Resonance Imaging Deformation Experiment. *J. Biomech. Eng.* **2013**, *135*, 061001. [CrossRef] [PubMed]

19. Morales-Orcajo, E.; Souza, T.R.; Bayod, J.; Casas, E.L. Non-linear finite element model to assess the effect of tendon forces on the foot-ankle complex. *Med. Eng. Phys.* **2017**, *49*, 71–78. [CrossRef] [PubMed]

20. Zhang, M.; Mak, A.F.T. In vivo friction properties of human skin. *Prosthet. Orthot. Int.* **1999**, *23*, 135–141. [CrossRef]

21. Cheung, J.T.-M.; Zhang, M.; An, K.-N. Effects of plantar fascia stiffness on the biomechanical responses of the ankle–foot complex. *Clin. Biomech.* **2004**, *19*, 839–846. [CrossRef] [PubMed]

22. Cheung, J.T.-M.; Zhang, M.; Leung, A.K.-L.; Fan, Y.-B. Three-dimensional finite element analysis of the foot during standing—A material sensitivity study. *J. Biomech.* **2005**, *38*, 1045–1054. [CrossRef] [PubMed]

23. Zhang, M.; Yu, J.; Cong, Y.; Wang, Y.; Cheung, J. Foot Model for Investigating Foot Biomechanics and Footwear Design. In *Computational Biomechanics of the Musculoskeletal System*, 1st ed.; Zhang, M., Fan, Y., Eds.; CRC Press (Taylor & Francis Group): Boca Raton, FL, USA, 2014; pp. 3–18.

24. Li, S.; Zhang, Y.; Gu, Y.; Ren, J. Stress distribution of metatarsals during forefoot strike versus rearfoot strike: A finite element study. *Comput. Biol. Med.* **2017**, *91*, 38–46. [CrossRef] [PubMed]

 *mathematics*

Article

Size Effects in Finite Element Modelling of 3D Printed Bone Scaffolds Using Hydroxyapatite PEOT/PBT Composites

Iñigo Calderon-Uriszar-Aldaca [1,2,*], Sergio Perez [1], Ravi Sinha [3], Maria Camara-Torres [3], Sara Villanueva [1], Carlos Mota [3], Alessandro Patelli [4], Amaia Matanza [5], Lorenzo Moroni [3] and Alberto Sanchez [1,*]

[1] TECNALIA, Basque Research and Technology Alliance (BRTA), Mikeletegi Pasealekua 2, 20009 Donostia-San Sebastián, Spain; sergio.perez@tecnalia.com (S.P.); sara.villanueva@tecnalia.com (S.V.)

[2] Department of Engineering, Public University of Navarra, 31006 Pamplona, Spain

[3] Complex Tissue Regeneration Department, MERLN Institute for Technology-Inspired Regenerative Medicine, University of Maastricht, 6211 LK Maastricht, The Netherlands; ravi.sinha@maastrichtuniversity.nl (R.S.); m.camaratorres@maastrichtuniversity.nl (M.C.-T.); c.mota@maastrichtuniversity.nl (C.M.); l.moroni@maastrichtuniversity.nl (L.M.)

[4] Department of Physics and Astronomy, Padova University, Via Marzolo 8, 35131 Padova, Italy; alessandro.patelli@unipd.it

[5] Materials Physics Center (MPC), Centro de Fisica de Materiales (CSIC, UPV/EHU), Paseo Manuel de Lardizabal 5, 20018 Donostia-San Sebastián, Spain; amatanza001@ehu.eus

[*] Correspondence: inigo.calderon@tecnalia.com (I.C.-U.-A.); alberto.sanchez@tecnalia.com (A.S.)

Citation: Calderon-Uriszar-Aldaca, I.; Perez, S.; Sinha, R.; Camara-Torres, M.; Villanueva, S.; Mota, C.; Patelli, A.; Matanza, A.; Moroni, L.; Sanchez, A. Size Effects in Finite Element Modelling of 3D Printed Bone Scaffolds Using Hydroxyapatite PEOT/PBT Composites. *Mathematics* **2021**, *9*, 1746. https://doi.org/10.3390/math9151746

Academic Editor: Mauro Malvè

Received: 2 June 2021
Accepted: 7 July 2021
Published: 24 July 2021

Publisher's Note: MDPI stays neutral with regard to jurisdictional claims in published maps and institutional affiliations.

Abstract: Additive manufacturing (AM) of scaffolds enables the fabrication of customized patient-specific implants for tissue regeneration. Scaffold customization does not involve only the macroscale shape of the final implant, but also their microscopic pore geometry and material properties, which are dependent on optimizable topology. A good match between the experimental data of AM scaffolds and the models is obtained when there is just a few millimetres at least in one direction. Here, we describe a methodology to perform finite element modelling on AM scaffolds for bone tissue regeneration with clinically relevant dimensions (i.e., volume > 1 cm^3). The simulation used an equivalent cubic eight node finite elements mesh, and the materials properties were derived both empirically and numerically, from bulk material direct testing and simulated tests on scaffolds. The experimental validation was performed using poly(ethylene oxide terephthalate)-poly(butylene terephthalate) (PEOT/PBT) copolymers and 45 wt% nano hydroxyapatite fillers composites. By applying this methodology on three separate scaffold architectures with volumes larger than 1 cm^3, the simulations overestimated the scaffold performance, resulting in 150–290% stiffer than average values obtained in the validation tests. The results mismatch highlighted the relevance of the lack of printing accuracy that is characteristic of the additive manufacturing process. Accordingly, a sensitivity analysis was performed on nine detected uncertainty sources, studying their influence. After the definition of acceptable execution tolerances and reliability levels, a design factor was defined to calibrate the methodology under expectable and conservative scenarios.

Keywords: finite element modelling; bone tissue engineering; 3D scaffold; additive manufacturing

1. Introduction

In order to manufacture patient-specific bone implants for tissue regeneration (Figure 1), it is important to keep in mind that any bone is a composite structure subjected to constant evolution normally composed of a high percentage of inorganic composition and a smaller organic fraction. The inorganic part is present intrafibrillarly in the bone and occupies up to 40% by volume filling with various degrees of space around the mineralized fibres, forming pore networks. Hence, human bones can be composed of high amounts of inorganics by volume, typically ranging from 50–60% up to 80–90% for some highly mineralized tissues, but always keeping some space for organics [1].

Consequently, with these boundary conditions, the engineering of patient-specific bone implants for tissue regeneration follows several approaches [2,3]. One of the most promising consists of the additive manufacturing of a temporary structure also known as a "scaffold", suitable for cell growth, with required porosity to stimulate osteogenesis, as stated by Jakus et al. [4], Karageorgiou and Kaplan [5] and Hutmacher [6]. These temporary scaffolds must meet several requirements, such as biocompatibility, biodegradability and suitable mechanical properties [2,3,7–13], but also printability [14,15]. The mechanical properties are of great importance, since the scaffolds should ideally have properties matching those of the surrounding tissue, ensuring its suitability to bear mechanical loading similar to the original tissue. The mechanical properties are mainly dictated by the bulk material properties. However, the geometry and porosity of the fabricated scaffolds also play an important role in the final mechanical properties of the implant [16,17].

There are diverse materials being investigated, spanning from synthetic and natural polymers, ceramics and combinations of these, with the research focused on finding an optimal formulation for a successful stimulation of bone healing [2,3], enabling the improvement of implants suitable for each patient condition in the future. Thereby, additive manufacturing of a bone scaffold requires a compatible biocomposite with reinforcing inorganic fillers to mimic natural bone composition and structure, while improving its mechanical behaviour to resist service life loadings [18,19]. Accordingly, there are many studies regarding material alternatives [20], such as silicate nanocomposites [21,22], graphene nanocomposites [23] and the extensive work on hydroxyapatite or nano-apatite [24–28]. Additionally, other studies focused on the mechanical optimization and performance of such material alternatives in the presence of micro and nano hydroxyapatite [29], or on bone regeneration and thermomechanical properties [30,31].

In this manuscript, a generalized procedure is developed for the numerical modelling of additive manufactured scaffolds for bone regeneration with clinically relevant dimensions (i.e., volume > 1 cm^3). The procedure is applicable to scaffolds made of any previously mechanically characterized material, and this procedure is then specifically validated with *poly(ethylene oxide terephthalate)/poly(butylene terephthalate)* (PEOT/PBT) with nano-hydroxyapatite fillers. This generalized procedure allows the scaffolds to be designed by function, fixing at first the requirements of geometry and the stresses to be supported and then selecting the optimal filler amount and porosity. A stiffer scaffold, for example, would require a greater amount of filler and less pores for stress to reduce the concentration spots. Consequently, this leads to design gradients and patterns for a single bone scaffold, which can be achieved using established topology optimization techniques.

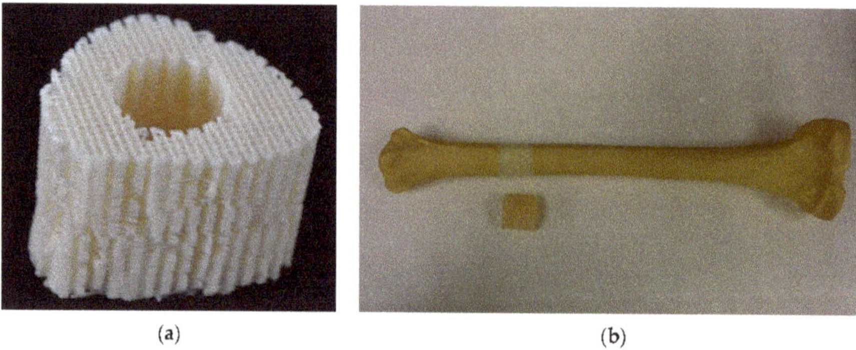

(a)   (b)

Figure 1. 3D printed bone scaffolds for tissue regeneration: (**a**) long bone defect-shaped scaffold printed with PEOT/PBT alone; (**b**) Scaffolds presenting longitudinal gradient patterns of plasma polymerized film deposition, visualized using methylene blue staining [32].

The mechanical properties of the bone scaffolds are a key aspect for the development of PEOT/PBT composites and a fast-complete characterization of these bulk materials can be carried out using monotonic tensile and compression tests according to standards. Nevertheless, the equivalent testing of non-standardized bone scaffolds made of such materials is not as easy, for the scaffolds need to be previously manufactured with different geometries and architectures, which slow down their rapid characterization. Therefore, a modelling procedure suitable to predict the mechanical performance of bone scaffolds with a certain pore architecture and bulk material mechanical properties has been developed.

Finally, a set of scaffolds has been produced and mechanically characterised as a way to confirm the suitability of the developed modelling protocol and its validation. Nevertheless, the 3D printing of clinically relevant scaffolds with the required micro-scale precision is a challenging task and the manufacturing process generally produces unintended imperfections causing uncertainty because of result scattering. Hence, a sensitivity analysis and statistical study have been conducted in order to derive some design factors taking into account these issues.

The final output is a finite element modelling (FEM) procedure for 3D-printed bone scaffolds able to forecast their mechanical behaviour depending on the bulk material properties. This will enable the optimization of the material development process and perform topology optimization of such scaffolds for bone tissues. This is not a topology optimization necessarily based on evolutionary structural optimization (ESO) algorithms, removing material where it is not working at certain stress level; nor it is based on additive evolutionary structural optimization (AESO) algorithms, simply adding materials where the stresses are above certain stress level, since the global shape of the bone implant needs to be kept "custom made", performing an external geometry for a specific patient. It is a topology optimization based on the scaffold properties' distribution within such a patient custom made geometry, i.e., manufacturing implants with stronger bone scaffolds at stress concentration "hot spots" by varying layer height, strand distance, fibre diameter (changing extrusion nozzle or printing speed), bulk material properties, plasma treatments, etc.

There are several studies regarding the FEM of 3D bone scaffolds for tissue regeneration. For instance, there are several bone scaffold aspects being currently analysed in terms of scaffold design [33], mechanical behaviour during failure [34], or covering the bone tissue regeneration and the progressive change in properties [35,36]. Nevertheless, the scope of such studies is focused on the mechanical behaviour of the 3D bone scaffold itself. Yet, there is little research on the holistic methodology connecting the bulk material properties with the scaffold mechanical behaviour within a 3D-printed bone, with mechanical properties depending on topology and size effect [37–39]. Moreover, this kind of predictive tool is important to forecast the expected mechanical behaviour of the scaffold on-site, using an optimizable geometry or bulk material.

In order to do it with FEM, there are two basic steps required. The first one is to represent every scaffold as an equivalent finite element, characterized by an approximately cubic shape with eight nodes, each being able to undergo a displacement in three degrees of freedom. The second step is to discretize the bone geometry in a compatible mesh, made of such solid finite elements, whose material properties are defined to match the mechanical behaviour of the scaffold (Figure 2A).

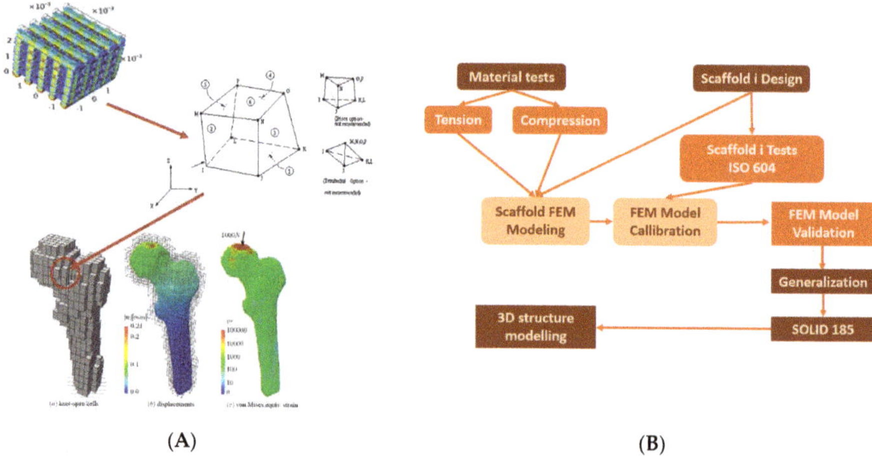

(A)   (B)

Figure 2. Finite element modelling of 3D-printed bone scaffolds with variable properties depending on bone topology: (**A**) Determination of equivalent finite element and bone discretization; (**B**) Flowchart of required steps for realistic finite element modelling of bone scaffolds within a bone.

Accordingly, a flowchart of the required works is presented in Figure 2B. It starts with the scaffold design at iteration i and the material tests for mechanical characterization under tension and compression. Then, an FEM of the bone scaffold is completed, and a monotonic compression test is simulated in that model. Thus, the results from the simulation can be compared to an actual scaffold compression test to calibrate and validate the FE model. This last step is required because the FEM is based on a theoretical or idealized geometry, whereas the real scaffolds manufactured at such scales present faults and uncertainties that need to be taken into account.

For the characterization of additive manufactured scaffolds, an FEM and support methodology has been carried out in this study, whose main objective is to be able to transfer the global mechanical properties of the scaffolds to solid eight-node finite elements, which will be used in FEM analyses. This will enable mechanical calculations to be addressed at a macroscopic level (complete bones, etc.), minimizing the computational costs, because it will not be necessary to take into account, in these models, the exact configuration of the scaffolds, but their equivalent mechanical properties. Additionally, this FEM of additive manufactured scaffolds has different properties, as equivalent finite elements can be further used to model mechanical property gradients simply by progressively changing such finite element material properties accordingly along a certain direction.

2. Materials and Methods

2.1. Materials

The bulk materials composing the scaffolds for bone tissue engineering were selected from amongst recently investigated new formulations based on combinations of a well characterized synthetic block copolymer of poly(ethylene oxide terephthalate)/poly(butylene terephthalate) (PEOT/PBT) with a diverse filler concentration [32,40,41]. For the improvement of the mechanical properties, the scaffolds produced from PEOT/PBT, including 45% of nano hydroxyapatite (nanoHA), were chosen due to their superior properties (Figure 3) and, therefore, were also used for the FE modelling. The mean mechanical behaviour obtained from the stress–strain curves was used (Figure 4). The stress–strain curves were obtained from the mechanical testing of the bulk materials according to the ASTM F2027-08 standard guide [42], prescribing ASTM D638-08 and ASTM D695-10 for tensile and compressive properties, respectively [43,44], and according also to international standard ISO 527 [45–47] (Figure 4). A Poisson's ratio of 0.4 was taken as usual practice for polymers, although there is some evidence that it could be higher in the case of PEOT/PBT [16]. The effect of the bulk material's Poisson's ratio on scaffold mechanical properties was studied in a sensitivity analysis.

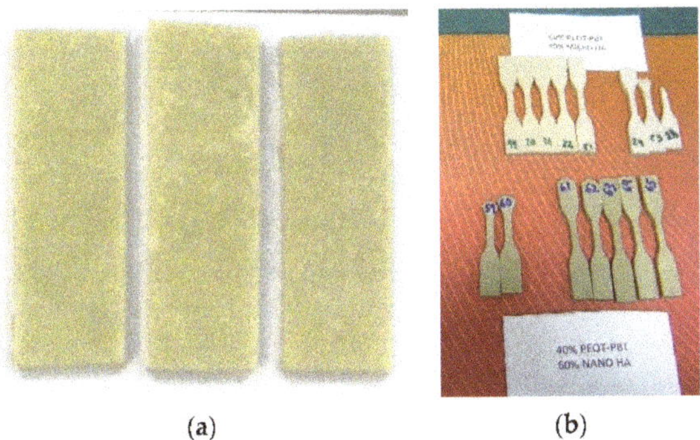

(a)   (b)

Figure 3. (a) Compression samples of PEOT/PBT 45% nano HA; (b) Tension samples made of different nano HA fillers concentration in PEOT/PBT.

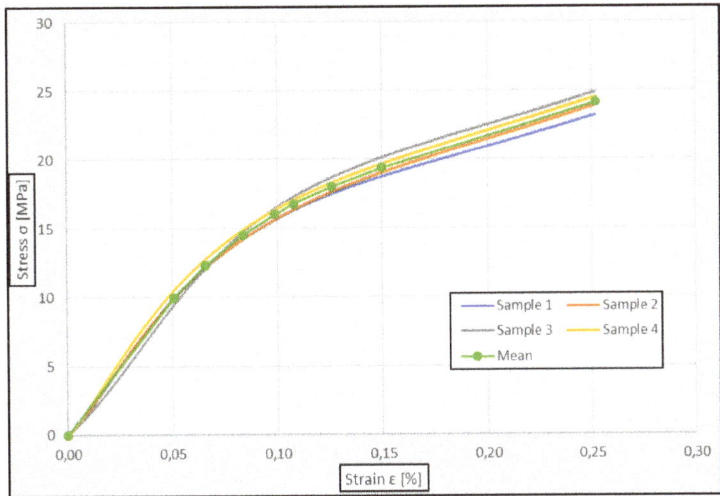

Figure 4. Compression stress–strain curve of PEOT/PBT with 45% nanoHA samples and mean behaviour considered for FEM.

For the scaffolds, monotonic compression tests, according to ASTM F2150-02 and ISO 604, were carried out on cylindrical scaffolds [48,49] to study the effect of several parameters, namely strand distance, the diameter of the extrusion needle (which determines the scaffold fibre diameter), layer height and the structure of the scaffold (external geometry, fibre position, scaffold diameter, scaffold height). Figure 5).

Figure 5. Standardized compression tests over 3D printed scaffold made of PEOT/PBT 45% nano HA.

Three variations of the scaffold were produced and tested, with three specimens each. The variations tested involved a decrease in the strand distance, meaning a denser, less porous scaffold, with a smaller inner volume of void spaces. The geometry of the different scaffold variations is summarized in Table 1.

Table 1. Geometrical properties of the tested scaffolds to be represented in the numerical models.

Scaffold Type	Fibre Diameter (mm)	Strand Distance (mm)	Layer Height (mm)	Scaffold Diameter (mm)]	Scaffold Height (mm)
G25_nanoHA 1.25	0.25	1.25	0.20	4.00	4.05
G25_nanoHA_1.00	0.25	1.00	0.20	4.00	4.05
G25_nanoHA_0.75	0.25	0.75	0.20	4.00	4.05

The monotonic compression tests of the scaffolds showed a good correspondence between samples of the same type within the elastic range, with a clear improvement in the mechanical properties as scaffold porosity decreases and scaffold density is, thus, increased (Figure 6). All scaffolds showed linear elastic behaviour under 10% strain and good reproducibility of stress–strain behaviour, supporting their suitability for load bearing bone tissue engineering, i.e., the intended use. Nevertheless, the differences in the strain–strain relationship of samples of the same type, evidenced in the curves, which were much more pronounced in the 3D-printed scaffolds than in the equivalent testing of bulk materials (Figure 4), indicate subsequent result scattering due to printing accuracy issues. Therefore, the only way to improve the result regularity on the scaffolds is to improve the manufacturing accuracy and quality control, where possible, whose only alternative is to deal with uncertainty, taking it into account during the design and calculation stage for a patient-customized scaffold.

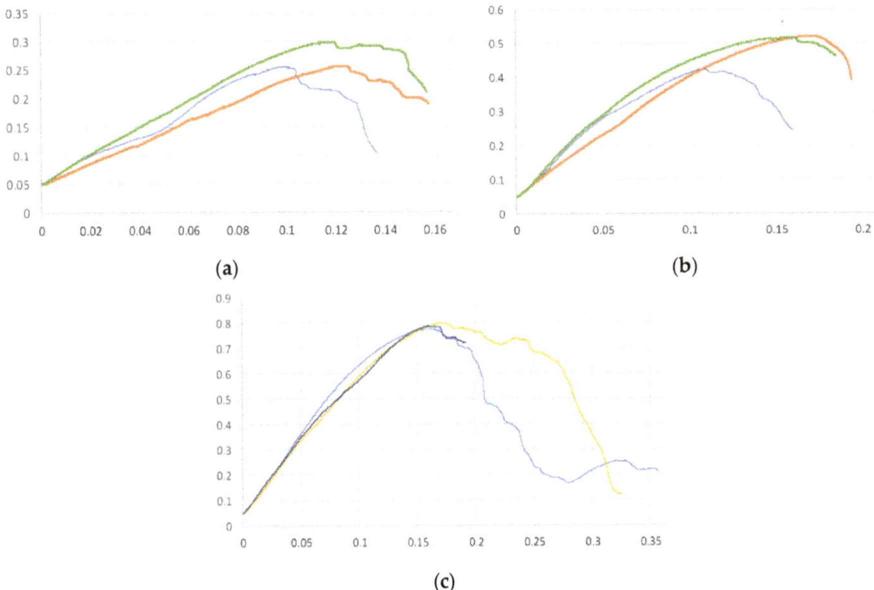

Figure 6. Curves of Stress (MPa)—Strain (%) relationship of three scaffold types (**a**) G25_nanoHA 1.25; (**b**) G25_nanoHA_1.00; (**c**) G25_nanoHA_0.75.

The most relevant values of the elastic range are the Young's modulus, defined as the slope of the initial linear relationship between the stress and strain, and the yield strength, defined as the stress where the linear range ends and the non-linear behaviour continues developing plasticity. Such values are summarized in Table 2, with corresponding Mean and Standard Deviation (SD) values from the tests of three replicates each.

Table 2. Results from monotonic compression tests of each sample and type, with corresponding mean value (Mean) and standard deviation (SD) of each type.

Sample	Young Modulus (MPa)	Yield Strength (MPa)	Elongation at Yield (%)
G25_nanoHA 1.25_1	2.33	0.25	9.80
G25_nanoHA 1.25_2	1.86	0.25	12.20
G25_nanoHA 1.25_3	2.56	0.29	11.60
G25_nanoHA_1.25 (Mean)	2.25	0.26	11.20
G25_nanoHA_1.25 (SD)	0.36	0.02	1.25
G25_nanoHA_1.00_1	5.54	0.42	10.90
G25_nanoHA_1.00_2	3.98	0.52	16.00
G25_nanoHA_1.00_3	5.88	0.50	13.50
G25_nanoHA_1.00 (Mean)	5.13	0.48	13.47
G25_nanoHA_1.00 (SD)	1.01	0.05	2.55
G25_nanoHA_0.75_1	6.93	0.78	15.90
G25_nanoHA_0.75_2	6.17	0.80	16.90
G25_nanoHA_0.75_3	5.90	0.79	16.40
G25_nanoHA_0.75 (Mean)	6.33	0.79	16.40
G25_nanoHA_0.75 (SD)	0.53	0.01	0.50

As a final comment, since this manuscript is about FEM of bone scaffolds, the results of monotonic compression tests on bulk materials and scaffolds are, respectively, just an input to set the material properties in a finite element model and a comparison basis for the output for validation. Thus, since these are boundary conditions of the subject matter of the manuscript, required to reproduce FEM results, they are presented in this section.

2.2. Methods

Several numerical models have been developed in order to simulate the behaviour of the three cylindrical scaffolds mentioned in Section 3.1 when subjected to a monotonic compression test. To this objective, ANSYS v17 finite elements software has been employed.

The numerical models developed within the frame of this study follow the geometrical characteristics defined in Table 2 in terms of fibre diameter (0.25 mm), strand distance (1.25/1.00/0.75 mm) and layer height (0.20 mm). A diameter of 4 mm has also been considered for the samples. Regarding the total height of the scaffolds, due to geometrical considerations to enable the idealized interlayer link modelling, there is little difference between the real height (4.00 mm) and the height considered in the numerical models (4.05 mm).

Considering the double symmetry, not only of the geometry of the scaffolds, but also of the loadings to be considered, just one quarter of the scaffolds has been represented in the numerical models, as shown in Figure 7.

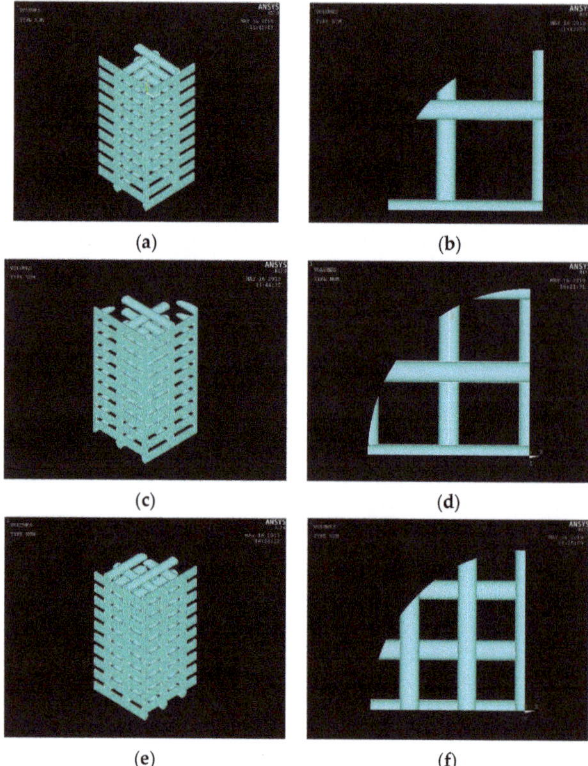

Figure 7. Geometry of scaffold numerical models: (**a**) Global view showing double symmetry of G25_nanoHA 1.25; (**b**) Cross-section of the resulting scaffold G25_nanoHA 1.25; (**c**) Global view showing double symmetry G25_nanoHA_1.00; (**d**) Cross-section of the resulting scaffold G25_nanoHA_1.00; (**e**) Global view showing double symmetry G25_nanoHA_1.00; (**f**) Cross-section of the resulting scaffold G25_nanoHA_0.75.

For the modelling of the fibres that conform to the scaffold, only one type of element (SOLID187) has been used. The SOLID187 element is a higher order 3D element with a quadratic displacement behaviour that is well suited to modelling irregular meshes. This element is defined by 10 nodes (tetrahedral element with nodes at vertices and mid edges) with three degrees of freedom at each node (translations in the nodal X, Y and Z directions). Moreover, for the modelling of the contacts, two additional types of elements have been automatically introduced by ANSYS, i.e., CONTA174 and TARGE170. CONTA174 is used to represent contact and sliding between 3D "target" surfaces and a deformable surface defined by this element, whereas the TARGE170 element is used to represent various 3D "target" surfaces for the associated CONTA174 contact elements. Hence, for demonstrative purposes, a detail of the meshing for G25_nanoHA_1.25 FEM with such finite elements is shown in Figure 8.

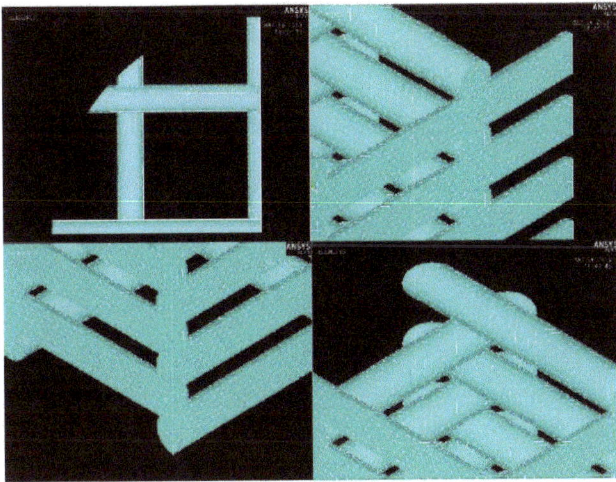

Figure 8. Detail of the meshing for G25_nanoHA_1.25 finite element model.

Regarding the boundary conditions, as the fibre diameter (0.25 mm) is larger than the layer height (0.20 mm), there will be stacking between perpendicular fibres. In these numerical models, it has been considered that there is a perfect bonding between fibres, as shown in Figure 9a. Additionally, a lower horizontal plane has been included in these numerical models to limit, by means of a contact, the vertical displacement of the lower fibres. This plane presenting infinite stiffness and its displacements are totally constrained, see Figure 9b. Finally, an upper horizontal plane has also been included in these numerical models in order to impose, by means of a contact, vertical displacement in the upper fibres. This plane presents infinite stiffness too and its lateral displacements are totally constrained (Figure 9c).

Finally, regarding the actions and type of analysis, in order to simulate a monotonic compression test, a vertical displacement has been imposed on the upper plane of the numerical models. This vertical displacement will be transferred to the scaffold by means of the contact defined between the upper plane and the upper fibres. Due to the negligible influence on the results, the self-weight of the scaffolds has not been considered in this study. Additionally, the need for evaluating the non-linearities related to both the material and the existence of contacts, has demanded the resolution of the numerical models developed in this study by means of a non-linear static analysis. In this resolution, the hypothesis of small displacements has also been taken into account.

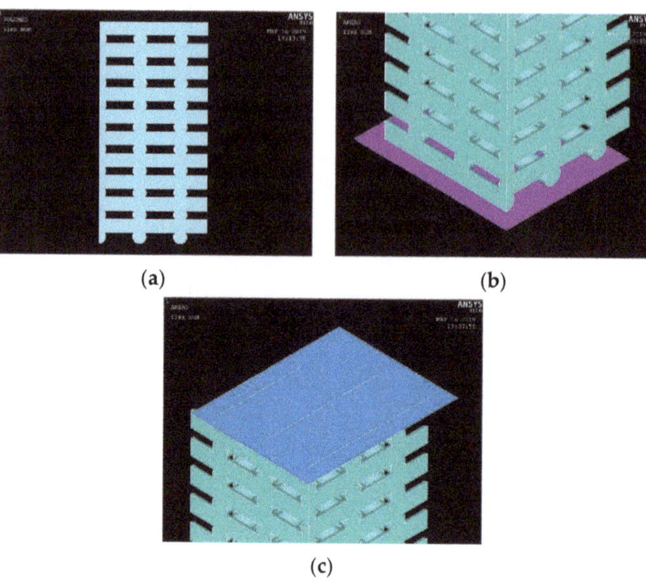

Figure 9. Boundary conditions considered for FEM, only G25_nanoHA_0.75 case is shown: (**a**) Detail of bonding between perpendicular fibres; (**b**) Detail of the lower; (**c**) Detail of the upper plane.

3. Results

3.1. G25_nanoHA 1.25

In Figure 10a, the stress–strain curve for the G25_nanoHA_1.25 scaffold is shown. The strain ($\varepsilon = \Delta l/l$) is then calculated as the displacement of the upper plane divided by the total height of the scaffold, 4.05 mm. Additionally, the stress is calculated as the force to be applied in order to obtain a certain displacement of the upper plane divided by the cross-sectional area of the scaffold ($\pi \cdot \varphi^2/4 = 4\pi$ mm^2), where φ is the diameter of the cylindrical scaffold. In this regard, this stress can be considered as an "apparent stress". This is a very important difference, because the stress–strain relationship derived by this way will differ from the actual stresses obtained using the FE model, depicting stresses of every prismatic solid FE according to a colour scale.

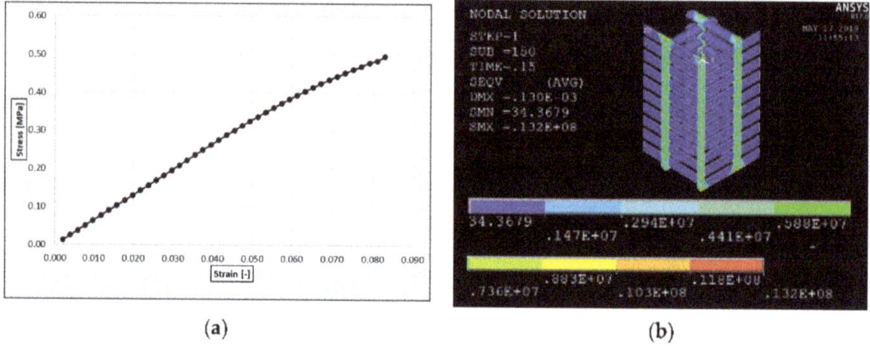

Figure 10. Results of FEM simulation of a monotonic compression test on a G25_nanoHA_1.25 Scaffold: (**a**) Stress–strain curve; (**b**) Nodal stresses (Von Mises equivalent stress, in Pascals).

Thus, defining the elastic modulus of the scaffold E as the stress divided by the strain for a strain value of 0.03 (within this range, the stress–strain relationship is almost linear) leads to the following Equation (1):

$$E = \frac{0.1955}{0.03} = 6.60 \text{ MPa} \tag{1}$$

Additionally, in the following Figure 10b, the Von Mises equivalent stresses for a scaffold strain of 0.03 are shown. Von Mises stress has been chosen for comparison because it is able to combine stresses in several axes in a single stress for comparison. Despite the fact there is a clearly dominant stress direction, since this is a simulated monotonic uniaxial compression test, it is combined with tangential and normal stresses in other directions at the spots where the fibres are crossing each other; therefore, simply plotting the principal stresses will underestimate the real triaxial stress state. As reflected in that figure, loads are mainly transferred through the columns formed by the intersections between perpendicular fibres.

It is noteworthy to remark the differences between the stresses of a stress–strain relationship (Figure 10a) and those depicted in the FEM (Figure 10b). The first going up to 0.5 MPa, while the second is going up to 13.2 MPa at substep 150. This is because, while the first is the quotient between the applied force divided by the cylinder area, the second is the actual stress at every SOLID 187 FE.

3.2. G25_nanoHA_1.00

Analogously, the same procedure is applied to the G25_nanoHA_1.00 numerical model. Accordingly, its stress–strain curve is shown in the following Figure 11a, with the corresponding Von Mises equivalent stress in Figure 11b. Moreover, the elastic modulus of the scaffold E, as the stress divided by the strain for a strain value of 0.03 leads to the following Equation (2):

$$E = \frac{0.2342}{0.03} = 7.90 \text{ MPa} \tag{2}$$

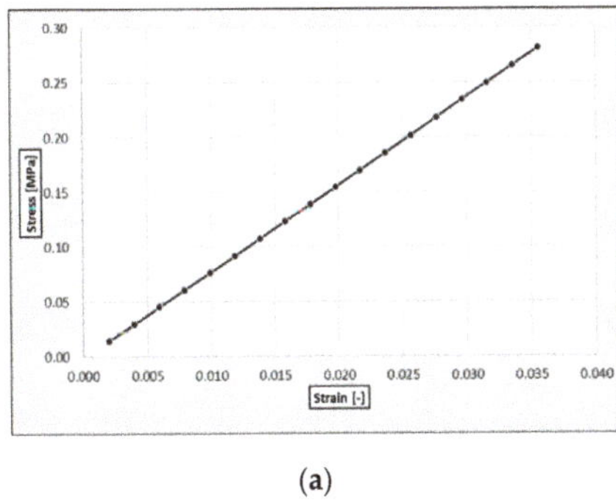

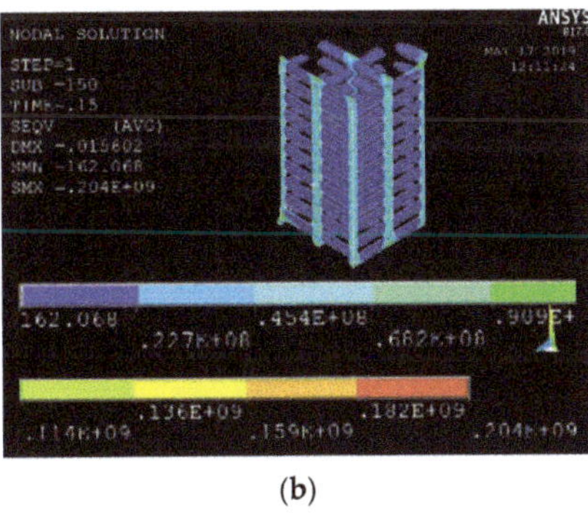

(a)　　　　　　　　　　(b)

Figure 11. Results of FEM simulation of a monotonic compression test on a G25_nanoHA_1.00 Scaffold: (**a**) Stress–strain curve; (**b**) Nodal stresses (Von Mises equivalent stress, in 10^{-1} Pascals).

It is noteworthy to remark the differences between the stresses in a stress–strain relationship (Figure 11a) and those depicted in the FEM (Figure 11b). The first going up to 0.28 MPa, while the second is going up to 20.4 MPa at substep 150. This is because, while the first is the quotient between the applied force divided by the cylinder area, the second is the actual stress at every SOLID 187 FE.

3.3. G25_nanoHA_0.75

Finally, the stress–strain curve of the G25_nanoHA_0.75 is shown in Figure 12a, while the corresponding equivalent Von Mises Stress is depicted in Figure 12b. The Young's modulus E is then calculated as the slope of the stress–strain curve at the point with 0.03 strain, as derived in the following Equation (3):

$$E = \frac{0.2342}{0.03} = 15.49 \text{ MPa} \tag{3}$$

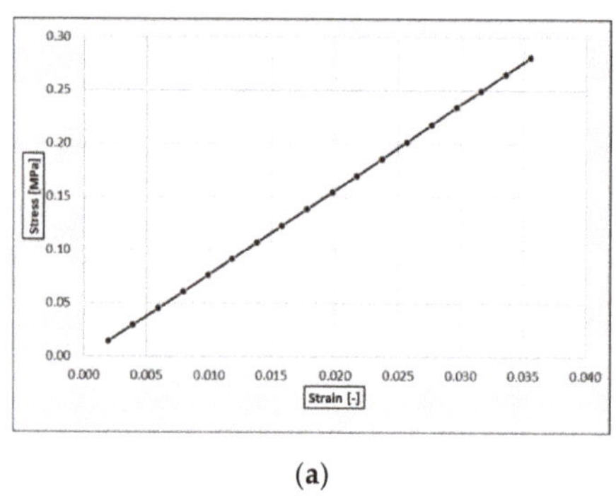
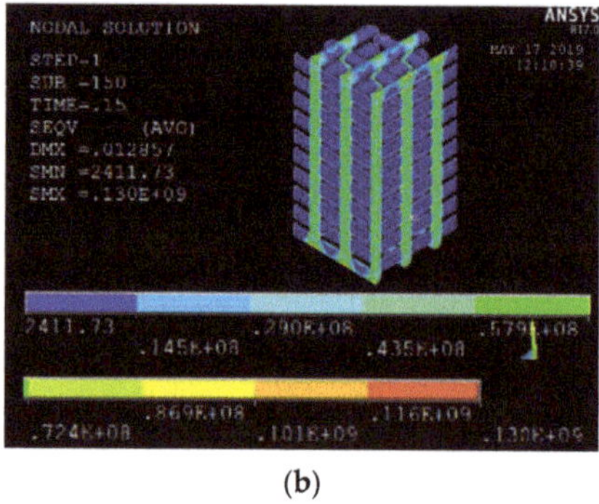

(a) (b)

Figure 12. Results of FEM simulation of a monotonic compression test on a G25_nanoHA_0.75 Scaffold: (**a**) Stress–strain curve; (**b**) Nodal stresses (Von Mises equivalent stress, in 10^{-1} Pascals).

It is noteworthy to remark the differences between the stresses in the stress–strain relationship (Figure 12a) and those depicted in the FEM (Figure 12b). The first going up to 0.28 MPa, while the second is going up to 13 MPa at substep 150. This is because, while the first is the quotient between the applied force divided by the cylinder area, the second is the actual stress at every SOLID 187 FE.

4. Discussion

The results obtained using an FEM of the different scaffolds G25 nanoHA 1.25, 1.00 and 0.75 are compared to the corresponding monotonic compression tests on the Young's modulus basis (Table 3). As can be seen in all the cases, the finite element model with idealized geometry showed a Young's modulus higher than the one measured using direct testing. Moreover, this improvement was higher than the mean value plus 2–3 times the standard deviation of each scaffold. In the case of G25 nanoHA 1.25, the relationship of the FEM value to the mean measured value was 6.60/2.25 = 2.93 times higher, while in the case of the other two, G25 nanoHA 1.00 and G25 nanoHA 0.75, it was 1.54 and 2.45, respectively.

Table 3. Comparison between the Young's modulus obtained using FEM and direct scaffold testing, with corresponding Mean and Standard Deviation values.

Scaffold		G25 nanoHA 1.25			G25 nanoHA 1.00			G25 nanoHA 0.75		
		Model	Mean	SD	Model	Mean	SD	Model	Mean	SD
Young's Modulus	(MPa)	6.60	2.25	0.36	7.90	5.13	1.01	15.49	6.33	0.53

Therefore, after repeating the FEM analysis several times, changing mesh sizes, contact definition and calculation steps and looking at the results, the difference was so high that it could not be explained by result scattering only. In fact, there are several assumptions on the FEM analysis on an idealized scaffold model that are not realistic at all, each contributing to these differences. For instance, a nominal fibre diameter, fibre position and straightness or strand distance are perfect in the model, but in the real world, with a 3D-printing process at such minimal scales, these parameters are impossible to be guaranteed and difficult to be controlled in real time. Thus, such imperfections are sources of uncertainty in terms of mechanical behaviour and the only way to deal with this is to set some execution tolerances and apply a design factor to the idealized geometry, as is common practice in structural engineering to deal with such execution imperfections.

Hence, the first step is to identify the main *prima facie* uncertainty sources. As such unintentional imperfections are impossible to reproduce, isolated from other errors and with accuracy in real bone scaffolds for testing, the best approach is to model and simulate such imperfections and see their influence on the results. In order to be able to simulate it, each uncertainty source is considered as an independent random variable, whose result is expected to be within a certain interval, with a reasonable grade of reliability, considering an according execution control procedure to guarantee some defined tolerances. Finally, all the variables need to be combined in a single design safety factor considering the global uncertainty, since each one will contribute to it to a certain degree.

As a last comment, a bone scaffold presents time-dependent properties in vivo. In fact, it undergoes deterioration processes related to fatigue, because a bone suffers from cyclic loadings and equivalent biological remodelling acting such as corrosion, since the scaffold material is progressively removed and substituted by bone tissue and this points for an initial deterioration up to a valley point where it starts to recover mechanical properties. Accordingly, the best approach is to develop a simplified model to take into account these two deterioration processes during service life, such as those presented by Calderon Uriszar-Aldaca et al. [50–52], developing an additional correction factor to take these into account.

4.1. Sensitivity Analysis

For the sensitivity analysis, the G25_nanoHA_1.25 numerical model under monotonic compression is taken as the base reference model, since it showed the highest difference between the tests and the idealized model, which is already described and whose results are presented in Section 3.1. Hence, the base characteristics and corresponding variations are presented in Table 4.

Table 4. Comparison between base G25_nanoHA_1.25 characteristics and developed variations for sensitivity analysis.

Source	Base	Variation 1	Variation 2
Poisson's Ratio	0.4	0.375	0.35
Position of fibres	At plane of sym.	Out plane of sym.	-
Fibre diameter (mm)	0.25	0.235	0.22
Strand distance (mm)	1.25	1.2	1.3
Layer height (mm)	0.2	0.215	0.23
Sample diameter (mm)	4	3.9	4.1
Deflection of fibres (mm)	0	0.06	0.12
Straightness of columns	Straight	Alternated	Cumulated
Existence of broken fibres	No broken fibres	breakage at crossing	far from crossing

4.1.1. Poisson's Ratio of the Base Material

As mentioned in Section 3.1 materials, a Poisson's coefficient of 0.4 has been considered for the base material as common practice for polymeric materials. Nevertheless, as no specific test has been carried out in the frame of the project in order to determine this value accurately, it can be considered as a source of uncertainty.

Accordingly, three simulations have been carried varying the Poisson's ratio with 0.4, 0.375 and 0.35 as inputs for G25_nanoHA_1.25 scaffold finite element model. Then, the stress–strain relationship and the elastic modulus has been obtained following the same procedure to see how this variation influences the results.

Thus, Figure 13 shows the stress–strain relationship when varying the Poisson's ratio and the elastic modulus obtained as the slope of such stress–strain relationship, the curves are so close that they stack over each other. Moreover, Table 5 summarize the results. According to these results, the influence is almost negligible, i.e., after increasing the Poisson's ratio of the base material, only a slight increase in the elastic modulus of the scaffold can be appreciated, lowering only up to 99.09% in stiffness.

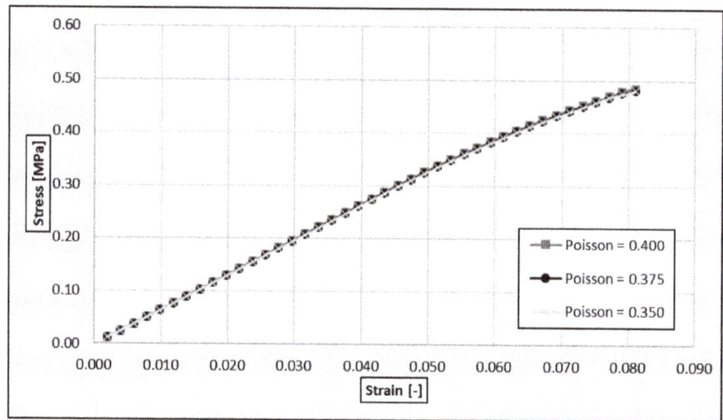

Figure 13. Influence of the Poisson's ratio on the stress–strain relationships from FEM.

Table 5. Compressive elastic modulus of a G25_nanoHA_1.25 scaffold for different values of the Poisson's ratio.

Poisson's Ratio (-)	Elastic Modulus (MPa)
0.400	6.60
0.375	6.57
0.350	6.54

4.1.2. Position of the Fibres within the Scaffold

According to Section 3 on results, the compression loads are mainly transferred to the basement through the columns formed by the intersections between the perpendicular fibres of the scaffold (Figure 10). Hence, due to the relatively low value of the sample diameter in comparison with the strand distance, the relative position of the fibres within the scaffold will affect the number and location of these columns, having some influence on the compressive stiffness of the scaffold, which can happen as a result of punching position variability.

As shown in Figure 14, two different configurations have been analysed. In the first one, at each layer of the scaffold, there is a fibre located in the relevant plane of symmetry. On the other hand, in the second configuration, at each layer of the scaffold, the relevant plane of symmetry is in the centre of the gap between two fibres.

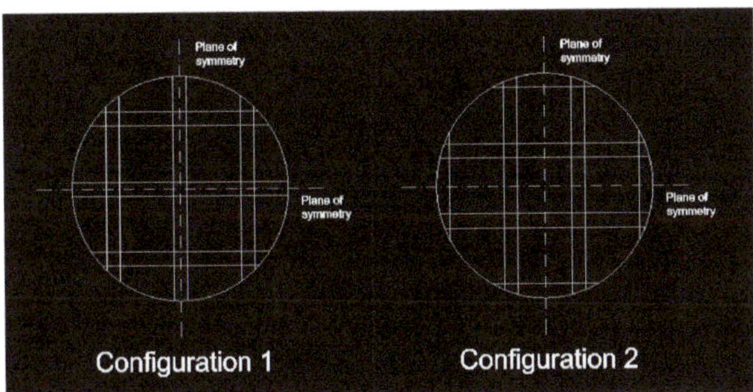

Figure 14. Configurations considered to analyse the influence of the position of the fibres on the compressive stiffness of the scaffold (G25_nanoHA 1.25).

Therefore, a detail of the geometry of the numerical models used for both configurations is shown in the following Figure 15:

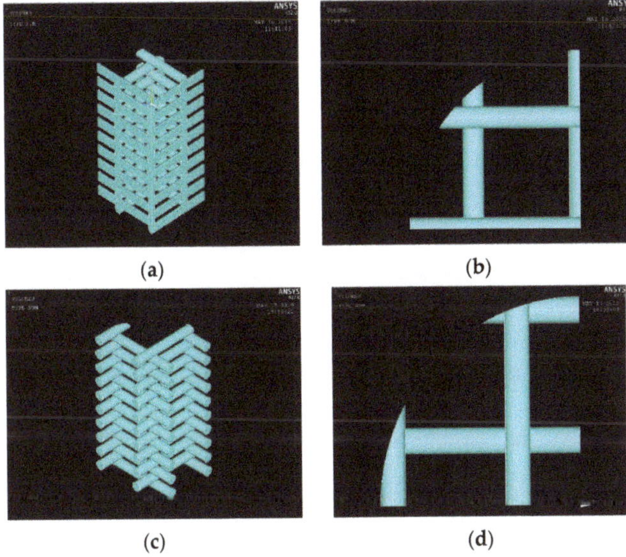

(a)   (b)

(c)   (d)

Figure 15. Geometry of the numerical models for configurations 1 and 2: (**a**) Global view of configuration 1; (**b**) Cross-section of configuration 1; (**c**) Global view configuration 2; (**d**) Cross-section of configuration 2.

The influence of the relative position of the fibres within the scaffold on the stress–strain relationship of the G25_nanoHA 1.25 scaffold subjected to a compression load has been analysed using FEM (Figure 16).

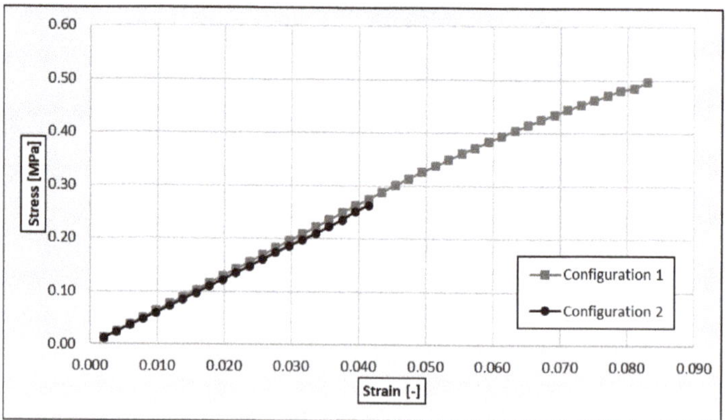

Figure 16. Influence of the relative position of the fibres within the scaffold G25_nanoHA_1.25 on the stress–strain curve.

The values of the compressive elastic modulus of a G25_nanoHA_1.25 scaffold of both configurations are shown in Table 6 (the compressive elastic modulus of the scaffold has been defined as the stress divided by the strain for a strain value of 0.03).

Table 6. Compressive elastic modulus of a G25_nanoHA_1.25 scaffold for different configurations in terms of the relative position of the fibres within the scaffold.

Poisson's Ratio (-)	Elastic Modulus (MPa)
0.400	6.60
0.375	6.26

According to these results, there is some influence of the relative position of the fibres within the scaffold on the compressive elastic modulus of a G25_nanoHA_1.25 scaffold, with configuration one being stiffer than configuration two, only reaching 94.85% of the stiffness. Finally, the Von Mises equivalent stresses for a scaffold strain of 0.03 are shown in Figure 17 for both the configurations.

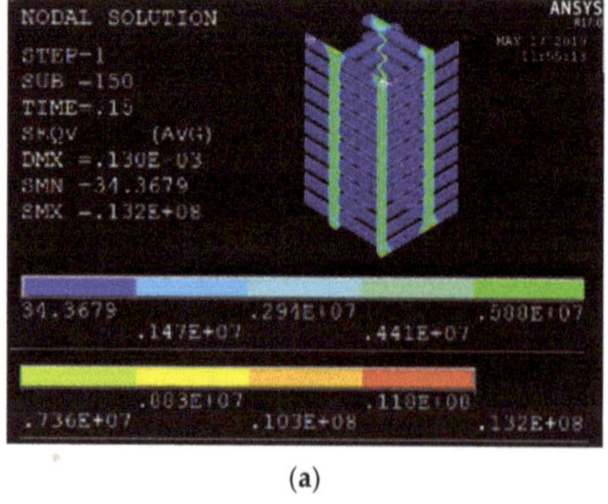

(a)

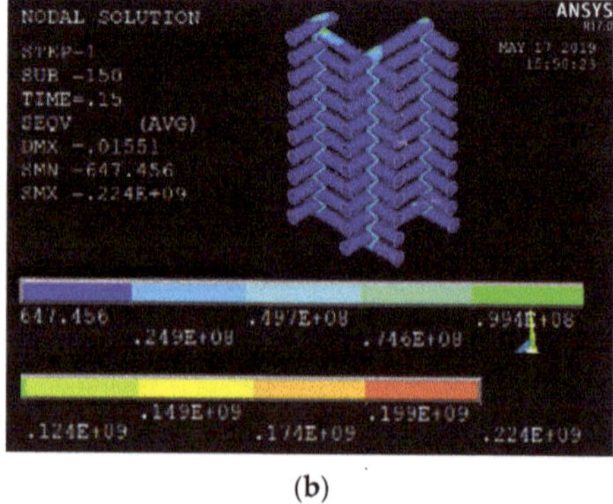

(b)

Figure 17. Nodal stresses of G25_nanoHA_1.25 (Von Mises equivalent stress): (**a**) Configuration 1, in Pascals; (**b**) Configuration 2, in 10^{-1} Pascals.

4.1.3. Fibre Diameter

As already disclosed in Table 1, scaffolds have been printed using a theoretical fibre diameter of 0.25 mm. Nevertheless, the accuracy of the printer in terms of deposition speed, positioning, temperature and material viscosity and variations in the material flow rate can lead to the real value of the fibre diameter being different to the theoretical one.

The influence of the fibre diameter on the stress–strain relationship of the G25_nanoHA_1.25 scaffold subjected to a compression load has been analysed by means of FEM (Figure 18).

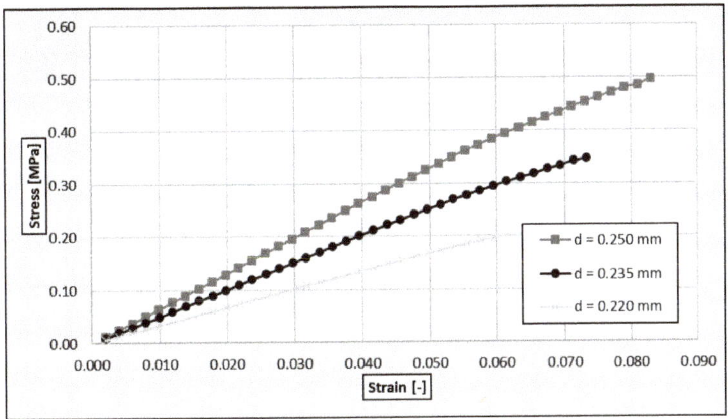

Figure 18. Influence of the fibre diameter on the stress–strain relationships from FEM.

The value of the compressive elastic modulus of a G25_nanoHA_1.25 scaffold for each of the three different values considered in this analysis for the fibre diameter is shown in Table 7.

Table 7. Compressive elastic modulus of a G25_nanoHA_1.25 scaffold for different values of the fibre diameter.

Fibre Diameter (mm)	Elastic Modulus (MPa)
0.250	6.60
0.235	5.08
0.220	3.39

The cross-sectional area of the columns formed by the intersections between perpendicular fibres of the scaffold will depend on the fibre diameter (the bigger the fibre diameter, the bigger the cross-sectional area of these columns). As the compression loads applied on the scaffold will be mainly transferred through these columns, the fibre diameter will have a great influence on the compressive elastic modulus of the scaffold, as can be seen in Table 7 and Figure 18.

For two different values of the fibre diameter ($\varphi_{\text{fibre}} = 0.235$ mm and $\varphi_{\text{fibre}} = 0.220$ mm), the Von Mises equivalent stresses for a scaffold strain of 0.03 are shown in Figure 19, for $\varphi_{\text{fibre}} = 0.250$ mm (see also Figure 10).

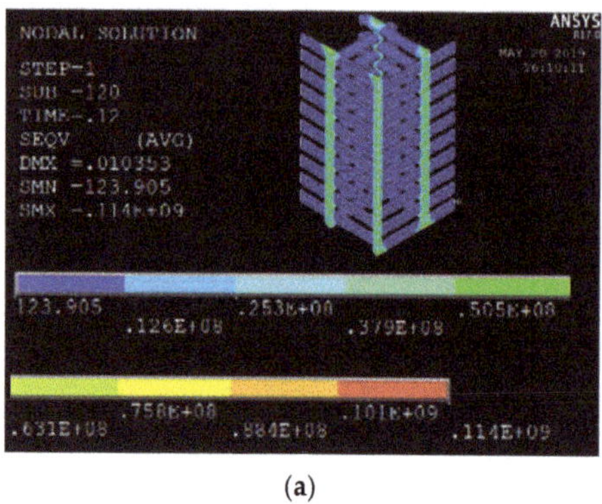 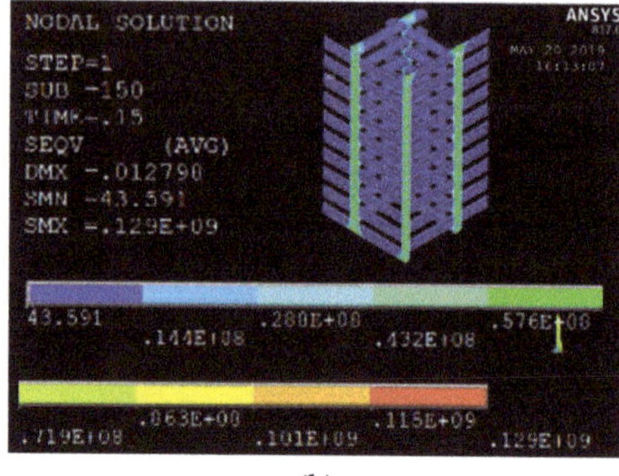

(a)   (b)

Figure 19. Nodal stresses of G25_nanoHA_1.25 (Von Mises equivalent stress): (**a**) φ_{fibre} = 0.235 mm, in 10^{-1} Pascals; (**b**) φ_{fibre} = 0.220 mm, in 10^{-1} Pascals.

4.1.4. Strand Distance

The G25_nanoHA_1.25 scaffold samples have been printed using a theoretical strand distance of 1.25 mm. Nevertheless, the accuracy of the positioning system of the printer can lead to a real value for the strand distance different from the theoretical one.

The influence of the strand distance on the stress–strain relationship of the G25_nanoHA_1.25 scaffold subjected to a compression load has been analysed using FEM and the results of this analysis are shown Figure 20.

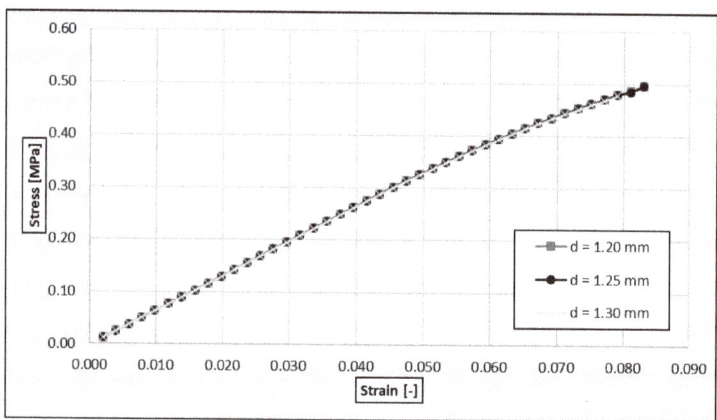

Figure 20. Influence of the strand distance on the stress–strain relationships from FEM.

The value of the compressive elastic modulus of a G25_nanoHA_1.25 scaffold for each of the three different values considered in this analysis for the strand distance is shown in Table 8.

Table 8. Compressive elastic modulus of a G25_nanoHA_1.25 scaffold for different values of the layer height.

Strand Distance (mm)	Elastic Modulus (MPa)
1.20	6.61
1.25	6.60
1.30	6.58

According to these results, when the variation of the strand distance is low, their influence on the stress–strain relationship of a G25_nanoHA_1.25 scaffold subjected to a compression load is almost negligible, i.e., increasing the strand distance causes only a slight decrease in the elastic modulus of the scaffold, up to 99.55% of the original stiffness.

As the compression loads applied on the scaffold will be mainly transferred through the columns formed by the intersections between the perpendicular fibres of the scaffold, variations in the strand distance will have a significant impact on the compressive elastic modulus of the scaffold, only when these variations lead to an increase/decrease in the number of those intersections.

For two different values of the strand distance (d = 1.20 mm and d = 1.30 mm), the Von Mises equivalent stresses for a scaffold strain of 0.03 are shown in Figure 21, for d = 1.25 mm (see also Figure 10).

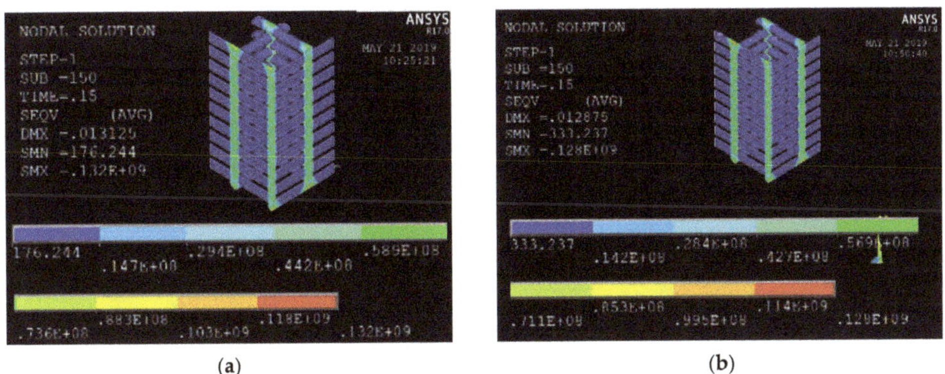

(a)   (b)

Figure 21. Nodal stresses of G25_nanoHA_1.25 (Von Mises equivalent stress): (**a**) d = 1.20 mm, in 10^{-1} Pascals; (**b**) d = 1.30 mm, in 10^{-1} Pascals.

4.1.5. Layer Height

The tested G25_nanoHA_1.25 scaffolds have been printed using a theoretical layer height of 0.20 mm. The influence of the layer height on the stress–strain relationship of the G25_nanoHA_1.25 scaffold subjected to a compression load has been analysed using FEM (Figure 22).

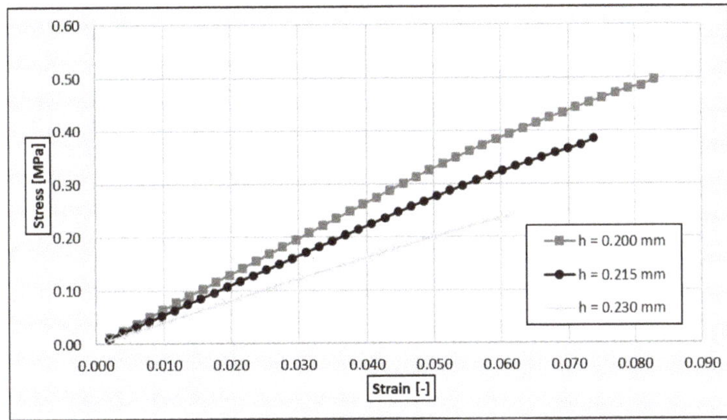

Figure 22. Influence of the layer height on the stress–strain relationships from FEM.

The value of the compressive elastic modulus of a G25_nanoHA_1.25 scaffold for each of the three different values considered in this analysis for the strand distance is shown in Table 9.

Table 9. Compressive elastic modulus of a G25_nanoHA_1.25 scaffold for different values of the layer height.

Layer Height (mm)	Elastic Modulus (MPa)
0.200	6.60
0.215	5.53
0.230	4.07

The cross-sectional area of the columns formed by the intersections between the perpendicular fibres of the scaffold will depend on the layer height; the bigger the layer height is, the lower the cross-sectional area of these columns is. As the compression loads applied on the scaffold will be mainly transferred through these equivalent columns, the layer height will have a great influence on the compressive elastic modulus of the scaffold, as can be seen in Table 9 and Figure 22. For two different values of the layer height (h = 0.215 mm and h = 0.230 mm), the Von Mises equivalent stresses for a scaffold strain of 0.03 are shown in Figure 23, for h = 0.200 mm (see Figure 10).

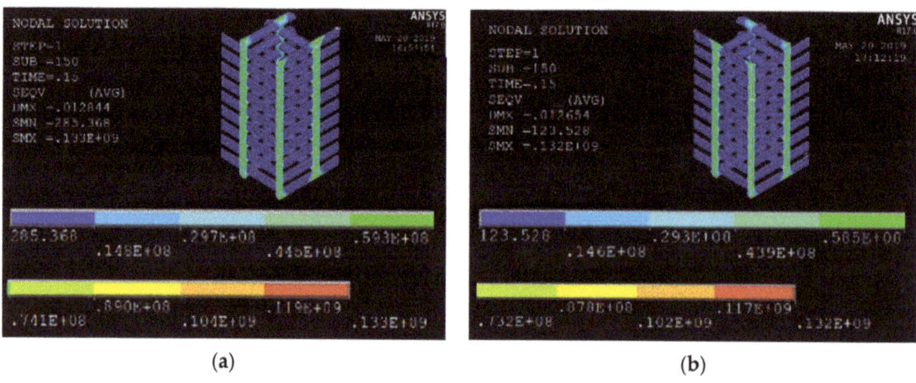

Figure 23. Nodal stresses of G25_nanoHA_1.25 (Von Mises equivalent stress): (**a**) h = 0.215 mm, in 10^{-1} Pascals; (**b**) h = 0.230 mm, in 10^{-1} Pascals.

4.1.6. Sample Diameter

In the frame of this study, monotonic compression tests and dynamic compression tests have been carried out on cylindrical samples with a theoretical diameter of 4 mm, extracted from G25_nanoHA_1.25 scaffolds. Nevertheless, the accuracy of the tools used to extract the samples from the printed scaffolds can lead to a real value for the sample diameter different from the theoretical one. The influence of the sample diameter on the stress–strain relationship of the G25_nanoHA_1.25 scaffold subjected to a compression load has been analysed using FEM (Figure 24).

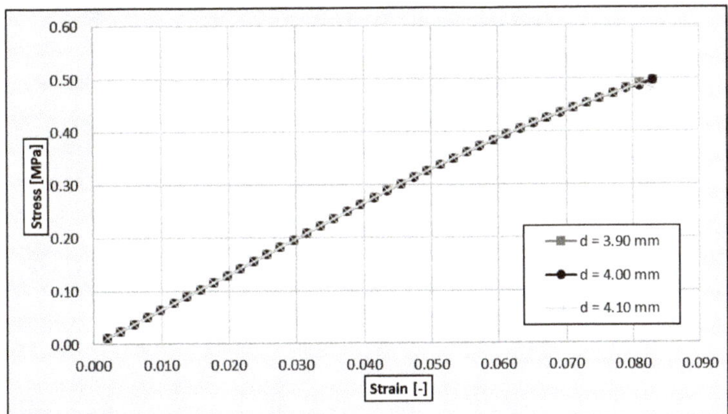

Figure 24. Influence of the sample diameter on the stress–strain relationships from FEM.

The value of the compressive elastic modulus of a G25_nanoHA_1.25 scaffold for each of the three different values considered in this analysis for the sample diameter is shown in Table 10.

Table 10. Compressive elastic modulus of a G25_nanoHA_1.25 scaffold for different values of the sample diameter. The theoretical value for the sample diameter has been used for calculating the stress.

Sample Diameter (mm)	Elastic Modulus (MPa)
3.90	6.61
4.00	6.60
4.10	6.62

According to these results, when the variation of the sample diameter is such that the number of columns formed by the intersections between perpendicular fibres of the scaffold is not affected, their influence on the stress–strain relationship of a G25_nanoHA_1.25 scaffold subjected to a compression load is negligible. On the other hand, if the actual sample diameter is used to determine the stress on the sample, a certain influence of this parameter on the stress–strain relationship of the G25_nanoHA_1.25 scaffold subjected to a compression load can be observed, as can be seen in Table 10 and Figure 24.

These results highlight that, due to the relative low value of the sample diameter in comparison with the strand distance, the stress–strain relationship of the G25_nanoHA_1.25 scaffold subjected to a compression load is very sensitive to the diameter of the tested sample. Nevertheless, this influence is expected to decrease as the sample diameter increases.

4.1.7. Deflection of Fibres during the Printing Process

Due to the low mechanical properties of the fused material, the fibres tend to bend and lose straightness during the printing process at middle span, between rigid or supporting nodes, such as the intersections between the perpendicular fibres, as can be appreciated in Figure 25.

Figure 25. Detail of printed scaffolds.

The influence of the deflection of the printed fibres on the stress–strain relationship of the G25_nanoHA_1.25 scaffold subjected to a compression load has been analysed using FEM (in Figure 26).

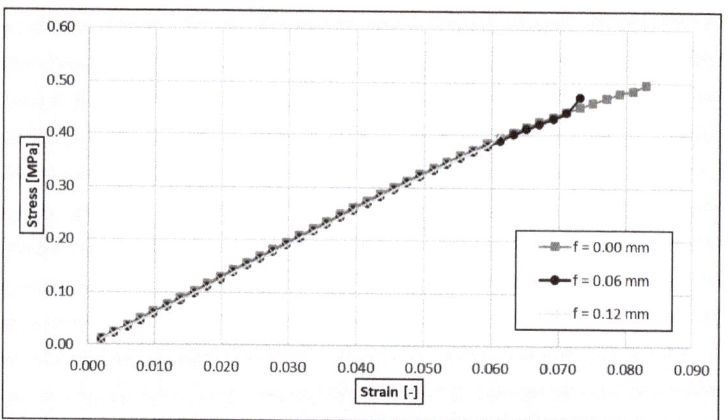

Figure 26. Influence of the deflection of the printed fibres on the stress–strain relationships from FEM.

The value of the compressive elastic modulus of a G25_nanoHA_1.25 scaffold for each of the three different values considered in this analysis for the deflection of the printed fibres is shown in Table 11.

Table 11. Compressive elastic modulus of a G25_nanoHA_1.25 scaffold for different values of the deflection of the printed fibres.

Deflection of the Printed Fibres (mm)	Elastic Modulus (MPa)
0.00	6.60
0.06	6.45
0.12	6.41

According to these results, the influence of the deflection of the printed fibres on the stress–strain relationship of a G25_nanoHA_1.25 scaffold subjected to a compression load is very low, i.e., increasing the deflection of the printed fibres means a slight decrease in the elastic modulus of the scaffold up to 97.12% of the original stiffness. It should be noted that, in this case, the compression load is parallel to the columns formed by the intersections between the perpendicular fibres. For compression loads parallel to any of the fibre directions, the influence of the deflection of the printed fibres on the stress–strain relationship of a G25_nanoHA_1.25 scaffold subjected to a compression load is expected to be much higher.

For two different values of the deflection of the printed fibres, f = 0.06 mm and f = 0.12 mm, the Von Mises equivalent stresses for a scaffold strain of 0.03 are shown in Figure 27; for no deflection, see Figure 10.

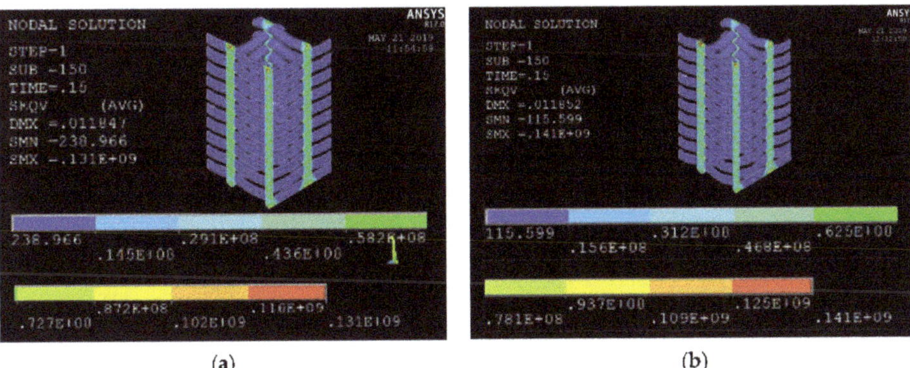

(a)   (b)

Figure 27. Nodal stresses of G25_nanoHA_1.25 (Von Mises equivalent stress): (**a**) f = 0.06 mm, in 10^{-1} Pascals; (**b**) f = 0.12 mm, in 10^{-1} Pascals.

4.1.8. Straightness of the Columns Responsible for Load Transmission

G25_nanoHA_1.25 scaffolds have been printed using a theoretical strand distance of 1.25 mm. However, the accuracy of the positioning system of the printer can lead to a real value for the strand distance different from the theoretical one.

Hence, the influence of the strand distance on the stress–strain relationship of the G25_nanoHA_1.25 scaffold subjected to a compression load was analysed in Section 4.1.4 on strand distance. In that case, it was supposed that there was some variation in the strand distance, but it was also considered that this variation remained constant within the whole scaffold. Under these circumstances, the columns formed by the intersections between the perpendicular fibres, which are responsible for the transmission of the compression load applied on the scaffold, remained vertically straight. Nevertheless, the x and y positioning in any printer present deviations; therefore, the real positions are x ± Δx and y ± Δy, meaning that some executions could present alternate positioning or show a displacement tendency.

In this section, on the other hand, the influence of the strand distance on the stress–strain relationship of the G25_nanoHA_1.25 scaffold subjected to a compression load has also been analysed. In this case, this variation in the strand distance affected the straightness of those columns.

Thus, the following two situations have been analysed:

- Alternate variations in the strand distance (situation one);
- Cumulative variations in the strand distance (situation two).

In the first situation, there is a certain eccentricity e between the theoretical location of the fibres and their actual location. In addition, it has been considered that the fibre corresponding to a certain layer is located on the opposite side, regarding the theoretical location of the fibres, of that of the fibres corresponding to the preceding and the following layers, see Figure 28.

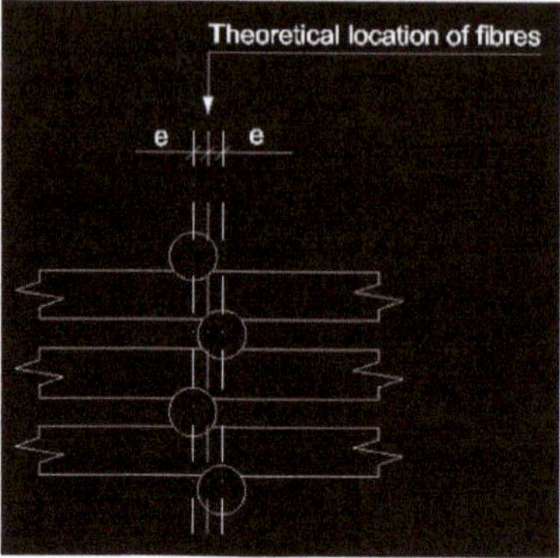

Figure 28. Location of fibres in situation 1.

A detail of the geometry of the numerical model used to simulate this situation is shown in Figure 29. In this situation, the following two values have been considered for the deviation of the fibres: e = 0.025 mm and e = 0.05 mm.

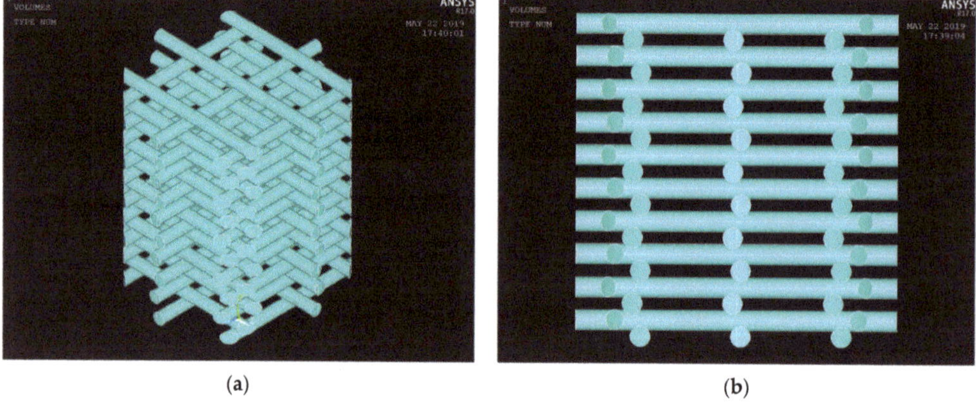

(a)   (b)

Figure 29. Geometry of the numerical model for situation 1 for G25_nanoHA 1.25: (**a**) Global view; (**b**) Vertical plane section.

In the second situation, there is also a certain eccentricity between the theoretical location of the fibres and their actual location. On the other hand, in this case, the deviation is cumulative, as shown in Figure 30.

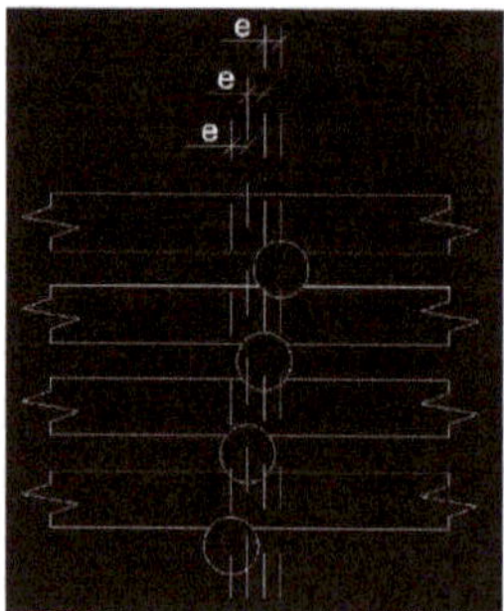

Figure 30. Location of fibres in situation 2.

A detail of the geometry of the numerical model used to simulate this situation is shown in Figure 31. In this situation, two values have also been considered for the deviation of the fibres, e = 0.005 mm and e = 0.009 mm.

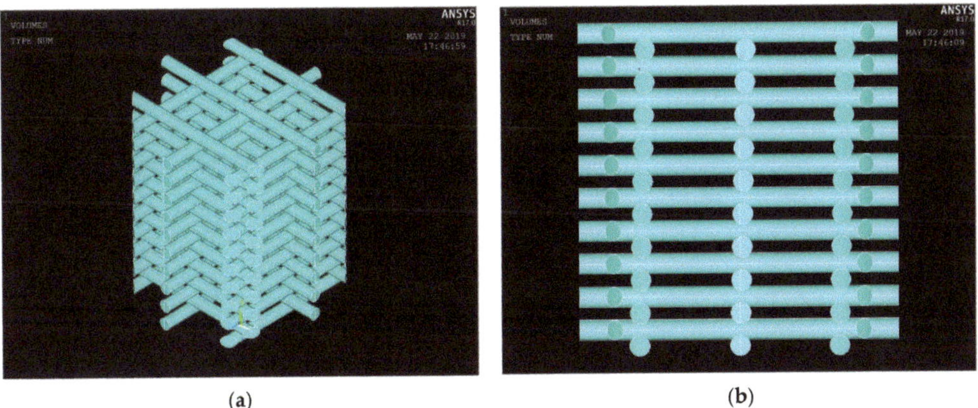

(a)   (b)

Figure 31. Geometry of the numerical model for situation 2 for G25_nanoHA 1.25: (a) Global view; (b) Vertical plane section.

In both situations, it has been considered that the deviation between the theoretical location of the fibres and their actual location affects the fibres oriented in both directions.

As can be seen in Figures 29 and 31, in these cases, there is no any symmetry in the geometry of the scaffolds; therefore, the whole scaffold has been represented in the numerical models. Therefore, in order to avoid lateral rigid body motions, which may impede the convergence of the numerical models, a friction coefficient of 0.3 has been considered in the contact between the upper plane and the upper fibres and the contact between the lower plane and the lower fibres.

The influence of the deviation between the theoretical location of the fibres and their actual location, affecting the straightness of the columns formed by the intersections between the perpendicular fibres, on the stress–strain relationship of the G25_nanoHA_1.25 scaffold subjected to a compression load has been analysed using FEM (Figure 32).

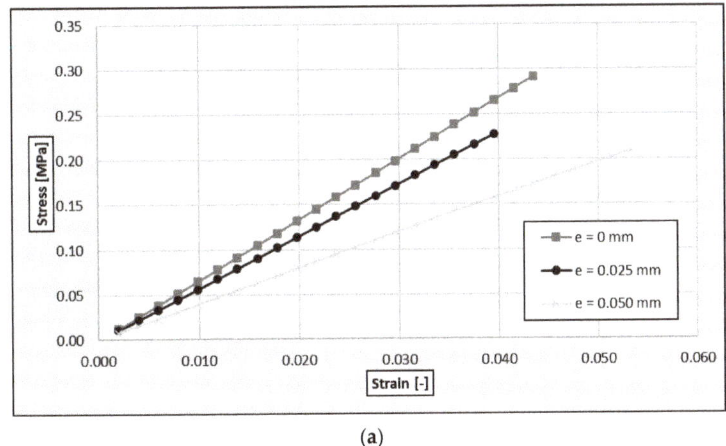

(a)

Figure 32. *Cont.*

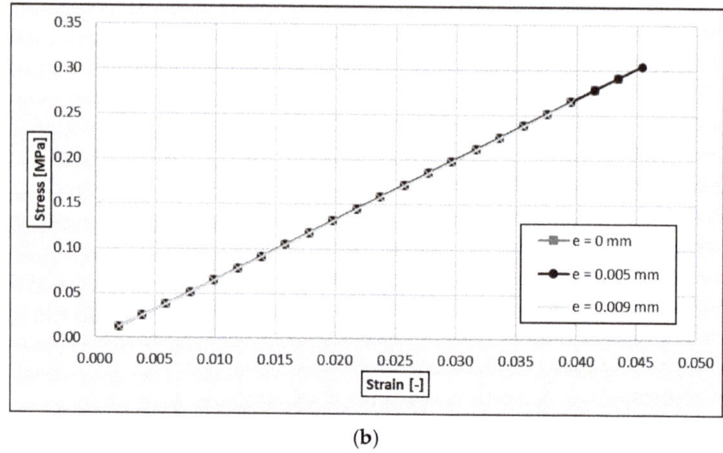

(b)

Figure 32. Influence of the straightness of the columns on the elastic modulus from FEM of G25_nanoHA 1.25: (**a**) Influence on the stress–strain relationships of configuration 1; (**b**) Influence on the stress–strain relationships of configuration 2.

Accordingly, the values of the compressive elastic modulus of a G25_nanoHA_1.25 scaffold for each of the three different values considered in this analysis, for the deviation between the theoretical location of the fibres and their actual location, are shown in Table 12 for situation one and Table 13 for situation two.

Table 12. Compressive elastic modulus of a G25_nanoHA_1.25 scaffold for different values of the eccentricity or deviation in situation 1.

Deviation (mm)	Elastic Modulus (MPa)
0.000	6.70
0.025	5.76
0.050	3.99

Table 13. Compressive elastic modulus of a G25_nanoHA_1.25 scaffold for different values of the eccentricity or deviation in situation 2.

Deviation (mm)	Elastic Modulus (MPa)
0.000	6.70
0.005	6.69
0.009	6.67

As can be seen in Tables 12 and 13, the compressive elastic modulus of the scaffold with no deviation (6.70 MPa) differs from the compressive elastic modulus of the reference scaffold considered in the previous sections (6.60 MPa). This difference lies in the fact that, as mentioned, the numerical models used in both cases differ regarding the friction coefficient used for the contacts between the upper and lower planes and the scaffold fibres.

According to these results, when errors are cumulative (situation two), there is no significant influence, for the range of values considered, of the deviation between the theoretical location of the fibres and their actual location on the stress–strain relationship of a G25_nanoHA_1.25 scaffold subjected to a compression load. However, when errors alternate with respect to the theoretical location of the fibres (situation one), the influence of the deviation between the theoretical location of the fibres and their actual location on the stress–strain relationship of a G25_nanoHA 1,25 scaffold subjected to a compression load may be considerable, because the straightness of the columns responsible for load transmission is much more affected in such situation.

4.1.9. Existence of Broken Fibres

The G25_nanoHA_1.25 scaffold samples for testing have been printed using a theoretical fibre diameter of 0.25 mm, leading to relatively fragile fibres that can suffer damage during the printing process, the cooling and polymerization process or the tooling of the samples.

Thus, the influence of the existence of broken fibres on the stress–strain relationship of the G25_nanoHA_1.25 scaffold subjected to a compression load has been analysed using FEM, having considered the following alternatives:

- No broken fibres within the scaffold—Situation one.
- One broken fibre within the scaffold (breakage located in one intersection between perpendicular fibres)—Situation two.
- Two broken fibres within the scaffold (breakages located in two different intersections between perpendicular fibres)—Situation three.
- One broken fibre within the scaffold (breakage located far from any intersection between perpendicular fibres)—Situation four.
- Two broken fibres within the scaffold (breakages located far from any intersection between perpendicular fibres)—Situation five.

All the fibre breakages considered in this study had a length of 0.15 mm and their location was randomly selected.

As in the previous section, in this case there is no symmetry in the geometry of the scaffolds; therefore, the whole scaffold has been represented in the numerical models. Therefore, in order to avoid lateral rigid body motions, which may impede the convergence of the numerical models, a friction coefficient of 0.3 has been considered in the contact between the upper plane and the upper fibres and the contact between the lower plane and the lower fibres. The results of this analysis are shown in Figure 33.

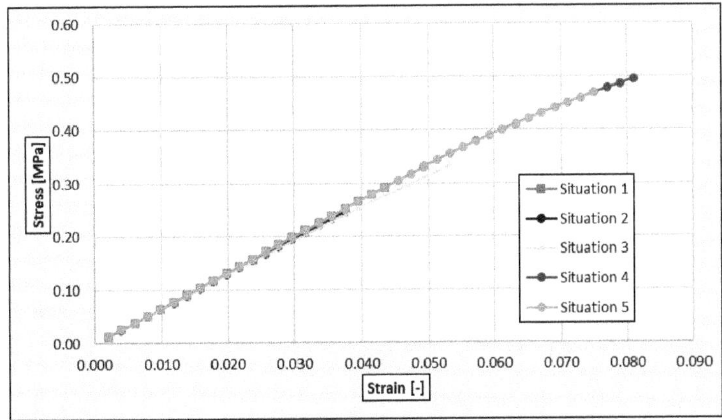

Figure 33. Influence of the existence of broken fibres on the stress–strain curve (G25_nanoHA 1.25).

The value of the compressive elastic modulus of a G25_nanoHA_1.25 scaffold for each of the five different situations considered in this analysis concerning the existence of broken fibres is shown in Table 14.

Table 14. Compressive elastic modulus of a G25_nanoHA_1.25 scaffold for different situations concerning the existence of broken fibres.

Situation (#)	Elastic Modulus (MPa)
Situation 1	6.70
Situation 2	6.57
Situation 3	6.43
Situation 4	6.70
Situation 5	6.70

As can be seen in Table 14, the compressive elastic modulus of the scaffold with no broken fibres (6.70 MPa) differs from the compressive elastic modulus of the reference scaffold considered in the previous sections (6.60 MPa). This difference lies in the fact that, as mentioned, the numerical models used in both cases differ regarding the friction coefficient used for the contacts between the upper and lower planes and the scaffold fibres.

According to these results, when the breakage of the fibres occurs far from any intersection between perpendicular fibres, there is no significant influence on the stress–strain relationship of a G25_nanoHA 1.25 scaffold subjected to a compression load, as the paths for the load transmission (columns formed by the intersections between perpendicular fibres) are not affected.

On the other hand, some influence on the stress–strain relationship of a G25_nanoHA_1.25 scaffold subjected to a compression load can be seen when the breakage of the fibres is located in the intersection between perpendicular fibres. In this case, it is expected that the longer the breakage is and the higher the number of affected fibres is, the bigger the influence is.

4.2. Design Safety Factor to Consider Uncertainty

In view of the Section 4.1 outcome, it is clear that the great differences in terms of mechanical behaviour summarized in Table 3 could be explained by the printing process uncertainty. Any model will be performed according to an idealized geometry, conditions and material properties but, in the end, assuming this idealized behaviour is unsafe because 3D printing bone scaffolds at such small-scale leads to several manufacturing mistakes, which become uncertainty sources.

Nevertheless, although it is not possible to forecast exact predictions on mechanical behaviour while unable to perform a perfect 3D-printed bone scaffold, it is possible to perform enough safe ones. This procedure is analogous to that used in purely structural or mechanical engineering for other non-medical applications, where the forecasts are performed running idealized models with reduction factors to consider execution mistakes, material properties uncertainties, lack of homogeneity, etc.

Thus, such reduction factors need to be balanced to be conservative enough but not too penalizing for the structure and the building process. Therefore, several tolerances are defined for each execution variable that need to be checked and controlled during structure execution or, analogously in this case, for bone scaffold 3D printing. Hence, each variable is defined by a mean-measured value, μ, and a standard deviation, σ, assuming a Gaussian distribution according to central limit theorem, since each variable is dependent on several uncontrolled factors, and defining the check and control intensity in a way that there is a small probability, typically 5%, of being under a lower boundary limit or exceeding an upper bound with a reliability level typically of 75%. As an additional comment, these fifth percentile values are known as characteristic values.

Accordingly, each uncertainty source is defined as a random variable x_1, x_2, ... , x_9 and, since the influence of each isolated random variable on the elastic modulus is linear dependent and biunivocal. the value of each random variable corresponds with an uncertainty factor, F_i, that is defined as the quotient between the actual elastic modulus and the theoretical one if 3D printed perfectly. For instance, the fibre diameter is the variable x_3, the nozzle diameter is 0.25 mm, and it theoretically prints at 0.25 mm but, depending on the printing speed, the fibre diameter could be lower. Thus, if the 3D printing process was perfect, then the fibre diameter would be 0.25 mm and the elastic modulus would be E = 6.6 MPa with a corresponding uncertainty factor F_3 = 6.6/6.6 = 1.0. However, if the printing speed was higher, then the fibre diameter would be lower instead and, in the case that it was even 0.22 mm, then the elastic modulus would be E = 3.39 MPa and the corresponding factor F_3 = 3.39/6.6 = 0.5136. Thus, because of the linear dependency, F_3 is a random variable with a Gaussian distribution matching x_3. Then, if we define a diameter tolerance and perform an execution control to guarantee that the fibre diameter is kept between 0.22 and 0.25 mm, with such an intensity that only 5% of the bone scaffolds present fibres below 0.22 mm and only 5% over 0.25 mm. Analogously, a mean, μ_i, and a standard deviation, σ_i, can be defined for each isolated uncertainty factor F_i, considering that in any Gaussian distribution, 90% of the probability is within the mean minus 1.6449 times the standard deviation and plus it. Hence, Table 15 summarizes the mean, μ_i, and standard deviation, σ_i, of each isolated uncertainty factor F_i, with corresponding limit values, $F_{i,min}$ and $F_{i,max}$, for the interval.

Table 15. Uncertainty factors F_i and corresponding mean μ_i and standard deviations σ_i, whose minimum and maximum values correspond to percentiles 5 and 95, respectively.

Variable (x_i)	Source (-)	$F_{i,min}$ (#)	$F_{i,max}$ (#)	μ_i (#)	σ_i (#)
x_1	Poisson's Ratio	0.9909	1.0000	0.9955	0.0028
x_2	Position of the fibres	0.9485	1.0000	0.9742	0.0157
x_3	Fibre Diameter	0.5136	1.0000	0.7568	0.1478
x_4	Strand Distance	0.9955	1.0000	0.9977	0.0014
x_5	Layer Height	0.6167	1.0000	0.8083	0.1165
x_6	Sample Diameter	0.9545	1.0530	1.0038	0.0299
x_7	Deflection of the fibres	0.9712	1.0000	0.9856	0.0088
x_8	Straightness of the columns	0.5955	1.0000	0.7978	0.1230
x_9	Existence of broken fibres	0.9597	1.0000	0.9799	0.0122

Thus, the next Figure 34 shows the probability density functions $f_i(x_i)$ of each Gaussian random variable thus defined. For the sake of simplicity, each variable x_i directly represents the uncertainty factor F_i = $E(x_i)/E(x_{theoretical})$ defined as the quotient between the actual elastic modulus and the theoretical one if 3D printed perfectly.

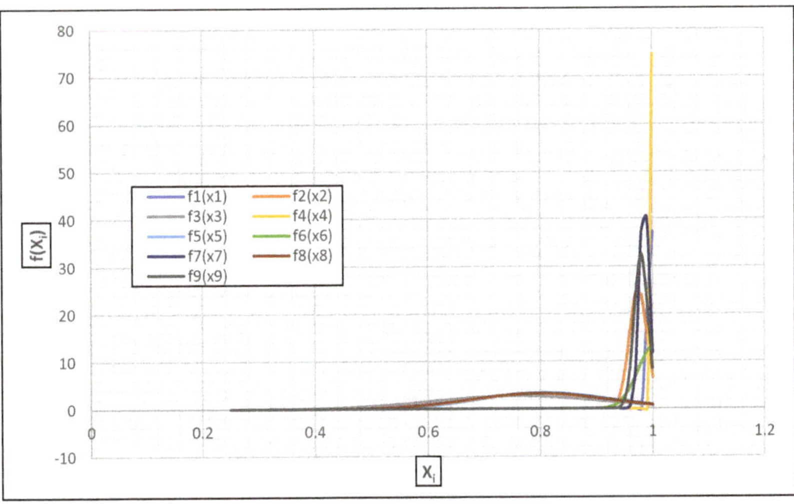

Figure 34. Gaussian density function of each uncertainty reduction factor.

The mathematical expression of each Gaussian probability density function corresponds to Equation (4), as follows:

$$f_i(x_i) = \frac{1}{\sigma_i \cdot \sqrt{2\pi}} \cdot e^{-\frac{(x_i - \mu_i)^2}{2 \cdot \sigma_i^2}} \tag{4}$$

Hence, considering each isolated uncertainty variable as independent from each other, the probability density function of the multiplication is the multiplication of the independent probability density functions, see Equation (5) as follows:

$$f(x_1 \cdot x_2 \cdots x_9) = f_1(x_1) \cdot f_2(x_2) \cdots f_9(x_9) \tag{5}$$

Therefore, to derive the probability density function of the multiplication it is required a variable change. Accordingly, such variable change is shown in Equation (6), where u is the multiplication of each N = 9 variables x_i and there are other N-1 variables z_i.

$$\begin{cases} u = x_1 \cdot x_2 \cdots x_9 \\ z_1 = x_1 \\ z_2 = x_2 \\ \vdots \\ z_8 = x_8 \end{cases} \tag{6}$$

Accordingly, the Jacobian determinant to execute the variable change is derived as the partial derivation of the variable multiplication by each variable, see Equation (7).

$$\frac{\partial(x_1 \cdot x_2 \cdots x_9)}{\partial(u, z_1 \cdot z_2 \cdots z_8)} = \begin{vmatrix} \frac{\partial x_1}{\partial u} & \frac{\partial x_1}{\partial z_1} & \cdots & \frac{\partial x_1}{\partial z_8} \\ \frac{\partial x_2}{\partial u} & \frac{\partial x_2}{\partial z_1} & \cdots & \frac{\partial x_2}{\partial z_8} \\ \vdots & \vdots & \vdots & \vdots \\ \frac{\partial x_9}{\partial u} & \frac{\partial x_9}{\partial z_1} & \cdots & \frac{\partial x_9}{\partial z_8} \end{vmatrix} = \begin{vmatrix} 0 & 1 & \cdots & \frac{\partial x_1}{\partial z_8} \\ 0 & 0 & \cdots & \frac{\partial x_2}{\partial z_8} \\ \vdots & \vdots & \ddots & \vdots \\ \frac{1}{z_1 \cdot z_2 \cdots z_8} & \frac{-u}{z_1^2 \cdot z_2 \cdots z_8} & \cdots & \frac{-u}{z_1 \cdot z_2 \cdots z_8^2} \end{vmatrix} = \frac{1}{z_1 \cdot z_2 \cdots z_8} \tag{7}$$

Thus, the probability density function of the nine variables, once changed, is shown in Equation (9), as follows:

$$g(u, z_1, \cdots, z_8) = f_1(z_1) \cdot f_2(z_2) \cdots f_9\left(\frac{u}{z_1 \cdot z_2 \cdots z_9}\right) \cdot \frac{1}{z_1 \cdot z_2 \cdots z_8} \tag{8}$$

In order to obtain the probability density function, the marginal distribution for variable u, which is obtained by the direct integration of the remaining variables z_1 to z_8, is required. See Equation (9) as follows:

$$g(u) = \iiint \cdots \int_{R^8} \left[f_1(z_1) \cdot f_2(z_2) \cdots f_9\left(\frac{u}{z_1 \cdot z_2 \cdots z_9}\right) \cdot \frac{1}{z_1 \cdot z_2 \cdots z_8} \right] \cdot dz_1 \cdot dz_2 \cdots dz_8 \tag{9}$$

As a last comment on Equation (9), it is worthy to remember that each $f_i(x_i)$ is a normal Gaussian distribution according to Equation (4) and that, despite being multiplied by the Jacobian fraction at the end, it is not directly integrable. Therefore, any solution can be obtained only by complex numerical means. Nevertheless, there is another simpler, more practical approach, considering that the distribution of the multiplication of Gaussian distributions is a Gaussian distribution itself. Then, it is possible to derive the mean and the standard deviation simply by taking into account their properties and their relationship with the expected value, E(x), and variance, Var(x).

4.2.1. Derivation of the Global Mean

The expected value corresponds with the first order moment of the probability distribution. Therefore, the expected value of the multiplication of variables is the ninth integration of the probability density function of the multiplication multiplied by each variable x_i, between $-\infty$ and ∞. See Equation (10), which is as follows:

$$E[\hat{x}_1 \cdot \hat{x}_2 \cdots \hat{x}_9] = \iiint \cdots \int_{R^9} \left[x_1 \cdot x_2 \cdots x_9 \cdot f_{\hat{x}_1 \cdot \hat{x}_2 \cdots \hat{x}_9}(x_1 \cdot x_2 \cdots x_9) \cdot dx_1 \cdot dx_1 \cdots dx_9 \right] \tag{10}$$

Nevertheless, as the probability density function of the multiplication for a set of independent variables is the product of the isolated probability density function of each variable, see Equation (5), then the Equation (10) can be developed into Equation (11), which is as follows:

$$E[\hat{x}_1 \cdot \hat{x}_2 \cdots \hat{x}_9] = \iiint \cdots \int_{R^9} \left[x_1 \cdot x_2 \cdots x_9 \cdot f_{\hat{x}_1}(x_1) \cdot f_{\hat{x}_2}(x_2) \cdots f_{\hat{x}_9}(x_9) \cdot dx_1 \cdot dx_1 \cdots dx_9 \right] \tag{11}$$

Accordingly, as every integral is independent from each other, then they can be regrouped as in the following Equation (12), presenting them as the product of integrals:

$$E[\hat{x}_1 \cdot \hat{x}_2 \cdots \hat{x}_9] = \int_{-\infty}^{\infty} x_1 \cdot f_{\hat{x}_1}(x_1) \cdot dx_1 \cdot \int_{-\infty}^{\infty} x_2 \cdot f_{\hat{x}_2}(x_2) \cdot dx_2 \cdots \int_{-\infty}^{\infty} x_9 \cdot f_{\hat{x}_9}(x_9) \cdot dx_9 \tag{12}$$

Thus, remembering that the mean is precisely the first order moment of a probability distribution, see Equation (13), which is as follows:

$$\int_{-\infty}^{\infty} x_i \cdot f_{\hat{x}_i}(x_i) \cdot dx_i = \mu_i \tag{13}$$

Then, by direct substitution, the mean of any variable multiplication is the product of the isolated variable means, see Equation (14), which is as follows:

$$E[\hat{x}_1 \cdot \hat{x}_2 \cdots \hat{x}_9] = \mu_1 \cdot \mu_2 \cdot \mu_3 \cdots \mu_9 \tag{14}$$

4.2.2. Derivation of the Global Standard Deviation

Regarding the variance of the variable multiplication, it corresponds to the expected value of the squared product minus the squared expected value of the product, see (15), which is as follows:

$$VAR(x_1 \cdot x_2 \cdot x_3 \cdots x_9) = E\left[(x_1 \cdot x_2 \cdot x_3 \cdots x_9)^2\right] - E([x_1 \cdot x_2 \cdot x_3 \cdots x_9])^2 \tag{15}$$

Now, making use of the expected value properties, the expected value of the product of variables is the multiplication of the expected value of such variables. Therefore, substituting in Equation (15) there it comes Equation (16), which is as follows:

$$VAR(x_1 \cdot x_2 \cdot x_3 \cdots x_9) = E\left[x_1^2\right] \cdot E\left[x_2^2\right] \cdots E\left[x_9^2\right] - (E[x_1])^2 \cdot (E[x_2])^2 \cdots (E[x_9])^2 \tag{16}$$

Now, considering that the expected value of a variable is the mean value $E[x_i] = \mu_i$, the expected value of the squared variable corresponds to Equation (17), which is as follows:

$$E\left[x_i^2\right] = \left(\mu_i^2 + \sigma_i^2\right) \tag{17}$$

Then, by a simple substitution in Equation (16), Equation (18) is then derived as follows:

$$VAR(x_1 \cdot x_2 \cdot x_3 \cdots x_9) = \left(\mu_1^2 + \sigma_1^2\right) \cdot \left(\mu_2^2 + \sigma_2^2\right) \cdots \left(\mu_9^2 + \sigma_9^2\right) - \mu_1^2 \cdot \mu_2^2 \cdots \mu_9^2 \tag{18}$$

4.2.3. Derivation of the Safety Factor

Thus, according to previous Equations (14) and (18), when substituting the mean and standard deviation values for the uncertainty factors, taken as isolated variables, which are summarized Table 15, the mean and standard deviation of the product of variables is derived. See Equations (19) and (20), which are as follows:

$$\mu = E[\hat{x}_1 \cdot \hat{x}_2 \cdots \hat{x}_9] \approx 0.4578 \tag{19}$$

$$\sigma = \sqrt{\mathrm{VAR}(x_1 \cdot x_2 \cdot x_3 \cdots x_9)} \approx 0.1346 \tag{20}$$

Therefore, with an execution control of each isolated uncertainty source, keeping 95% of the bone scaffolds within the required intervals for the fibre diameter, layer height, etc., then the combined outcome in terms of the global uncertainty factor will show a Gaussian distribution (Figure 35). In this figure, the mean value is 0.4578, while the characteristic value corresponding to fifth percentile (shadowed area) is 0.2364.

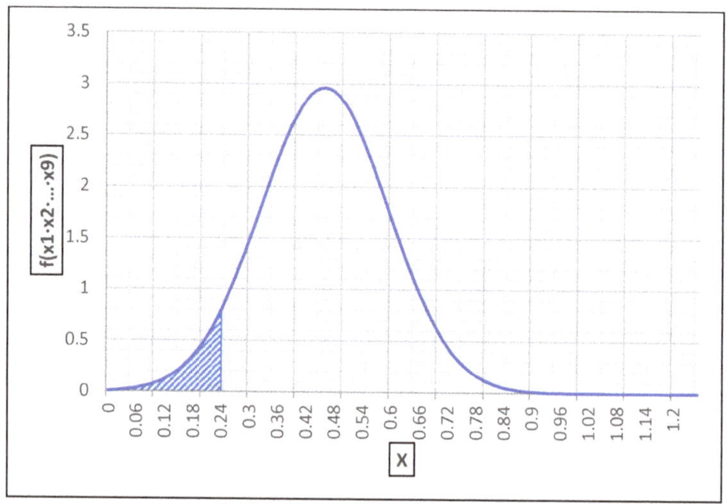

Figure 35. Gaussian density function of global uncertainty factor derived from variable product.

Consequently, there are two reduction factors derived from uncertainty to be applied to the elastic modulus directly obtained from the bulk material as correction factors for the scaffolds modelling. The first is derived from the mean value and shall be used for FEM simulations where the accuracy in terms of deformations and displacements are the main outcome, and the second is derived from the characteristic value for conservative forecasts that shall be used for FEM simulations where the main outcome is a safe enough prediction in terms of stress or an upper bound of deformations. Thus, the design value for the global elastic or Young's modulus after printing E_d can be derived from the elastic modulus obtained from models with idealized or theoretical conditions, E_M, and material input obtained from bulk material testing, but with a reduction material printing factor as a design safety factor γ_{MP}, see Equation (21). Thus, for accurate simulations regarding deformations it takes 2.2 as value and for safer conservative simulations regarding stresses or the upper bound of local deformations it takes 4.25 instead, see Equations (22) and (23), which are as follows:

$$E_d = \frac{E_M}{\gamma_{MP}} \tag{21}$$

$$E_d = 0.4578 \cdot E_M \approx \frac{E_M}{2.2} \tag{22}$$

$$E_d = 0.2364 \cdot E_M \approx \frac{E_M}{4.25} \tag{23}$$

Additionally, this design safety factor can be lowered if the execution control becomes more intense; thus 99% of the bone scaffolds presents the variables within the defined tolerance limits instead 95%, or if the intensity remains the same but the tolerance limits are lowered to ensure that geometry deviations are mitigated and derived, the uncertainty factors are lowered accordingly. The fibre diameter, layer height and straightness of the columns are key uncertainty sources affecting the bone scaffold behaviour. Therefore, the best strategy to improve the resulting scaffold behaviour is to improve the execution control and reduce the tolerances in these three variables.

4.2.4. Use Case

For instance, let us have a use case analogous to that presented in Figure 1, where a hollow cylindric scaffold made of G25 nanoHA 1.25, G25 nanoHA 1.00 or G25 nanoHA 0.75 is substituting some damaged length of a femur for tissue regeneration, as representative of a uniaxial loading case. The specific patient body weight is around 75 kg; therefore, the scaffold section must face F = 735 N of loading peaks. If the cylindric scaffold length was 20 mm, the external diameter was 50 mm and the inner diameter was 10 mm, then the cross-sectional area would be A = 1885 mm^2. Thus, the axial apparent stress of the scaffold would be σ_x = F/A = −0.39 MPa, with σ_y = σ_z = 0 Mpa, since this is a monoaxial loading and a cylindric loading state, and the axial strain ε_x is, according to Hooke's Equation (24) and taking into account the Poisson coefficients μ_{yx} and μ_{zx}, as follows:

$$\varepsilon_x = \frac{\sigma_x}{E_x} - \mu_{yx} \cdot \frac{\sigma_y}{E_y} - \mu_{zx} \cdot \frac{\sigma_z}{E_z} = \frac{\sigma_x}{E_x} \tag{24}$$

Now, using the apparent young modulus obtained from the idealized models $E_{A,FEM}$ already presented in Table 3 without any correction factor, the mean young modulus, $E_{m,d}$, does not present any change. Thus, the mean strain ε_m is obtained by direct application of Equation (24). Then, the displacement δ is obtained by multiplying the strain by the scaffold length. Moreover, the real young modulus, E_{FEM}, taking into account the real stresses of solid 187 elements at the scaffold, multiplied by the strain ε_m considered as uniform, gives the real foreseeable stress, σ_{FEM}. This calculation process for each scaffold type is presented in Table 16.

Table 16. Derivation of displacement δ and stress σ at hollow cylindric scaffold within a femur bone without correction factor (γ_{MP} = 1).

	$E_{A,FEM}$	$E_{m,d}$	ε_m	δ	E_{FEM}	σ_{FEM}
	(MPa)	(MPa)	(-)	(mm)	(MPa)	(MPa)
G25 nanoHA 1.25	6.6	6.6	0.059	1.182	566.2	33.4
G25 nanoHA 1.00	7.9	7.9	0.049	0.987	730.7	36.1
G25 nanoHA 0.75	15.49	15.5	0.025	0.503	563.1	14.2

Now, if the design limits were σ_{max} = 40 MPa and δ_{max} = 1.5 mm, then the selected scaffold type could be G25 nanoHA 0.75 or G25 nanoHA 1.25, as they are the only ones fulfilling both requirements regarding the maximum displacement and stress. Nevertheless, if the scaffold was finally made of such materials, considering that the realistic scaffold will present lower stiffness due to several uncertain sources, then the calculation needs to be readjusted, taking into account the mean expectable behaviour of the scaffold (Table 17). This can be made by applying the reduction factor γ_{MP} = 2.2 to derive a mean Young's modulus for design $E_{m,d}$ and repeating the calculations with this adjusted elasticity. As can be seen, by calculating it this way, the only scaffold type left fulfilling both requirements is G25 nanoHA 0.75.

Table 17. Derivation of displacement δ and stress σ at hollow cylindric scaffold within a femur bone with mean correction factor ($\gamma_{MP} = 2.2$).

	$E_{A,FEM}$	$E_{m,d}$	ε_m	δ	E_{FEM}	σ_{FEM}
	(MPa)	(MPa)	(-)	(mm)	(MPa)	(MPa)
G25 nanoHA 1.25	6.6	3.0	0.130	2.600	566.2	73.6
G25 nanoHA 1.00	7.9	3.6	0.109	2.172	730.7	79.3
G25 nanoHA 0.75	15.49	7.0	0.055	1.108	563.1	31.2

However, while conducting the calculations this way is a correct procedure to forecast displacements depending on a mean scaffold behaviour, it is not a safe enough procedure to consider the mechanical failure due to stress, since a crack starting at certain spot of the scaffold due to a local weakness could propagate, causing a global structural failure. This is the reason why the calculations on the resistance take a characteristic value instead, i.e., the fifth percentile or the value that is exceeded 19 times of every 20 measurements. This guarantees that there is a certain amount of resistance left to compensate for local weaknesses. Moreover, it can be made by applying the reduction factor $\gamma_{MP} = 4.25$ instead to derive a characteristic Young's modulus for design $E_{k,d}$ and repeating the calculations with this adjusted elasticity. As can be seen, calculating it this way discards the three initial designs, since the three scaffolds present a stress over 40 MPa (Table 18).

Table 18. Derivation of displacement δ and stress σ at hollow cylindric scaffold within a femur bone with characteristic correction factor ($\gamma_{MP} = 4.25$).

	$E_{A,FEM}$	$E_{k,d}$	ε_k	δ	E_{FEM}	σ_{FEM}
	(MPa)	(MPa)	(-)	(mm)	(MPa)	(MPa)
G25 nanoHA 1.25	6.6	1.6	0.251	5.022	566.2	142.2
G25 nanoHA 1.00	7.9	1.9	0.210	4.195	730.7	153.3
G25 nanoHA 0.75	15.49	3.6	0.107	2.140	563.1	60.2

The outcome of this evidence is that, if we cannot reduce the loadings, we need to increase the effective cross-sectional area to face them. This can be made by increasing the external diameter or reducing the inner diameter of the scaffold, but most of the time it is not possible in a patient customized geometry; therefore, the only alternatives are as follows:

- Increasing the execution control, enabling to reduce the factor γ_{MP};
- Improving the material, pushing forward the stress limits and the global stiffness;
- Increasing the fibre diameter and/or reducing the strand distance, thus increasing the column area.

Reducing the strand distance from 0.75 to 0.5 leads to an increase in the amount of "resisting columns", from 21 to 45; this results in an expectable stress of 37.47 MPa, taking into account the previous results in columns made of spatially crossed cylinders of the same geometry. Moreover, increasing the fibre diameter instead, to pass from 60.2 to 40 MPa, will require a cross sectional augmentation by a scale factor equal to the square root of the quotient $F = (60.2/40)^{0.5}$, giving a 0.307-mm diameter. Nevertheless, the diameter augmentation reduces the fibre cross-sectional curvature, increasing contact in a non-linear basis; therefore, it should be enough to increase the diameter up to 0.3 mm.

Thus, this patient will require a G25 nanoHA 0.75 with a fibre diameter of 0.3 mm and a G25 nanoHA 0.5 with fibre diameter of 0.25 mm, both produced from 55% PEOT/PBT + 45% nanoHA material or, finally, a G25 nanoHA 0.75 with 0.25 mm fibre diameter produced from an improved material.

5. Conclusions

Thus, comparing the results of finite element modelling to the bone scaffold compression tests according to considered methodology, and once the uncertainty sources derived from manufacturing process are analysed during discussion, the following conclusions can be summarized:

A methodology to perform FEM of 3D-printed bone scaffolds of any initial bulk material properties and geometry is defined. Moreover, it is experimentally calibrated and validated by direct testing of the scaffolds. The bone scaffold is then represented as an equivalent finite element to enable the FEM of bone tissue discretized with a compatible mesh, enabling the required characteristics to be studied, depending on bone topology.

The methodology is executed for a 55% PEOT/PBT + 45% nanoHA bulk material and three different scaffold geometries. The results coming from the FEM and the monotonic compression test of the bone scaffolds show significative differences, being that the idealized simulation is always better than the actual bone scaffolds. Considering the tests' mean values and standard deviations, the differences cannot be explained by result scattering only.

A deep look at the scaffolds show several fabrication imprecisions, meaning a resulting geometry far different from the theoretical one considered for FEM, such as different strand distance, layer height, fibre diameter, Poisson's ratio, fibre position, sample diameter, straightness of the columns, existence of broken fibres. This is due to the small imprecisions adding up layer-by-layer as large clinically relevant scaffolds (i.e., volume > 1 cm^3) are being built. Thus, there are up to nine uncertainty sources detected in scaffolds coming from different variables.

Consequently, a sensitivity analysis is performed for each uncertainty source, deriving the influence or uncertainty factor and reducing the expected behaviour of each isolated source in terms of scaffold stiffness. Then, some tolerances are defined for each variable to be controlled during execution, with such an intensity that every variable presents a 95% probability of being within the selected interval in a Gaussian distribution.

The two main effects that can give strong changes to the elastic modulus are the strand overlapping and the fibre diameter. The first can be accommodated by design planning the shifting of the planes while printing, and the second may be modulating the printing speed, these will contribute to palliate the effect and minimize the corresponding design factor. A global design safety factor is defined in a predictive and conservative scenario, combining each isolated uncertainty factor in a single value taken from the mean value and fifth percentile of the Gaussian distribution of the product of each reducing uncertainty factor.

Thus, the aprioristic FEM of a bone tissue region with the inclusion of bone scaffold volume made of a certain bulk material becomes possible simply by using solid 185 equivalent finite elements to mesh it, assigning as material properties the ones derived from the FEM simulation of the monotonic tests of the scaffold, divided by the global design safety factor. This enables the further topological optimization of the scaffold along the bone tissue and the material selection, depending on the requirements.

Author Contributions: Conceptualization and methodology, I.C.-U.-A.; finite element simulations, S.P.; research on bulk materials and selection, A.S. and S.V.; experimental tests and validation, A.S. and S.V.; formal analysis and investigation, I.C.-U.-A. and A.M; providing resources and scaffolds for testing, R.S. and M.C.-T.; data curation, A.S., S.V. and S.P.; writing—original report, S.P.; writing— manuscript, I.C.-U.-A.; writing—review and editing, I.C.-U.-A., R.S. and L.M.; visualization, A.M. and C.M.; supervision, I.C.-U.-A.; project administration, A.S., A.P. and L.M.; funding acquisition, C.M., A.P. and L.M. All authors have read and agreed to the published version of the manuscript.

Funding: This research was funded by the European Union, represented by the European Commission, grant number 685825-FAST-H2020-NMP-2014-2015/H2020-NMP-PILOTS-2015.

Institutional Review Board Statement: Not applicable.

Informed Consent Statement: Not applicable.

Data Availability Statement: The data presented in this study are available on request from the corresponding authors.

Acknowledgments: Authors would like to acknowledge Miguel Calderon Felices for his help and scientific advice regarding the statistical development of the design safety factor.

Conflicts of Interest: The authors declare no conflict of interest.

References

1. Stevens, M.M. Biomaterials for Bone Tissue Engineering. *Mater. Today* **2008**, *11*, 18–25. [CrossRef]
2. Fernandez-Yague, M.A.; Abbah, S.A.; McNamara, L.; Zeugolis, D.I.; Pandit, A.; Biggs, M.J. Biomimetic Approaches in Bone Tissue Engineering: Integrating Biological and Physicomechanical Strategies. *Adv. Drug Deliv. Rev.* **2015**, *84*, 1–29. [CrossRef]
3. Rezwan, K.; Chen, Q.Z.; Blaker, J.J.; Boccaccini, A.R. Biodegradable and Bioactive Porous Polymer/Inorganic Composite Scaffolds for Bone Tissue Engineering. *Biomaterials* **2006**, *27*, 3413–3431. [CrossRef]
4. Jakus, A.E.; Rutz, A.L.; Jordan, S.W.; Kannan, A.; Mitchell, S.M.; Yun, C.; Koube, K.D.; Yoo, S.C.; Whiteley, H.E.; Richter, C.-P.; et al. Hyperelastic "Bone": A Highly Versatile, Growth Factor–Free, Osteoregenerative, Scalable, and Surgically Friendly Biomaterial. *Sci. Transl. Med.* **2016**, *8*, 358ra127. [CrossRef]
5. Karageorgiou, V.; Kaplan, D. Porosity of 3D Biomaterial Scaffolds and Osteogenesis. *Biomaterials* **2005**, *26*, 5474–5491. [CrossRef]
6. Hutmacher, D.W. Scaffolds in Tissue Engineering Bone and Cartilage. *Biomaterials* **2000**, *21*, 2529–2543. [CrossRef]
7. Lee, W.C.; Lim, C.H.Y.X.; Shi, H.; Tang, L.A.L.; Wang, Y.; Lim, C.T.; Loh, K.P. Origin of Enhanced Stem Cell Growth and Differentiation on Graphene and Graphene Oxide. *ACS Nano* **2011**, *5*, 7334–7341. [CrossRef] [PubMed]
8. Chung, C.; Kim, Y.-K.; Shin, D.; Ryoo, S.-R.; Hong, B.H.; Min, D.-H. Biomedical Applications of Graphene and Graphene Oxide. *Acc. Chem. Res.* **2013**, *46*, 2211–2224. [CrossRef] [PubMed]
9. Raucci, M.G.; Giugliano, D.; Longo, A.; Zeppetelli, S.; Carotenuto, G.; Ambrosio, L. Comparative Facile Methods for Preparing Graphene Oxide-Hydroxyapatite for Bone Tissue Engineering. *J. Tissue Eng. Regen. Med.* **2017**, *11*, 2204–2216. [CrossRef]
10. Badar, M.; Rahim, M.I.; Kieke, M.; Ebel, T.; Rohde, M.; Hauser, H.; Behrens, P.; Mueller, P.P. Controlled Drug Release from Antibiotic-Loaded Layered Double Hydroxide Coatings on Porous Titanium Implants in a Mouse Model. *J. Biomed. Mater. Res. A* **2015**, *103*, 2141–2149. [CrossRef] [PubMed]
11. Ai, J.; Chen, Y.; Jing, H.; Gao, X.; Liu, J.; Ma, K.; Suo, J.; Yu, S.; Song, S. Dynamic Release of Antibiotic Drug Gentamicin Sulfate from Novel Zirconium Phosphate Nano-Platelets. *Sci. Adv. Mater.* **2014**, *6*, 2603–2610. [CrossRef]
12. Rodriguez-Losada, N.; Wendelbo, R.; Garcia-Fernandez, M.; Pavia, J.; Martinez-Montañez, E.; Lara-Muñoz, J.P.; Arenas, E.; Aguirre-Gomez, J.A. Graphene Derivative Scaffolds Facilitate in Vitro Cell Survival and Maturation of Dopaminergic SN4741 Cells. *Acta Physiol.* **2014**, *212*, 69. [CrossRef]
13. Fiorillo, M.; Verre, A.F.; Iliut, M.; Peiris-Pagés, M.; Ozsvari, B.; Gandara, R.; Cappello, A.R.; Sotgia, F.; Vijayaraghavan, A.; Lisanti, M.P. Graphene Oxide Selectively Targets Cancer Stem Cells, across Multiple Tumor Types: Implications for Non-Toxic Cancer Treatment, via "Differentiation-Based Nano-Therapy". *Oncotarget* **2015**, *6*, 3553–3562. [CrossRef]
14. Butscher, A.; Bohner, M.; Roth, C.; Ernstberger, A.; Heuberger, R.; Doebelin, N.; von Rohr, P.R.; Müller, R. Printability of Calcium Phosphate Powders for Three-Dimensional Printing of Tissue Engineering Scaffolds. *Acta Biomater.* **2012**, *8*, 373–385. [CrossRef]
15. Butscher, A.; Bohner, M.; Hofmann, S.; Gauckler, L.; Müller, R. Structural and Material Approaches to Bone Tissue Engineering in Powder-Based Three-Dimensional Printing. *Acta Biomater.* **2011**, *7*, 907–920. [CrossRef] [PubMed]
16. Moroni, L.; de Wijn, J.R.; van Blitterswijk, C.A. 3D Fiber-Deposited Scaffolds for Tissue Engineering: Influence of Pores Geometry and Architecture on Dynamic Mechanical Properties. *Biomaterials* **2006**, *27*, 974–985. [CrossRef]
17. Woodfield, T.B.F.; Malda, J.; de Wijn, J.; Péters, F.; Riesle, J.; van Blitterswijk, C.A. Design of Porous Scaffolds for Cartilage Tissue Engineering Using a Three-Dimensional Fiber-Deposition Technique. *Biomaterials* **2004**, *25*, 4149–4161. [CrossRef] [PubMed]
18. Tjong, S.C. Structural and Mechanical Properties of Polymer Nanocomposites. *Mater. Sci. Eng. R Rep.* **2006**, *53*, 73–197. [CrossRef]
19. Paul, D.R.; Robeson, L.M. Polymer Nanotechnology: Nanocomposites. *Polymer* **2008**, *49*, 3187–3204. [CrossRef]
20. Schieker, M.; Seitz, H.; Drosse, I.; Seitz, S.; Mutschler, W. Biomaterials as Scaffold for Bone Tissue Engineering. *Eur. J. Trauma* **2006**, *32*, 114–124. [CrossRef]
21. Pavlidou, S.; Papaspyrides, C.D. A Review on Polymer–Layered Silicate Nanocomposites. *Prog. Polym. Sci.* **2008**, *33*, 1119–1198. [CrossRef]
22. Alexandre, M.; Dubois, P. Polymer-Layered Silicate Nanocomposites: Preparation, Properties and Uses of a New Class of Materials. *Mater. Sci. Eng. R Rep.* **2000**, *28*, 1–63. [CrossRef]
23. Potts, J.R.; Dreyer, D.R.; Bielawski, C.W.; Ruoff, R.S. Graphene-Based Polymer Nanocomposites. *Polymer* **2011**, *52*, 5–25. [CrossRef]
24. Liu, Q.; de Wijn, J.R.; van Blitterswijk, C.A. Composite Biomaterials with Chemical Bonding between Hydroxyapatite Filler Particles and PEG/PBT Copolymer Matrix. *J. Biomed. Mater. Res.* **1998**, *40*, 490–497. [CrossRef]
25. Liu, Q.; de Wijn, J.R.; de Groot, K.; van Blitterswijk, C.A. Surface Modification of Nano-Apatite by Grafting Organic Polymer. *Biomaterials* **1998**, *19*, 1067–1072. [CrossRef]
26. Liu, Q.; de Wijn, J.R.; van Blitterswijk, C.A. Nano-Apatite/Polymer Composites: Mechanical and Physicochemical Characteristics. *Biomaterials* **1997**, *18*, 1263–1270. [CrossRef]

27. Liu, O.; Du Wijn, J.R.; Van Blitterswijk, C.A. Intermolecular complexation between peg/pbt block copolymer and polyelectrolytes polyacrylic acid and maleic acid copolymer. *Eur. Polym. J.* **1997**, *33*, 1041–1047. [CrossRef]
28. Liu, Q.; De Wijn, J.R.; Bakker, D.; Van Blitterswijk, C.A. Surface Modification of Hydroxyapatite to Introduce Interfacial Bonding with PolyactiveTM 70/30 in a Biodegradable Composite. *J. Mater. Sci. Mater. Med.* **1996**, *7*, 551–557. [CrossRef]
29. Munarin, F.; Petrini, P.; Gentilini, R.; Pillai, R.S.; Dirè, S.; Tanzi, M.C.; Sglavo, V.M. Micro- and Nano-Hydroxyapatite as Active Reinforcement for Soft Biocomposites. *Int. J. Biol. Macromol.* **2015**, *72*, 199–209. [CrossRef]
30. Nandakumar, A.; Cruz, C.; Mentink, A.; Birgani, Z.T.; Moroni, L.; van Blitterswijk, C.; Habibovic, P. Monolithic and Assembled Polymer—Ceramic Composites for Bone Regeneration. *Acta Biomater.* **2013**, *9*, 5708–5717. [CrossRef] [PubMed]
31. Fouad, H.; Elleithy, R.; Alothman, O.Y. Thermo-Mechanical, Wear and Fracture Behavior of High-Density Polyethylene/Hydroxyapatite Nano Composite for Biomedical Applications: Effect of Accelerated Ageing. *J. Mater. Sci. Technol.* **2013**, *29*, 573–581. [CrossRef]
32. Sinha, R.; Cámara-Torres, M.; Scopece, P.; Falzacappa, E.V.; Patelli, A.; Moroni, L.; Mota, C. A Hybrid Additive Manufacturing Platform to Create Bulk and Surface Composition Gradients on Scaffolds for Tissue Regeneration. *bioRxiv* **2020**, 165605. [CrossRef]
33. Olivares, A.L.; Marsal, È.; Planell, J.A.; Lacroix, D. Finite Element Study of Scaffold Architecture Design and Culture Conditions for Tissue Engineering. *Biomaterials* **2009**, *30*, 6142–6149. [CrossRef]
34. Miranda, P.; Pajares, A.; Guiberteau, F. Finite Element Modeling as a Tool for Predicting the Fracture Behavior of Robocast Scaffolds. *Acta Biomater.* **2008**, *4*, 1715–1724. [CrossRef]
35. Sandino, C.; Planell, J.A.; Lacroix, D. A Finite Element Study of Mechanical Stimuli in Scaffolds for Bone Tissue Engineering. *J. Biomech.* **2008**, *41*, 1005–1014. [CrossRef] [PubMed]
36. Milan, J.-L.; Planell, J.A.; Lacroix, D. Computational Modelling of the Mechanical Environment of Osteogenesis within a Polylactic Acid–Calcium Phosphate Glass Scaffold. *Biomaterials* **2009**, *30*, 4219–4226. [CrossRef]
37. Almeida, H.A. Design of Tissue Engineering Scaffolds Based on Hyperbolic Surfaces: Structural Numerical Evaluation. *Med. Eng.* **2014**, *8*, 1033–1040. [CrossRef] [PubMed]
38. Dinis, J.C.; Morais, T.F.; Amorim, P.H.J.; Ruben, R.B.; Almeida, H.A.; Inforçati, P.N.; Bártolo, P.J.; Silva, J.V.L. Open Source Software for the Automatic Design of Scaffold Structures for Tissue Engineering Applications. *Procedia Technol.* **2014**, *16*, 1542–1547. [CrossRef]
39. De Amorim Almeida, H.; da Silva Bártolo, P.J. Virtual Topological Optimisation of Scaffolds for Rapid Prototyping. *Med. Eng. Phys.* **2010**, *32*, 775–782. [CrossRef]
40. Cámara-Torres, M.; Duarte, S.; Sinha, R.; Egizabal, A.; Álvarez, N.; Bastianini, M.; Sisani, M.; Scopece, P.; Scatto, M.; Bonetto, A.; et al. 3D Additive Manufactured Composite Scaffolds with Antibiotic-Loaded Lamellar Fillers for Bone Infection Prevention and Tissue Regeneration. *Bioact. Mater.* **2021**, *6*, 1073–1082. [CrossRef]
41. Bastianini, M.; Scatto, M.; Sisani, M.; Scopece, P.; Patelli, A.; Petracci, A. Innovative Composites Based on Organic Modified Zirconium Phosphate and PEOT/PBT Copolymer. *J. Compos. Sci.* **2018**, *2*, 31. [CrossRef]
42. F04 Committee. *ASTM F2027-08, Standard Guide for Characterization and Testing of Raw or Starting Biomaterials for Tissue-Engineered Medical Products*; ASTM International: West Conshohocken, PA, USA, 2008.
43. D20 Committee. *ASTM D638-08, Standard Test Method for Tensile Properties of Plastics*; ASTM International: West Conshohocken, PA, USA, 2008.
44. D20 Committee. *ASTM D695-10, Standard Test Method for Compressive Properties of Rigid Plastics*; ASTM International: West Conshohocken, PA, USA, 2010.
45. Technical Committee: ISO/TC 61/SC 2 Mechanical Behavior. *ISO 527-1:2019 Plastics—Determination of Tensile Properties—Part 1: General Principles*; IOS: Geneva, Switzerland, 2019; p. 26.
46. Technical Committee: ISO/TC 61/SC 2 Mechanical Behavior. *ISO 527-2:2012 Plastics—Determination of Tensile Properties—Part 2: Test Conditions for Moulding and Extrusion Plastics*; IOS: Geneva, Switzerland, 2012; p. 11.
47. Technical Committee: ISO/TC 61/SC 13 Composites and Reinforcement Fibres. *ISO 527-4:1997 Plastics—Determination of Tensile Properties—Part 4: Test Conditions for Isotropic and Orthotropic Fibre-Reinforced Plastic Composites*; IOS: Geneva, Switzerland, 1997; p. 11.
48. F04 Committee. *ASTM F2150-02e1, Standard Guide for Characterization and Testing of Biomaterial Scaffolds Used in Tissue-Engineered Medical Products*; ASTM International: West Conshohocken, PA, USA, 2002.
49. Technical Committee: ISO/TC 61/SC 2 Mechanical Behavior. *ISO 604:2002 Plastics—Determination of Compressive Properties*; IOS: Geneva, Switzerland, 2002; p. 18.
50. Calderón-Urízar-Aldaca, I.; Biezma, M.V.; Matanza, A.; Briz, E.; Bastidas, D.M. Second-Order Fatigue of Intrinsic Mean Stress under Random Loadings. *Int. J. Fatigue* **2020**, *130*, 105257. [CrossRef]
51. Calderon-Uriszar-Aldaca, I.; Briz, E.; Biezma, M.V.; Puente, I. A Plain Linear Rule for Fatigue Analysis under Natural Loading Considering the Coupled Fatigue and Corrosion Effect. *Int. J. Fatigue* **2019**, *122*, 141–151. [CrossRef]
52. Calderon-Uriszar-Aldaca, I.; Biezma, M.V. A Plain Linear Rule for Fatigue Analysis under Natural Loading Considering the Sequence Effect. *Int. J. Fatigue* **2017**, *103*, 386–394. [CrossRef]

Article

Obtaining Expressions for Time-Dependent Functions That Describe the Unsteady Properties of Swirling Jets of Viscous Fluid

Eugene Talygin * and Alexander Gorodkov

Bakulev National Medical Research Center for Cardiovascular Surgery, 121552 Moscow, Russia; euegene.talygin@skmenergy.com
* Correspondence: skalolom@gmail.com

Abstract: Previously, it has been shown that the dynamic geometric configuration of the flow channel of the left heart and aorta corresponds to the direction of the streamlines of swirling flow, which can be described using the exact solution of the Navier–Stokes and continuity equations for the class of centripetal swirling viscous fluid flows. In this paper, analytical expressions were obtained. They describe the functions $C_0(t)$ and $\Gamma_0(t)$, included in the solutions, for the velocity components of such a flow. These expressions make it possible to relate the values of these functions to dynamic changes in the geometry of the flow channel in which the swirling flow evolves. The obtained expressions allow the reconstruction of the dynamic velocity field of an unsteady potential swirling flow in a flow channel of arbitrary geometry. The proposed approach can be used as a theoretical method for correct numerical modeling of the blood flow in the heart chambers and large arteries, as well as for developing a mathematical model of blood circulation, considering the swirling structure of the blood flow.

Keywords: potential swirling flow; Navier–Stokes equations; unsteady swirling flow; tornado-like jets

Citation: Talygin, E.; Gorodkov, A. Obtaining Expressions for Time-Dependent Functions That Describe the Unsteady Properties of Swirling Jets of Viscous Fluid. *Mathematics* **2021**, *9*, 1860. https://doi.org/10.3390/math9161860

Academic Editor: Mauro Malvè

Received: 23 June 2021
Accepted: 29 July 2021
Published: 5 August 2021

Publisher's Note: MDPI stays neutral with regard to jurisdictional claims in published maps and institutional affiliations.

1. Introduction

In our previous work, we investigated the dynamic geometry of the left heart and aorta to find consistency between the configuration of the flow channel and the directions of the swirling streamlines of the TLJ (Tornado-Like Jets) class [1–3]. The found correspondence allowed us to assume that the blood flow in the heart and large vessels belong to the TLJ class. These flows were described using the explicit solutions [4], which determine the conditions for a swirling jet of Newtonian fluid in space to appear and evolve.

It has been experimentally proved that swirling flows of this class, under certain conditions, are formed on a concave surface streamlined by the viscous medium. Apparently, an important role in the formation of such a swirling flow is played by the vortex boundary layer arising on the concave surface. This layer should consist of some small-scale vortex structures such as Taylor–Görtler vortices and differ in their properties from the classical shear boundary layer of L. Prandtl. These vortices are cylindrical structures, the axis of rotation of which is parallel to the incoming flow. This shape allows the swirling flow to rely on these vortex structures, conjugating with them only at one point. In this case, the movement of the swirling flow relative to the concave surface causes the appearance of rolling stresses, which are much less than the shear stresses in the boundary layer of L. Prandtl. The geometric shape of the generatrix of the concave surface determines the shape of the streamlines of the swirling flow. In this case, the vortex boundary layer is able to change its thickness along the concave surface dynamically. This allows it to compensate for local inconsistencies between the real surface and the geometric shape built along streamlines. Such compensation is necessary when the swirling blood flow moves into a section of the vascular bed with pathological geometry disorders.

Obviously, blood has complex rheology and, in the general case, cannot be regarded as a Newtonian fluid. However, the available experimental data allow us to assert that in large-caliber vessels at relatively high speeds, the dynamic viscosity of blood hardly changes, and blood can be considered a Newtonian fluid.

Calculating the Reynolds number for the blood flow in the heart and great vessels, we obtain a value that allows us to make an unambiguous conclusion about the turbulent nature of the flow under consideration. However, a turbulent flow is characterized by active mixing of the liquid volume and spontaneous vortex formation. It is obvious that these properties of turbulent flows do not correspond to the physiological characteristics of blood circulation. Considering the blood flow as TLJ, the flow structure can be preserved without noticeable swirls and perturbations at such Reynolds numbers that, for other principles of flow organization, would lead to flow turbulization. However, it was not possible to take into account the nonstationarity blood flow because, in the exact solution, it is determined by arbitrary time-dependent functions that do not have a formal analytical record. At the same time, it is obvious that the real blood flow is strictly unsteady. A formal description of the unsteady properties of the swirling blood flow in terms of analytical functions would make it possible to construct a model of blood circulation with a higher degree of accuracy.

Earlier it was found that the geometric characteristics of the bloodstream (if we assume that they correspond to the geometric characteristics of the flow channel) are described with a high degree of accuracy by quasi-stationary solutions [1–4]. In these solutions, the non-stationary properties of the flow are determined by the behavior of the time-dependent functions $C_0(t)$ and $\Gamma_0(t)$. These functions could be determined experimentally if measuring the vector field of flow velocities. However, so far, making such a measurement with sufficient accuracy is rather difficult. The task is complicated by the fact that the bloodstream at all stages of its evolution interacts with the movable walls of the flow channel, taking the direction of movement given by the instantaneous geometric configuration of the channel. At the same time, the bloodstream is a submerged stream, and its structure changes upon interaction and merging with residual blood volume that retains a certain movement after the previous cardiac cycle.

In previous studies, we considered the swirling blood flow in the heart and great vessels as quasi-stationary—a continuous set of values of the functions $C_0(t)$ and $\Gamma_0(t)$ on a time interval was replaced by a discrete subset of values, each of which corresponded to a certain configuration of the flow channel. This made it possible to replace the functions $C_0(t)$ and $\Gamma_0(t)$ with a set of constants that reflect the nonstationarity behavior of the flow. The analysis of the dynamic geometry of the left heart and aorta confirmed this hypothesis; the values of some parameters of the swirling blood flow in this segment of blood circulation were calculated. However, the accuracy of the obtained results was limited by the resolution of discretization.

Obtaining formal relations for the functions $C_0(t)$ and $\Gamma_0(t)$ will make it possible to unambiguously relate the structure of an unsteady swirling flow with the conditions that form it—the dynamics of the inflowing blood flow, the geometry of the flow channel, and the interaction with residual volumes at each stage of the evolution of the bloodstream both in time and along the length of the flow channel.

Therefore, the aim of this work was to obtain formal mathematical expressions for the functions $C_0(t)$ and $\Gamma_0(t)$, which describe in general form the unsteadiness of the swirling flow. Due to the calculation difficulties, we could not validate the proposed equations with the experimental results. This work will be conducted in the next stage of our study.

The presented paper has the following structure:

- Introduction;
- Materials and Methods;
- Results;
- Discussion;
- Conclusion;

- References.

2. Materials and Methods

According to [1–4], velocity components of the TLJ in a cylindrical coordinate system are expressed by the following relations:

$$\begin{cases} u_z = 2C_0(t)z \\ u_r = -C_0(t)r \\ u_\varphi = \frac{\Gamma_0(t)}{2\pi r} * \left(1 - e^{-\frac{C_0(t)r^2}{2v}}\right) \end{cases} \tag{1a}$$

where u_z, u_r, and u_φ are the velocity components, v is the kinematic viscosity, and the unsteadiness and evolution of the flow are determined by the behavior of the time-dependent functions $C_0(t)$ and $\Gamma_0(t)$. $C_0(t)$ is an arbitrary function of time, which, by its meaning, represents the gradient of the longitudinal component of the velocity (sec^{-1}); $\Gamma_0(t)$ is an arbitrary function of time, corresponding to the physical meaning of the circulation of the medium (m^2/sec).

It can be seen from the presented evidence that the unsteady properties of the flow under study are determined by the behavior of the time-dependent functions $C_0(t)$ and $\Gamma_0(t)$. While obtaining relations for the swirling flow velocity (1a) from the original system of Navier–Stokes equations and continuity, the functions $C_0(t)$ and $\Gamma_0(t)$ were considered as arbitrary functions of time.

These ratios were analyzed using the methods of linear differential equations solving and standard methods of differentiation for chain functions with several arguments.

3. Results

To obtain an analytical form of the functions $C_0(t)$ and $\Gamma_0(t)$, relations (1a) were substituted into the Navier–Stokes equations.

For this, the equations for the three velocity components were considered separately in a cylindrical coordinate system.

Azimuthal velocity component of the Navier–Stokes equation can be expressed as follows [5]:

$$\frac{u_r u_\varphi}{r} + \frac{\partial u_\varphi}{\partial r}u_r + \frac{1}{r}\frac{\partial u_\varphi}{\partial \varphi}u_\varphi + \frac{\partial u_\varphi}{\partial z}u_z + \frac{\partial u_\varphi}{\partial t} = F_\varphi - \frac{1}{\rho r}\frac{\partial p}{\partial \varphi} + v\left(\Delta u_\varphi - \frac{u_\varphi}{r^2} + \frac{2}{r^2}\frac{\partial u_r}{\partial \varphi}\right)$$
$$\Delta = \frac{\partial^2}{\partial r^2} + \frac{1}{r}\frac{\partial}{\partial r} + \frac{1}{r^2}\frac{\partial^2}{\partial \varphi^2} + \frac{\partial^2}{\partial z^2} \tag{1b}$$

In the case of axial symmetry of the considered swirling flow (this condition means that all partial derivatives with respect to φ must be equal to 0), the Equation (1b) has the following form:

$$\frac{u_r u_\varphi}{r} + \frac{\partial u_\varphi}{\partial r}u_r + \frac{\partial u_\varphi}{\partial z}u_z + \frac{\partial u_\varphi}{\partial t} = v\left(\frac{\partial^2 u_\varphi}{\partial r^2} + \frac{1}{r}\frac{\partial u_\varphi}{\partial r} + \frac{\partial^2 u_\varphi}{\partial z^2} - \frac{u_\varphi}{r^2}\right) \tag{1c}$$

It is seen from relations (1a) that u_φ does not depend on z. This means that the partial derivative u_φ with respect to z equals 0. Then Equation (1c) can be changed as follows:

$$\frac{u_r u_\varphi}{r} + \frac{\partial u_\varphi}{\partial r}u_r + \frac{\partial u_\varphi}{\partial t} = v\left(\frac{\partial^2 u_\varphi}{\partial r^2} + \frac{1}{r}\frac{\partial u_\varphi}{\partial r} - \frac{u_\varphi}{r^2}\right) \tag{1d}$$

We substitute relations (1a) into Equation (1d), writing out each term separately:

$$\frac{\partial u_\varphi}{\partial t} = \frac{d\Gamma_0(t)}{dt} * \frac{1}{2\pi r} * \left(1 - e^{-\frac{C_0(t)r^2}{2v}}\right) + \frac{\Gamma_0(t)r}{4\pi v} * \frac{dC_0(t)}{dt} * e^{-\frac{C_0(t)r^2}{2v}}$$

$$\frac{u_r u_\varphi}{r} = -\frac{C_0(t)\Gamma_0(t)}{2\pi r} * \left(1 - e^{-\frac{C_0(t)r^2}{2v}}\right)$$

$$\frac{\partial u_\varphi}{\partial r} u_r = -C_0(t)r * \left(\frac{C_0(t)\Gamma_0(t)}{2\pi v} * e^{-\frac{C_0(t)r^2}{2v}} - \frac{\Gamma_0(t)}{2\pi r^2} * \left(1 - e^{-\frac{C_0(t)r^2}{2v}}\right)\right)$$

$$\frac{\partial^2 u_\varphi}{\partial r^2} = \frac{\Gamma_0(t)}{\pi r} * \left[\frac{1 - e^{-\frac{C_0(t)r^2}{2v}}}{r^2} + \frac{C_0(t) * e^{-\frac{C_0(t)r^2}{2v}} * \left(1 - e^{-\frac{C_0(t)r^2}{2v}}\right)}{2v} - \frac{C_0(t) * e^{-\frac{C_0(t)r^2}{2v}}}{v}\right]$$

$$\frac{1}{r}\frac{\partial u_\varphi}{\partial r} = \frac{C_0(t)\Gamma_0(t)}{2\pi v r} * e^{-\frac{C_0(t)r^2}{2v}} - \frac{\Gamma_0(t)}{2\pi r^3} * \left(1 - e^{-\frac{C_0(t)r^2}{2v}}\right)$$

$$\frac{u_\varphi}{r^2} = \frac{\Gamma_0(t)}{2\pi r^3} * \left(1 - e^{-\frac{C_0(t)r^2}{2v}}\right)$$

Substituting the obtained relations into the original Equation (1d), the left-hand side is transformed as follows:

$$\frac{d\Gamma_0(t)}{dt} * \frac{1}{2\pi r} * \left(1 - e^{-\frac{C_0(t)r^2}{2v}}\right) + \frac{\Gamma_0(t)r}{4\pi v} * \frac{dC_0(t)}{dt} * e^{-\frac{C_0(t)r^2}{2v}} - \frac{C_0(t)\Gamma_0(t)}{2\pi r} * \left(1 - e^{-\frac{C_0(t)r^2}{2v}}\right) - C_0(t)r$$

$$* \left(\frac{C_0(t)\Gamma_0(t)}{2\pi v} * e^{-\frac{C_0(t)r^2}{2v}} - \frac{\Gamma_0(t)}{2\pi r^2} * \left(1 - e^{-\frac{C_0(t)r^2}{2v}}\right)\right)$$

$$= \frac{d\Gamma_0(t)}{dt} * \frac{1}{2\pi r} * \left(1 - e^{-\frac{C_0(t)r^2}{2v}}\right) + \frac{\Gamma_0(t)r}{4\pi v} * \frac{dC_0(t)}{dt} * e^{-\frac{C_0(t)r^2}{2v}} - \frac{C_0^2(t)\Gamma_0(t)r}{2\pi v} * e^{-\frac{C_0(t)r^2}{2v}}$$

Similarly, the right-hand side of Equation (1d) can be transformed as follows:

$$\frac{\Gamma_0(t)v}{\pi r} * \left[\frac{1 - e^{-\frac{C_0(t)r^2}{2v}}}{r^2} + \frac{C_0(t) * e^{-\frac{C_0(t)r^2}{2v}} * \left(1 - e^{-\frac{C_0(t)r^2}{2v}}\right)}{2v} - \frac{C_0(t) * e^{-\frac{C_0(t)r^2}{2v}}}{v}\right] + \frac{C_0(t)\Gamma_0(t)}{2\pi r} * e^{-\frac{C_0(t)r^2}{2v}}$$

$$- \frac{\Gamma_0(t)v}{2\pi r^3} * \left(1 - e^{-\frac{C_0(t)r^2}{2v}}\right) - \frac{\Gamma_0(t)v}{2\pi r^3} * \left(1 - e^{-\frac{C_0(t)r^2}{2v}}\right) = -\frac{C_0(t)\Gamma_0(t)}{2\pi r} e^{-\frac{C_0(t)r^2}{v}}$$

The final Equation (1d) will be written as follows:

$$\frac{d\Gamma_0(t)}{dt}\frac{1}{2\pi r}\left(1 - e^{-\frac{C_0(t)r^2}{2v}}\right) + \frac{dC_0(t)}{dt}\frac{\Gamma_0(t)r}{4\pi v}e^{-\frac{C_0(t)r^2}{2v}} - \frac{C_0^2(t)\Gamma_0(t)r}{2\pi v}e^{-\frac{C_0(t)r^2}{2v}} =$$
$$= -\frac{C_0(t)\Gamma_0(t)}{2\pi r}e^{-\frac{C_0(t)r^2}{v}} \tag{1e}$$

The Navier-Stokes equation for calculating the radial velocity component is written as follows [5]:

$$\frac{\partial u_r}{\partial t} + u_r\frac{\partial u_r}{\partial r} - \frac{u_\varphi}{r}\frac{\partial u_r}{\partial \varphi} - \frac{u_\varphi^2}{r} + u_z\frac{\partial u_r}{\partial z} = -\frac{1}{\rho}\frac{\partial p}{\partial r} + v\left(\frac{\partial^2 u_r}{\partial r^2} + \frac{1}{r}\frac{\partial u_r}{\partial r} - \frac{u_r}{r^2} + \frac{\partial^2 u_r}{\partial z^2}\right) \tag{2a}$$

Considering the axisymmetric flow, we have:

$$\frac{\partial u_r}{\partial t} + u_r\frac{\partial u_r}{\partial r} - \frac{u_\varphi^2}{r} + u_z\frac{\partial u_r}{\partial z} = -\frac{1}{\rho}\frac{\partial p}{\partial r} + v\left(\frac{\partial^2 u_r}{\partial r^2} + \frac{1}{r}\frac{\partial u_r}{\partial r} - \frac{u_r}{r^2} + \frac{\partial^2 u_r}{\partial z^2}\right)$$

Substituting the expressions for the velocity Components (1a) into this equation, we get:

$$-r\frac{dC_0(t)}{dt} + C_0^2(t)r - \frac{\Gamma_0(t)^2}{4\pi^2 r^3}\left(1 - e^{-\frac{C_0(t)r^2}{2v}}\right)^2 = -\frac{1}{\rho}\frac{\partial p}{\partial r} \tag{2b}$$

The Navier-Stokes equation for calculating the longitudinal velocity component will then be written as follows [5]:

$$\frac{\partial u_z}{\partial t} + u_r \frac{\partial u_z}{\partial r} + \frac{u_\varphi}{r}\frac{\partial u_z}{\partial \varphi} + u_z \frac{\partial u_z}{\partial z} = -\frac{1}{\rho}\frac{\partial p}{\partial z} + \nu\left(\frac{\partial^2 u_z}{\partial r^2} + \frac{1}{r}\frac{\partial u_z}{\partial r} + \frac{1}{r^2}\frac{\partial^2 u_z}{\partial \varphi^2} + \frac{\partial^2 u_z}{\partial z^2}\right) \quad (3a)$$

In the case of axial symmetry of the considered swirling flow, Equation (3a) is written as follows:

$$\frac{\partial u_z}{\partial t} + u_r\frac{\partial u_z}{\partial r} + u_z\frac{\partial u_z}{\partial z} = -\frac{1}{\rho}\frac{\partial p}{\partial z} + \nu\left(\frac{\partial^2 u_z}{\partial r^2} + \frac{1}{r}\frac{\partial u_z}{\partial r} + \frac{\partial^2 u_z}{\partial z^2}\right)$$

Substituting relations (1a) into Equation (3a), writing out each term separately, we get:

$$2z\frac{dC_0(t)}{dt} + 4C_0^2(t)z = -\frac{1}{\rho}\frac{\partial p}{\partial z} \quad (3b)$$

Thus, finding the functions $C_0(t)$, $\Gamma_0(t)$ is reduced to solving the following system of differential equations:

$$\begin{cases} \frac{d\Gamma_0(t)}{dt}\frac{1}{2\pi r}\left(1 - e^{-\frac{C_0(t)r^2}{2\nu}}\right) + \frac{\Gamma_0(t)r}{4\pi\nu}e^{-\frac{C_0(t)r^2}{2\nu}}\left(\frac{dC_0(t)}{dt} - 2C_0^2(t)\right) = -\frac{C_0(t)\Gamma_0(t)}{2\pi r}e^{-\frac{C_0(t)r^2}{\nu}} \\[2mm] -r\frac{dC_0(t)}{dt} + C_0^2(t)r - \frac{\Gamma_0(t)^2}{4\pi^2 r^3}\left(1 - e^{-\frac{C_0(t)r^2}{2\nu}}\right)^2 = -\frac{1}{\rho}\frac{\partial p}{\partial r} \\[2mm] 2z\frac{dC_0(t)}{dt} + 4C_0^2(t)z = -\frac{1}{\rho}\frac{\partial p}{\partial z} \end{cases} \quad (4)$$

It should be noted that the equations in the written system (4), in addition to the sought-for time-dependent functions, also contain the derivatives of pressure with respect to coordinates and these coordinates. Thus, the conditions for which the exact solutions were derived are satisfied if and only if any function of the partial derivatives of pressure with respect to coordinates and of the coordinates themselves is a function only of time.

The last equation of system (4) contains only one unknown function—$C_0(t)$; therefore, such an equation is most conveniently solved using the method of Lie groups. The solution to this equation gives the following expression for the desired function:

$$C_0(t) = -\frac{\alpha_1}{2\tanh(\alpha_1 * (C_1 - t))} \quad (5a)$$

where C_1 is the constant of integration and can be taken as 0.

Then the resulting expression for $C_0(t)$ was substituted into the second equation of the system (4). After this, the equation is a differential equation from one unknown function $\Gamma_0(t)$. Its solution gives the following result:

$$\Gamma_0(t) = \frac{\pi\sqrt{\dfrac{\alpha_3 * \tanh^2(\alpha_1 * (C_1 - t)) + \alpha_4}{\exp\left(\dfrac{\alpha_2}{4\nu * \tanh(\alpha_1 * (C_1 - t))}\right) - 1}}}{\tanh(\alpha_1 * (C_1 - t))} \quad (5b)$$

The following notations are introduced in Expressions (5a) and (5b):

$$\begin{cases} \alpha_1 = \sqrt{-\frac{1}{\rho z} * \frac{\partial p}{\partial z}} \quad \left[\frac{1}{s}\right] \\[2mm] \alpha_2 = r^2 * \sqrt{-\frac{1}{\rho z} * \frac{\partial p}{\partial z}} \quad \left[\frac{m^2}{s}\right] \\[2mm] \alpha_3 = \frac{4r^3}{\rho}\frac{\partial p}{\partial r} - \frac{2r^4}{\rho z} * \frac{\partial p}{\partial z} \quad \left[\frac{m^4}{s^2}\right] \\[2mm] \alpha_4 = \frac{3r^4}{\rho z} * \frac{\partial p}{\partial z} \quad \left[\frac{m^4}{s^2}\right] \end{cases} \quad (5c)$$

If the pressure change along the longitudinal and radial coordinates is constant, the parameters $(\alpha_1, \ldots, \alpha_4)$ show only the change in the geometric configuration of the jet over time. If the jet is enclosed in a flow channel that takes its shape, then the parameters $(\alpha_1, \ldots, \alpha_4)$ can be determined experimentally by measuring the dimensions of the channel.

In [6,7], the empirical dependences on time of $\Gamma_0(t)$ and of $C_0(t)/\Gamma_0(t)$ were obtained. Comparing these dependencies, the approximate order of values of the constants was determined as $(\alpha_1, \ldots, \alpha_4) = (10^{-1}, 10^{-6}, 10^{-4}, 10^{-2})$.

Using the suggested values of the parameters, the graphs of the dependence of the functions $C_0(t)$ and $\Gamma_0(t)$ on time were plotted (Figures 1 and 2).

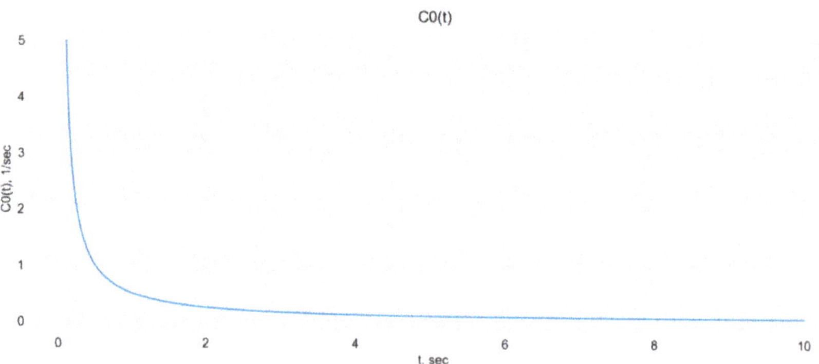

Figure 1. Plot of the $C_0(t)$ function.

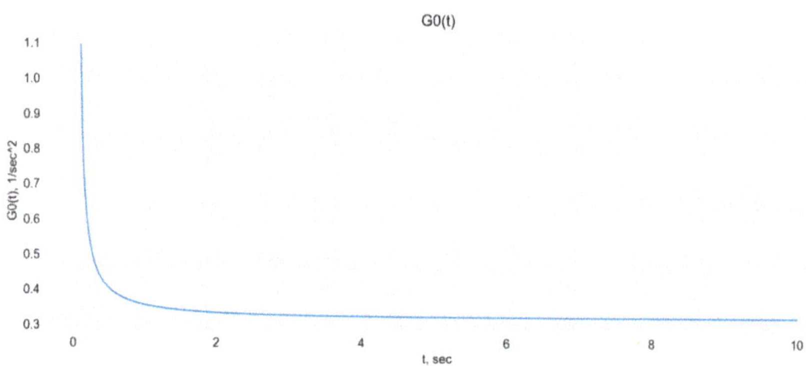

Figure 2. Plot of the $\Gamma_0(t)$ function.

4. Discussion

As a result, we managed to obtain a general analytical form of the functions $C_0(t)$ and $\Gamma_0(t)$. The ratios for these functions contain several parameters, the exact value of which should be determined within the framework of a specific problem. As can be seen from the graphs, the studied functions $C_0(t)$ and $\Gamma_0(t)$ are strictly positive and decrease monotonically with time. This behavior expresses the fulfillment of the energy conservation law for the considered swirling flow. Indeed, solution (1a) shows that if the values of the functions under study could increase in absolute value over time, the kinetic energy of the swirling flow would have increased in the absence of external action, which is impossible. The dynamics of the functions $C_0(t)$ and $\Gamma_0(t)$ approximately corresponds to the experimental curves plotted, based on studies of the blood flow structure in the left ventricle [7].

Analysis of the constants $(\alpha_1, \ldots, \alpha_4)$ allows the following statements:

The dimensions of the constants correspond to the dimensions of the studied functions $C_0(t)$ and $\Gamma_0(t)$. In particular, the dimension α_1 coincides with the dimension $C_0(t)$, the dimension α_2 coincides with the dimension $\Gamma_0(t)$ and the dimensions α_3, α_4 coincide with the dimensions $\Gamma_0(t)^2$.

The swirling flow belongs to the TLJ class and can be described by exact solutions (1a) if and only if the form of the dynamic pressure dependence on the channel coordinates does not change with time.

Two swirling flows belonging to the TLJ class and having different geometric characteristics can be determined by the same functions $C_0(t)$ and $\Gamma_0(t)$ and, accordingly, have an identical dynamic structure if the set of constants $(\alpha_1, \ldots, \alpha_4)$ is the same for both.

Therefore, the obtained relations (5a) and (5b) make it possible to describe in a general form the unsteady swirling flow of a viscous fluid, the velocity components of which are expressed by solution (1a). Considering the obtained expressions for the functions $C_0(t)$ and $\Gamma_0(t)$, the exact solution can be used for numerical modeling of the swirling blood flow in the heart and large vessels without considering the non-Newtonian properties of blood.

Attempts to formally describe the unsteady blood flow have been undertaken earlier by many researchers [8–11]. Computational fluid dynamics methods were mainly used in these works. With the development of computing power, it has become possible to simulate the geometry of heart cavities using neural networks [12]. However, despite great advances in the field of numerical modeling, these models turn out to be rather complex and lose the necessary clarity. This is even more true for models based on neural networks.

In this work, based on the well-studied non-stationary equations of hydrodynamics, relatively simple relations were obtained that describe the non-stationary properties of the swirling blood flow in the heart and great vessels. The use of the obtained ratios in CFD models and for calculating initial approximations of the geometry of flow channels for training neural networks will enable the development of relatively simple and effective blood circulation models.

5. Conclusions

Most modern circulatory models are based on numerical modeling of blood flow and based on methods of computational hydrodynamics and deep learning [8–12]. This approach simplifies the calculations but does not allow derivation of the physical and mathematical meaning of the functional dependencies, which are included in the ratios for the components of the blood flow velocity. So, it was possible to obtain analytical relations for functions reflecting the unsteady properties of the swirling blood flow. These relations, on the one hand, make it possible to perform a more accurate analysis of the physical characteristics of an unsteady swirling blood flow. On the other hand, they allow formulating characteristic flow parameters that can be verified by experiment. Thus, the results of this study allow us to deepen the physical understanding of the peculiarities of the evolution of the swirling blood flow, as well as to provide convenient characteristic parameters for validating experimental data.

It should be noted that the obtained ratios do not consider the pulsating nature of the blood flow in the heart and large vessels. The necessary improvements are planned to be added to the model at subsequent stages of the study.

The next steps in the development of this study will be to carry out experimental measurements of the field of velocities and pressures and to validate the obtained results using the functional relationships obtained in this work.

Author Contributions: Development of the concept of the article and collection of the material, E.T.; Development of a research algorithm, finding suitable mathematical methods for solving the assigned tasks, E.T. and A.G.; Carrying out mathematical transformations, solving equations, E.T., Description of the physiological side of the study, A.G.; Supervision and Project Administration, A.G.; Writing, E.T. and A.G. Both authors have read and agreed to the published version of the manuscript.

Funding: The study was supported by the Russian Scientific Foundation (grant No. 16-15-00109).

Institutional Review Board Statement: Not applicable.

Informed Consent Statement: Not applicable.

Data Availability Statement: Not applicable.

Conflicts of Interest: The authors declare no conflict of interest.

References

1. Talygin, E.; Kiknadze, G.; Agafonov, A.; Gorodkov, A. Application of the Tornado-Like Flow Theory to the Study of Blood Flow in the Heart and Main Vessels: Study of the Potential Swirling Jets Structure in an Arbitrary Viscous Medium. In Proceedings of the ASME 2019 International Mechanical Engineering Congress and Exposition, Salt Lake City, UT, USA, 11–14 November 2019. [CrossRef]
2. Talygin, E.A.; Zhorzholiani, S.T.; Agafonov, A.V.; Kiknadze, G.I.; Gorodkov, A.Y.; Bokeriya, L.A. Quantitative Evaluation of Disorders of the Swirled Blood Flow Structure in the Aorta with Pathological Alteration of Its Channel Geometry Using Numerical Simulation of the Aorta. *Hum. Physiol.* **2019**, *45*, 527–535. [CrossRef]
3. Zhorzholiani, S.T.; Talygin, E.A.; Krasheninnikov, S.V.; Tsigankov, Y.M.; Agafonov, A.V.; Gorodkov, A.Y.; Kiknadze, G.I.; Chvalun, S.N.; Bokeria, L.A. Elasticity Change along the Aorta is a Mechanism for Supporting the Physiological Self-organization of Tornado-like Blood Flow. *Hum. Physiol.* **2018**, *44*, 532–540. [CrossRef]
4. Kiknadze, G.I.; Krasnov, Y.K. Evolution of a spout-like flow of viscous fluid. *Sov. Phys. Dokl.* **1986**, *31*, 799–801.
5. Sedov, L.I. Solid Mechanics. *Nauka. Moscow.* **1973**, *1*, 178–180. (In Russian)
6. Bockeria, L.; Kiknadze, G.I.; Gachechiladze, I.A.; Gorodkov, A.Y. Application of Tornado-Flow Fundamental Hydrodynamic Theory to the Study of Blood Flow in the Heart: Further Development of Tornado-Like Jet Techology. In Proceedings of the ASME IMECE, Denver, CO, USA, 11–17 November 2011. [CrossRef]
7. Talygin, E.A.; Zazybo, N.A.; Zhorzholiany, S.T.; Krestinich, I.M.; Mironov, A.A.; Kiknadze, G.I.; Bokerya, L.A.; Gorodkov, A.Y.; Makarenko, V.N.; Alexandrova, S.A. Quantitative Evaluation of Intracardiac Blood Flow by Left Ventricle Dynamic Anatomy Based on Exact Solutions of Non-Stationary Navier-Stocks Equations for Selforganized tornado-Like Flows of Viscous Incompresssible Fluid. *Uspekhi Fiziologicheskikh Nauk* **2016**, *47*, 48–68. [PubMed]
8. Doost, S.N.; Ghista, D.; Su, B.; Zhong, L.; Morsi, Y.S. Heart blood flow simulation: A perspective review. *Biomed. Eng. Online* **2016**, *15*, 1–28. [CrossRef] [PubMed]
9. Khalafvand, S.S.; Ng, E.Y.-K.; Zhong, L.; Hung, T.-K. Three-dimensional diastolic blood flow in the left ventricle. *J. Biomech.* **2017**, *50*, 71–76. [CrossRef] [PubMed]
10. Pedrizzetti, G.; Domenichini, F. The Long Way from Theoretical Models to Clinical Applications. *Ann. Biomed. Eng.* **2014**, *40*. [CrossRef]
11. Habibi, M.; D'Souza, R.M.; Dawson, S.T.; Arzani, A. Integrating multi-fidelity blood flow data with reduced-order data assimilation. *Comput. Biol. Med.* **2021**, *135*, 104566. [CrossRef] [PubMed]
12. Liao, X.; Qian, Y.; Chen, Y.; Xiong, X.; Wang, Q.; Heng, P.-A. MMTLNet: Multi-Modality Transfer Learning Network with adversarial training for 3D whole heart segmentation. *Comput. Med Imaging Graph.* **2020**, *85*, 101785. [CrossRef] [PubMed]

Article

Continuum Scale Non Newtonian Particle Transport Model for Hæmorheology

Torsten Schenkel [1,*] and Ian Halliday [2]

1 Department of Engineering and Mathematics, Sheffield Hallam University, Sheffield S1 1WB, UK
2 Department of Infection, Immunity and Cardiovascular Disease, University of Sheffield, Sheffield S10 2RX, UK; i.halliday@sheffield.ac.uk
* Correspondence: t.schenkel@shu.ac.uk

Abstract: We present a continuum scale particle transport model for red blood cells following collision arguments, in a diffusive flux formulation. The model is implemented in FOAM, in a framework suitable for haemodynamics simulations and adapted to multi-scaling. Specifically, the framework we present is able to ingest transport coefficient models to be derived, prospectively, from complimentary but independent meso-scale simulations. For present purposes, we consider modern semi-mechanistic rheology models, which we implement and test as proxies for such data. The model is verified against a known analytical solution and shows excellent agreement for high quality meshes and good agreement for typical meshes as used in vascular flow simulations. Simulation results for different size and time scales show that migration of red blood cells does occur on physiologically relevany timescales on small vessels below 1 mm and that the haematocrit concentration modulates the non-Newtonian viscosity. This model forms part of a multi-scale approach to haemorheology and model parameters will be derived from meso-scale simulations using multi-component Lattice Boltzmann methods. The code, `haemoFoam`, is made available for interested researchers.

Keywords: haemorheology; blood flow modelling; particle transport; numerical fluid mechanics

Citation: Schenkel, T.; Halliday, I. Non Newtonian Particle Transport Model for Hæmorheology. *Mathematics* **2021**, *9*, 2100. https://doi.org/10.3390/math9172100

Academic Editors: Victor Mitrana and Alexander Zeifman

Received: 29 May 2021
Accepted: 25 August 2021
Published: 30 August 2021

Publisher's Note: MDPI stays neutral with regard to jurisdictional claims in published maps and institutional affiliations.

1. Introduction

Blood is a non-Newtonian fluid with very complex behaviour deriving from a meso-scopic composition which—minimally described—is a dense, mono-disperse suspension of deformable vesicles suspended in incompressible plasma. Accordingly, blood rheology is dominated by the interaction of cells, with a multitude of models having been proposed to account for such meso-scale effects as deformation, aggregation, and rouleaux formation which underline emergent macroscopic flow properties like concentration dependant viscosity and shear thinning. The authors are currently developing a multi-scale approach, explicitly modelling meso-scale effects using Lattice Boltzmann Models (LBM), in which erythrocyte mechanics are fully resolved, while describing the macro-scale rheology using particle transport modelling and quasi-mechanistic non-Newtonian rheology models described using traditional Eulerian numerics. We give the rationale of this pairing of methodologies shortly, in the next sub-section. In essence, though, it is intended that the latter model will eventually be parameterised using LBM data. Here, we present the continuum mechanical part of this multi-scale modelling approach, which is designed to facilitate the simulation of realistic vessel geometries and spatially complex flow patterns.

Multi-scale models are necessary in many applications. In meteorology, the Eulerian grids of the macroscopic simulation are characterised by a 1 km spatial resolution. Clearly it is necessary to account for more local variation of field variables and macroscopic atmospheric simulations are parameterised based on data gathered from, e.g., semi-analytic meso-scale models of convection [1]. (Note, we use the term meso-scale (macroscale) to identify the shortest (largest) length scale considered. We use the term continuum-scale to describe a scale where particle dynamics are neglected and integral bulk material parameters are used—within the scope of this paper macroscale is treated as continuum-scale).

Similarly, in the simulation of turbulent flow fields, temporal or spatial averages are used to derive macroscopic turbulence properties, like an isotropic turbulent viscosity, which describes the statistical effect of the inherently chaotic, and anisotropic nature of meso-scale and micro-scale turbulence. While the micro-scale can be simulated (Direct Numerical Simulation), this approach is limited by size, Reynolds number and computational resource and thus it is mostly used in fundamental research and in the development and parametrisation the macro-scale model.

In a high particle load suspension like blood, many types of mechanical interactions between particles and carrier fluid occur [2]. A variety of models are used from 1 to d reduced order models, through meso-scale models to continuum-scale models [3]. Meso-scale modelling, e.g., Particle Dynamics Methods [4], or the multi-component Lattice Boltzmann Method [5] has a widely acknowledged facility for Lagrangian particulate flows [6–10] and has been employed to describe these interactions and the dynamics of the collisions in detail. Very recently, we have developed a single framework, three-dimensional methodology for capturing detailed, particle-scale interactions between neutrally buoyant suspended vesicles (i.e., erythrocytes), using our novel chromodynamics multi-component Lattice Boltzmann method variant [11]. We have previously shown that this same essential model is able to capture detailed hydrodynamic interactions, lubrication effects and ballistic collisions between transported particles, in two dimensional simulations containing O(1000) liquid drops at high volume faction [12]. Current work aims to scale these models up, to allow extraction, through statistical averaging and measuring mechanical dissipation, of relevant transport coefficients, e.g.,

$$\nu = \nu(\phi, \dot{\gamma}), \quad D = D(\phi, \dot{\gamma}). \tag{1}$$

Above, ν and D denote kinematic viscosity and diffusion co-efficient and ϕ and $\dot{\gamma}$ denote particle density and local shear rate.

As with direct numerical simulation in turbulence modelling, finite computational resource means that these detailed explicit particulate models are limited to small volumes containing relatively few particles in their simulation domain (an the order of magnitude of hundreds to thousands at the time of writing). To address the much greater scales of medical significance, it is, therefore, necessary to develop macro or continuum scale models, which encapsulate an integrated effect of these interactions, without explicitly resolving them. Crucially, these continuum models must be amenable to parametrisation, in the present context using meso-scale information such as that encapsulated in the data of Burgin [12].

Based on the work of Leighton and Acrivos [13], and Phillips et al. [14], we present here a macro-scale, or continuum-scale model for haematocrit transport, which, together with modern formulations of "quasi-Newtonian" (i.e., not considering viscoelastic and viscoplastic effects) non-Newtonian model, allows for the simulation of macroscopic flows, accounting for the integral effect of meso-scopic phenomena.

Currently, the models which have been proposed for continuum-scale haematocrit transport models can, roughly, be divided [15] into suspension balance models [13,16] and diffusive flux models [14].

Suspension balance models use an Euler–Euler mixture modelling approach, where the carrier fluid and the particle load are represented as separate species, with a transport Equation (typically convection-diffusion) and physical transport properties for each species; in diffusive flux models, the suspension is modelled as a single species, with the particle volume fraction being modelled as a scalar property, which influences the bulk transport coefficients.

Our macroscopic model is a particle transport model after Phillips [14]. It follows the collision arguments by Leighton and Acrivos [13] by describing the particle migration based on the gradients of shear strain, concentration and viscosity. The local concentration of haematocrit is then used to establish the local effective viscosity.

The haematocrit distribution emerges, locally, from a balance of the competing effects of diffusion-driven flux down the local concentration gradient (which, for steady, developed flow, lies in the radial direction) and shear-induced migration (which, for steady, developed, flows, would be in the negative radial direction). Clearly, the problem of flow-concentration coupling is made still more complex since the diffusion co-efficient varies with local shear rate and particle concentration.

Previous attempts to implement this class of particle transport model in a finite volume code were either limited to a single viscosity model (due to the necessity of linearising the source terms [17]) or they had to introduce damping terms (to suppress instability [18], mostly due to the fact that they attempted to implement the non-linear transport equation for haematocrit into commercial software packages (Fluent), which does not allow a fully implicit additional transport equation in user subroutines).

The framework we present in this work expands on those commonly used in the simulation of clinically relevant vascular flows in the following important aspects:

- It includes a transport equation for the most important group of suspended particles based on continuum-scale, semi-mechanistic modelling.
- It includes more sophisticated non-Newtonian viscosity models that couple to local haematocrit concentration in addition to shear rate in the viscosity function.
- The implementation does not require linearisation of the viscosity source term in the haematocrit transport equation, which, crucially, allows for other viscosity models to be implemented eventually.

It is implemented using the *Field Operation And Manipulation* (FOAM) framework. FOAM, or OpenFOAM, is an open source library which allows easy implementation of Finite Volume Method (FVM) solvers and achieves stability without artificial damping by implementing the non-linear terms in a hybrid explicit-implicit fashion which allows stable simulations even in realistic physiological geometries at high rates.

2. Materials and Methods

2.1. Particle Migration Model

In a particle transport model, the transport of haematocrit is dominated by advection —following the bulk flow—variations in concentration are evened by diffusive processes, and the migration within the bulk is driven by a migration pressure. This migration pressure is the result of two phenomena: (1) spatial variation of collision (interaction) frequency and (2) spatial variation of viscosity [14]

A detailed treatise on the rationale behind the compression arguments can be found in Leighton and Acrvios, and Phillips [13,14], we only give a brief outline at this point.

2.1.1. Spatial Variation of Collision Frequency

Particles that are moving relative to each other in neighbouring shear surfaces will experience collisions. The frequency of these collisions is proportional to the shear rate $\dot{\gamma}$, the particle concentration ϕ, and the particle collision radius a. In a field of constant concentration and constant shear, $\dot{\gamma}\phi = $ const, the collisions are in equilibrium either side of the shear surface, and no net migration will occur. In the presence of gradients of shear rate or concentration, the imbalance of collisions will lead to a "migration pressure" down the gradient. This collision driven migration pressure can be described as a function of $a\nabla(\dot{\gamma}\phi))$. Using a proportionality factor of K_c and assuming a displacement proportional to the particle radius a, the migratory flux N_c due to variations in collision frequency can be expressed as (using the chain rule):

$$N_c = -K_c a^2 (\phi^2 \nabla \dot{\gamma} + \phi \dot{\gamma} \nabla \phi) \tag{2}$$

2.1.2. Spatial Variation of Viscosity

The displacement of particles after a collision is moderated by viscous effects. In a constant viscosity field the displacement is isotropic and thus balanced with no net migration effects. In a viscosity gradient the displacement will be less damped in direction of the lower viscosity, leading to a net migration effect down the viscosity gradient.

The displacement velocity is proportional to the relative change in viscosity over a distance that is of order a: $a(1/\mu)\nabla\mu$. With the displacement frequency scaling with $\dot{\gamma}\phi$, and a proportionality factor of K_μ, the migratory flux due to viscosity gradient can be described as (flux is proportional to ϕ):

$$N_\mu = -K_\mu\dot{\gamma}\phi^2\left(\frac{a^2}{\mu}\nabla\mu\right) \tag{3}$$

The scalar transport equation for haematocrit, ϕ, is then (neglecting molecular diffusion, Brownian motion), where D/Dt is the total differential:

$$\frac{D\phi}{Dt} = \nabla\cdot\left(N_c + N_\mu\right) \tag{4}$$

$$\begin{aligned}
\frac{D\phi}{Dt} = \;& \nabla\cdot\left(a^2 K_c\phi\dot{\gamma}\nabla\phi\right) \\
& + \; a^2 K_c\nabla\cdot\left(\phi^2\nabla\dot{\gamma}\right) \\
& + \; a^2 K_\mu\nabla\cdot\left(\dot{\gamma}\phi^2\frac{1}{\mu}\nabla\mu\right),
\end{aligned} \tag{5}$$

with a, particle radius, $\dot{\gamma}$, shear strain rate magnitude, μ, dynamic viscosity, K_c and K_μ, collision parameters.

Typically, the viscosity is $\mu = f(\dot{\gamma},\phi)$, which makes the last source term non-linear, which can, in turn, make the solution of this transport equation difficult.

Previous attempts to solve this problem analytically or implement this type of migration model in a numerical model used linearisation of this source term, which involves the derivative of μ in both $\dot{\gamma}$ and ϕ, and thus limits the model to a specific viscosity model, for which it has been implemented [17,19]. Our current implementation deals with the non-linear viscosity source term in a way that leaves the viscosity gradient term intact and is thus agnostic to the rheology model used.

2.2. Rheology Models

It is obvious from the third RHS term in Equation (5), that the particle transport strongly depends on the rheology model it is coupled with. This model implementation aims to be independent of the rheology model. The draw-back of this approach is that errors present in the rheology model, which influence the particle transport, cannot be calibrated out with the parameters of the migration model alone, but the combined set of model parameters will need to be found for any new rheology model that is to be implemented.

Typically, only the shear thinning effects are taken into account, when modelling the non-Newtonian properties of blood in CFD. Common models are of the Carreau and Casson types. In these models, the haematocrit concentration is only used as a bulk parameter in the parametrisation, if at all. Our framework, incorporating the transport of haematocrit, allows the rheological model to take the local particle concentration into account when calculating the local, effective viscosity.

The rheology models that have been implemented and tested in this study are the concentration dependent Krieger–Dougherty model [20], the Quemada model [21–23] with modification by Das [24] (and a new parameter set, which avoids the singularity problem commonly associated with this model), an extended Krieger model, accommodating shear thinning and aggregation effects [25], a Casson model with haematocrit dependence following Merril et al. [24,26], and a modified Carreau type model, proposed by Yeleswarapu [27]. All model parameters have been fitted to the experimental data of Brooks [28] (Figure 1).

2.2.1. Krieger–Dougherty Model

The traditional Krieger–Dougherty model [20] was developed to describe the rheology of high volume ratio suspensions of rigid spherical particles. Rigid, spherical particles do not exhibit shear-thinning behaviour, so the Krieger–Dougherty model is only dependent on the haematocrit concentration ϕ. It shows a singularity for $\phi = \phi^*$, where ϕ^* is the haematocrit concentration for which the suspension does stop to behave like a fluid. For rigid spheres $\phi^* = 0.68$ [20], while for blood it can go up to $\phi^* = 0.98$, which is usually attributed to the deformability of the erythrocytes [25].

$$\mu = \mu_P \left(1 - \frac{\phi}{\phi^*}\right)^{-n}.$$ (6)

The parameter $n = k\phi^*$ is often set to $n = 2$, but more commonly to the high shear limit of $n = 1.82$ for $\phi^* = 0.68$ [22,29], which is also the value used in this work to allow comparison with the results from Phillips and others [14,17,19]. μ_P is the Newtonian viscosity of the liquid phase (plasma).

In this study, the Krieger–Dougherty model is not used as for modelling blood viscosity but as a reference model for verification and validation.

2.2.2. Quemada Model

The Quemada model is based on "optimisation of viscous dissipation" [21]. In its original form it is formulated as a Newtonian, concentration dependent viscosity:

$$\mu = \mu_P(1 - k\phi)^{-2},$$ (7)

with k being related to the packing concentration and (for the high shear limit) given as: $k = 2/\phi^*$. In this form it is closely related to the Krieger–Dougherty model (Equation (6)).

In its non-Newtonian form k is expressed as [22,23]:

$$k = \frac{k_0 + k_\infty \sqrt{\dot{\gamma}/\dot{\gamma}_c}}{1 + \sqrt{\dot{\gamma}/\dot{\gamma}_c}},$$ (8)

where k_0 and k_∞ are the intrinsic viscosities at zero and infinite shear, respectively, and $\dot{\gamma}_c$ is a critical shear rate.

The shear rate magnitude $\dot{\gamma}$ is defined as

$$\dot{\gamma} := \sqrt{2D : D},$$ (9)

with D, the symmetric part of the velocity gradient tensor.

Different parameter fits have been proposed for $k_0, k_\infty, \dot{\gamma}_c$. Cokelet [26,30] proposed:

$$
\begin{aligned}
k_0 &= \exp(a_0 + a_1\phi + a_2\phi^2 + a_3\phi^3) & (10)\\
k_\infty &= \exp(b_0 + b_1\phi + b_2\phi^2 + b_3\phi^3) & (11)\\
\dot{\gamma}_c &= \exp(c_0 + c_1\phi + c_2\phi^2 + c_3\phi^3). & (12)
\end{aligned}
$$

Das [24] noted that Cokelet us parameter set causes the viscosity to be non-monotonous over haematocrit concentration for low shear, and exhibits singularities for zero shear. Das changed the parameter fit for k_0 to

$$k_0 = a_0 + \frac{2}{a_1 + \phi},$$ (13)

which results in a monotonous behaviour for low shear (the lowest shear measured in the Brooks dataset is around $\dot{\gamma} = 0.15$ s^{-1}), but still shows a singularity for $\phi = 80.4\%$. While this is outside the haematocrit values typically encountered in clinical practice, it can still pose a problem if cell migration is taken into account, which will concentrate cells in the core region. In order to overcome this problem, a new parameter set, based on Das's formulation, is derived in this work, which does not show a singularity. Figure 2 shows viscosity over shear rate for low shear rate ($\dot{\gamma} = 0.15$ s^{-1}) and zero shear rate. While all the curves show a good fit with the data, the new parameter set does show monotonous behaviour throughout and no singularity below the critical haematocrit.

2.2.3. Modified 5 Parameter Krieger Model

Hund et al. [25] proposed and developed a quasi-mechanistic extension to the Krieger–Dougherty model.

Starting from the traditional formulation of the Krieger–Dougherty model:

$$\mu = \mu_P \left(1 - \frac{\phi}{\phi^*}\right)^{-n}, \tag{14}$$

describing the haematocrit dependence, the shear-thinning behaviour is introduced by a variable exponent n:

$$n = n_\infty + \begin{cases} 0, \; \phi < \phi_{st} \\ n_{st}, \phi > \phi_{st}, \end{cases} \tag{15}$$

where ϕ_{st} is the threshold haematocrit concentration below which no shear-thinning is observed. Based on Brooks [28], this threshold is around $\phi = 0.15$, and n_∞ is modelled using a exponential dependency on ϕ:

$$n_\infty = a + b \, exp(-c \, \phi). \tag{16}$$

Hund's [25] shear-thinning exponent n_{st} comprises contributions of red blood cell aggregation and deformability:

$$n_{st} = n_{agg} + n_{def}, \tag{17}$$

where each component is described by a power law:

$$n_{agg/def} = \beta_{agg/def} \, \gamma'_{agg/def}{}^{-\nu_{agg/def}}, \tag{18}$$

with the empirical coefficient β and ν, and the non-dimensional shear rate $\gamma' = 1 + (\lambda\dot{\gamma})^{\nu_g}$, as defined by Carreau and Yasuda [31], with a time constant λ, and $\nu_g = 2$. This formulation ensures finite n_{st} at zero shear.

In the 5-component form the aggregation and deformation influences on the shear-thinning exponent are combined into a single power law, due to the limited data on these effects:

$$n_{st} = \beta\gamma'^{-\nu}. \tag{19}$$

The model proposed by Hund et al. allows for inclusion of the influence of large molecule concentration (proteins polysacharides, lipids), as well as fibrinogen, and temperature on the constitutive model. Due to a lack of data these are not included in the 5-parameter model.

2.2.4. Yeleswarapu-Wu Model

This model is based on a visco-elastic Oldroyd-B model developed by Yeleswarapu et al. [27,32]. In this study, the visco-elastic effects are neglected, only the shear-thinning behaviour and haematocrit dependency are implemented. The shear-thinning behaviour follows a modified Carraeu-type model based on a mixture model by Jung et al. [33].

The model is based on a mixture model and thus the viscosity is decribed as a function of plasma viscosity μ_P and red blood cell viscosity μ_{rbc} [32]:

$$\mu_{mix} = (1 - \phi)\mu_P + \phi\mu_{rbc}, \tag{20}$$

where the red blood cell viscosity is described as:

$$\mu_\infty(\phi) + (\mu_0(\phi) - \mu_\infty(\phi))\frac{1 + ln(1 + k\dot{\gamma})}{1 + k\dot{\gamma}}, \tag{21}$$

where, in this implementation, k is a constant model parameter, and μ_0 and μ_∞ are modelled as third order polynomials of ϕ:

$$\mu_0 = a_1\phi + a_2\phi^2 + a_3\phi^3 \tag{22}$$
$$\mu_\infty = b_1\phi + b_2\phi^2 + b_3\phi^3 \tag{23}$$

2.2.5. Casson-Merrill Model

The Casson model [34] is a classical non-Newtonian model in which the viscosity is modelled as:

$$\mu = \left(\sqrt{\mu_\infty} + \sqrt{\frac{\tau_0}{\dot{\gamma}}}\right)^2, \tag{24}$$

where μ_∞ is the Casson viscosity (asymptote at high shear rate) and τ_0 is the yield stress. The yield effect means that this model has a singularity at zero shear, leading to infinite viscosity. While there is an argument that blood does exhibit yield at slow time scales and low shear, this effect will typically make this type of model unsuited for numerical simulation within a generalised Newtonian approach with a local effective viscosity due to numerical instability.

For blood, Merill et al. gave the expressions for μ_∞ and τ_0 as [24,26]

$$\mu_\infty = \left(\frac{\mu_{pl}}{(1 - \phi)^\alpha}\right) \tag{25}$$

$$\tau_0 = \beta^2\left[\left(\frac{1}{1 - \phi}\right)^{\alpha/2} - 1\right]^2, \tag{26}$$

with the fitting parameters α and β.

2.2.6. Characteristics of Rheology Models

All viscosity model parameters were fitted to experimental data for varying levels of haematocrit in ADC plasma reported by Brooks [28]. While these data are for steady state shear only, it is still considered on of the best datasets for blood rheology data and is used in the majority of work on blood rheology. The parameters were fitted using a Levenberg–Marquardt least squares fit, implemented in Scientific Python (SciPy), using the MINPACK library. Table 1 shows the parameter sets for the different models, Figure 1 shows the comparison of model results and experimental data. All models show a good fit to the experimental data in the range were experimental data is available ($\dot{\gamma} > 0.15\ \mathrm{s}^{-1}$), while the behaviour for low shear stress varies between the models. The Casson model shows a singularity for zero shear (yield stress behaviour), while the other models all have finite viscosity for zero shear. However, the values at low shear vary widely. For $\dot{\gamma} = 10^{-2}\ \mathrm{s}^{-1}$, the range of relative viscosity is between $\mu/\mu_P = 71.4$ to 936. This variation will heavily influence the behaviour at low shear rate, e.g., on the axis of the flow.

Figure 2 shows the parameter fit for the Quemada model, where the classical Cokelet fit [30] exhibits singularities at 12.2%, 18%, 73.1%, and 85.6% for zero shear. The Das variation [24] improves on this, but the original parameter set by Das still shows a singularity for 80.4% haematocrit. The new parameter fit performed in this study removes the singularities completely and shows monotonous behaviour for the whole range of haematocrit concentrations and shear rates.

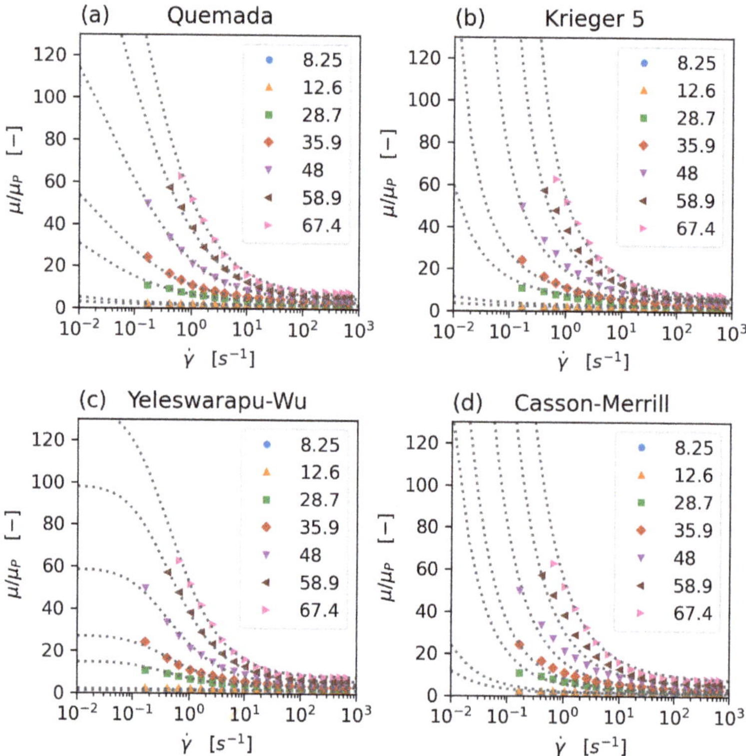

Figure 1. Comparison of non-Newtonian rheology models: Quemada (**a**), modified 5-parameter Krieger (**b**), Yeleswarapu-Wu (**c**), and Casson-Merril (**d**) model. All model parameters have been fitted to Brooks' data. Dots: experimental data (Brooks), dotted lines: model equations, data series: haematocrit concentration.

Table 1. Viscosity model parameters. Levenberg–Marquardt least squares fit (SciPy, MINPACK), to Brooks's data (all viscosities calculated in Pa s), $\mu_P = 1.23 \times 10^{-3}$ Pa s.

Quemada	MKM5	Yeleswarapu	Casson
-	-	-	-
a0: 0.06108	-	a1: −0.02779	-
a1: 0.04777	-	a2: 1.012	-
-	-	a3: −0.636	-
b0: 1.803	b: 8.781	b1: 0.0749	α: 1.694
b1: −3.68	c: 2.824	b2: −0.1911	β: 0.01197
b2: 2.608	β: 16.44	b3: 0.1624	-
b3: −0.001667	λ: 1296	-	-
-	-	k: 8.001	-
c0: −7.021	ν: 0.1427	-	-
c1: 34.45	-	-	-
c2: −39.94	-	-	-
c3: 14.09	-	-	-

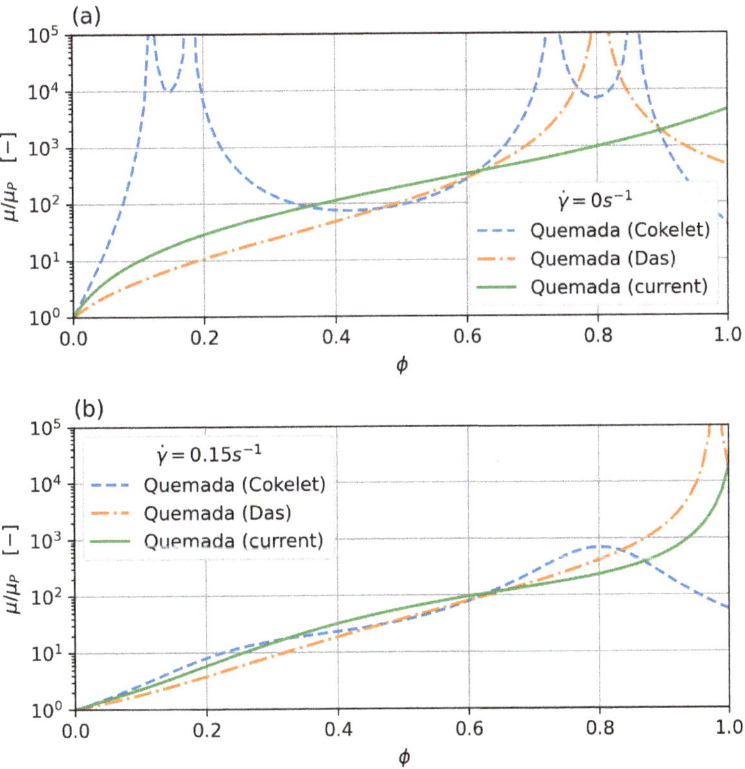

Figure 2. Comparison of Quemada parameterisation for zero (**a**) and low (0.15 s^{-1}) (**b**) shear rate. The classic Cokelet parameter set shows singularities at 12.2%, 18%, 73.1%, 85.6%, the modified parameterisation by Das improves on this, but still shows a singularity for 80.4% haematocrit. The current parameter set removes the singularity and shows monotonous behaviour.

2.3. Implementation

The fundamental equations for mass and momentum conservation were implemented using the SIMPLE (Semi-Implicit Method for Pressure-Linked Equations) [35] method for steady state, and the PISO (Pressure-Implicit with Splitting of Operators) [36] and PIMPLE (combining PISO and SIMPLE) methods for transient simulations.

Discretision is typically second order in space and time. The code supports all discretisation methods that are supported in the FOAM library.

The haematocrit transport Equation (5) is implemented as a scalar transport Equation (Listing 1), solved outside of the SIMPLE loop. The Laplacians in ϕ are implemented implicitly (*fvm*) as diffusion terms, while the source terms in $\dot{\gamma}$ and μ are calculated explicitly (*fvc*).

Listing 1. Haematocrit transport equation as implemented in Openfoam.

```
gammaDot = pow(2,0.5)*mag(symm(fvc::grad(U)));

sourceC  = fvc::laplacian(Kc*sqr(a)*sqr(H), gammaDot);

sourceV  = fvc::laplacian(Kmu*sqr(a)
                            * gammaDot*sqr(H)/laminarTransport.nu(),
                            laminarTransport.nu());

fvScalarMatrix HEqn
            (
            fvm::ddt(H)
            + fvm::div(phi, H)
            - fvm::laplacian(Kc*sqr(a)*H*gammaDot, H)
            ==
            sourceC
            + sourceCnonlin
            + sourceV
            );
```

For steady state (SIMPLE) and transient cases with the PIMPLE algorithm, underrelaxation is required. Typically the underrelaxation factor that is required can be estimated from the order of magnitude of the ratio between collision radius and vessel radius. Stable simulation has been achieved for relaxation factors of $0.1 \log(O(a/R))$, e.g., a radius $R = 50$ µm and collision radius of 3.5 µm will require an underrelaxation factor of $\approx$0.1 with no underrelaxation for the final iteration. The PISO algorithm does not use underrelaxation and requires a time estimated as the smaller of (a) timestep estimated from the Courant number ($Co < 1$), and (b) time step calculated based on a Courant number scaled with the migration velocity (instead of the convective velocity).

The discretisation schemes used in the calculations presented in this paper are: second order Euler backward in time and second order (Gauss linear, and Gauss linear upwind for advective terms) in space, gradients are approximated using the least squares theme.

Rheology models are implemented as quasi-Newtonian, with calculation of local cell viscosity based on the shear rate and haematocrit value in the cell from the previous iteration/time step. The new rheology models that are implemented at the time of writing are the standard Krieger–Dougherty, the modified 5-parameter Krieger, the Yeleswarapu–Wu, and the Quemada model.

3. Results

All results shown in this paper are for fully developed pipe flow, with periodic boundary conditions between outlet and inlet, with prescribed average velocity. The radius of the pipe varies between 50 µm and 5 mm, to represent typical vessel diameters. The pipe length is two diameters.

3.1. Verification and Influence of Mesh Type

The verification case for the implementation is a pipe of radius 50 µm, average velocity $V = 0.0065$ m s^{-1}. The rheology model used in the verification case is the Krieger–Dougherty model to allow comparison to the analytical solution [14] (no analytical solution available for the non-linear terms in the shear-stress and concentration dependent models). Model parameters for the Krieger–Dougherty model are $K_c = 0.41$, $K_\mu = 0.62$, $\phi^* = 0.68$, and $n = 1.82$.

The simulation was performed for different meshes, Figure 3, (a) an axisymmetric (2D) wedge with 50 cells resolution in radial direction, (b) a hexahedral, block structured mesh—50 cells radial, and (c) a polyhedral mesh with boundary layer inflation with ~60 cells across the diameter—this type of mesh is common in the simulation of vascular flow in patient specific geometries. The given resolutions were chosen based on a mesh convergence study and realistic mesh resolutions as typically used in vascular simulations. The migration model requires a mesh that is of similar resolution as meshes that aim at resolving wall shear stress (WSS) and WSS derived metrics.

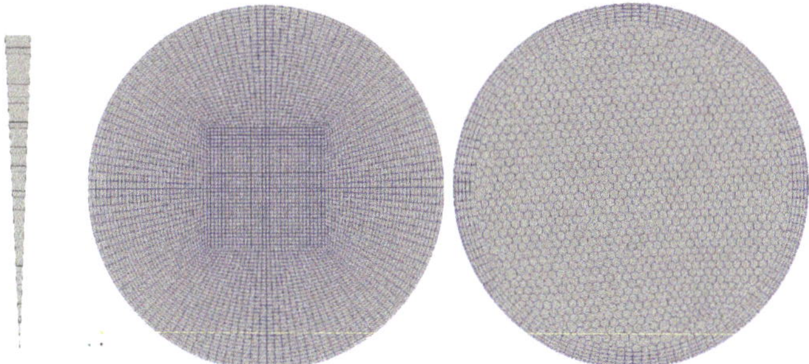

Figure 3. Mesh topology for the verification of the model: axisymmetric wedge, 50 cells radial; hexahedral, block-structured, 50 cells radial; polyhedral with boundary layer extrusion, 60 cells diameter.

Figure 4 shows the results for the different meshes in comparison to the analytical solution of the migration model with the Krieger–Dougherty model. The axisymmetric two-dimensional and the hexahedral three-dimensional meshes show excellent agreement, with only a slight rounding of the peaked analytical solution at the axis. The polyhedral three-dimensional mesh also shows good agreement, but the additional numerical diffusion blunts the profile at the axis, the concentration close to the wall is well represented.

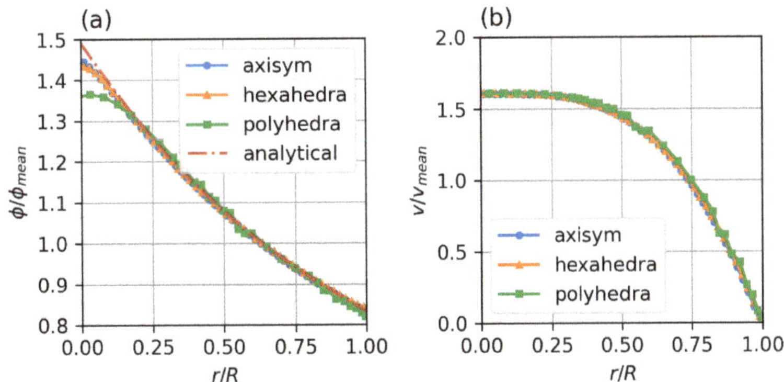

Figure 4. Steady state particle distribution (**a**) and velocity profiles (**b**) for different mesh types, compared with analytical solution for particle distribution by Krieger et al. Parameters: fully developed pipe flow, $R = 50$ µm, $V = 0.0065$ m s^{-1}, $K_c = 0.41$, $K_\mu = 0.62$, $n = 1.82$, $\phi^* = 0.68$, Standard Krieger–Dougherty Model.

3.2. Length and Time Scale Dependency

3.2.1. Wall Shear Strain Scaling

The parabolic velocity profile for a Newtonian flow is given as:

$$v = -2V\left(\frac{r^2}{R^2} - 1\right),$$ (27)

where V is the average velocity.

Therefore, the velocity gradient in radial direction is:

$$\frac{\partial v}{\partial r} = -\frac{4Vr}{R^2}.$$ (28)

So the gradient at the wall ($r = R$) scales with V and R^{-1}. The velocity is, therefore, scaled with R, such that the wall velocity gradient is constant. The Reynolds number scales with R^2. For the given values of $R = 0.05, 0.5, 5$ mm, $V = 0.0065,\ 0.065,\ 0.65$ m s^{-1}, the wall velocity gradient is constant at $\dot{\gamma}_w \approx 650$ s^{-1}, to cover the significant three decades of shear strain magnitude for shear-thinning non-Newtonian blood models.

The steady state particle distribution profile is independent of the length scale and the diameter ratio. It will only depend on the ratio of K_c / K_μ. Figure 5 shows steady state profiles for a range of diameters from 0.1 to 10 mm. The computational effort for the particle migration model, however, scales with R^2 / a^2, with R, the vessel radius, and a, the particle collision radius. While the small diameter $D = 0.1$ mm case is fully converged after around 10^4 iterations, the $D = 10$ mm case requires 10^6 iterations. This corresponds to the diffusion timescales.

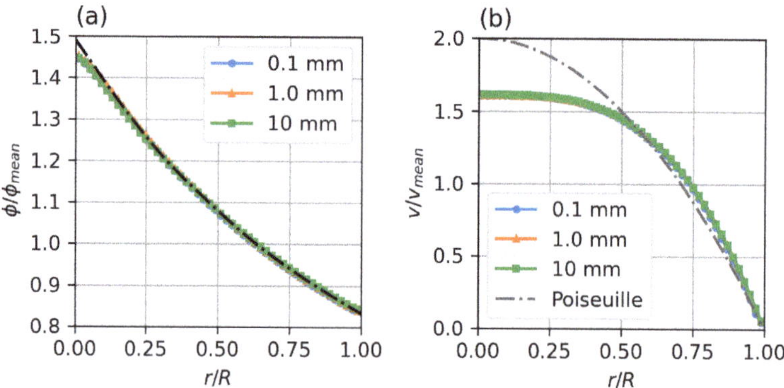

Figure 5. Steady state particle distribution (**a**) and velocity profiles (**b**) for different diameters. Parameters: fully developed pipe flow, $R = 0.05,\ 0.5,\ 5$ mm, $V = 0.0065,\ 0.065,\ 0.65$ m s^{-1}, $K_c = 0.41$, $K_\mu = 0.62$, $n = 1.82$, $\phi^* = 0.68$, Standard Krieger–Dougherty Model.

3.2.2. Kinematic and Particle Migration Timescales

Blood flow with particle migration is governed by several different time scales for flow kinematics and particle migration. The timescale for the development of the velocity profile (kinematic timescale) is

$$\tau_k = \frac{R^2}{\nu}.$$ (29)

The timescales for the development of the particle migration profile can be derived from the particle migration flux diffusion terms as:

$$\tau_{c\phi} = \frac{R^2}{K_c a^2 \phi \dot{\gamma}},$$ (30)

$$\tau_{c\dot{\gamma}} = \frac{R^2}{K_c a^2 \phi^2},$$ (31)

$$\tau_\mu = \frac{R^2}{K_\mu a^2 \dot{\gamma} \frac{\partial(\ln \mu)}{\partial \phi}}.$$ (32)

The kinematic timescale scales with R^2/v, while the particle migration timescales scale with the square diameter ratio R^2/a^2, where R is the pipe radius, and a is the particle (collision) radius.

The kinematic viscosity, $v \approx 3 \times 10^{-6}$ m^2 s^{-1}, while for an average collision radius of red blood cells of $a = 3.5$ μm, the particle migration diffusion coefficients are of the order of 10^{-9}–10^{-11} m^2 s^{-1}. This means the particle migration happens on timescales that are three orders of magnitude greater than the kinematic timescales.

Figure 6 shows the temporal development of the particle distribution and non-Newtonian velocity profile. The flows were initialised with a fully developed parabolic velocity profile and a uniform particle distribution of $\phi = 0.45$ volume fraction. The 0.1 mm case has reached steady state conditions within 0.5 s, the 1.0 mm case shows significant particle migration after physiologically relevant times, while the 10 mm case does show only minimal migration after 10 s. It can be seen that temporal scaling follows the predicted R^2/a^2 scaling factor.

3.3. Variation of Rheology Model and Collision Parameter Ratio

As is obvious from Equation (5), the particle migration is strongly dependent on the viscosity model and the balance between collision and viscosity driven migration, as expressed in the model parameters K_c and K_μ.

While the magnitude of K_c and K_μ controls the magnitude of the migration pressures and thus the temporal response of the system, the concentration profile only depends on the balance between collision and viscosity driven fluxes. This balance is expressed by the ratio between the parameters K_c/K_μ. Figure 7 shows the haematocrit profiles as they develop for different viscosity models—Krieger–Dougherty (K-D), Quemada (Q), Yeleswarapu–Wu (Y), modified 5-parameter Krieger model (K5), and varying K-ratios $K_c/K_\mu = 0.4$ to 0.75.

Compared to the verification K-D case with K-ratio of 0.66, it can be seen that a shift in the balance to higher influence of the collision frequency (higher K-ratio) steepens the profile, while a lower K-ratio, i.e., a shift of the balance to the resistive influence of the viscosity increase in the low shear region causes a flatter profile.

Comparing the different viscosity models clearly shows the main difference in the core region, where the strong variation in the low shear behaviour, discussed earlier, leads to a strong variation in the relative viscosity gradient (last term in Equation (5)). It is obvious that there is a need for further study and comparison with experimental or meso-scale modelling data to find realistic parameters for each of the potential viscosity models. Especially the modified Krieger model (K5) shows a, most likely unrealistic, double-bump profile at the axis.

Based on these results, the Quemada model with a K-ratio of between 0.5 and 0.6 seems to be the most promising candidate for a semi-mechanistic rheology model for blood.

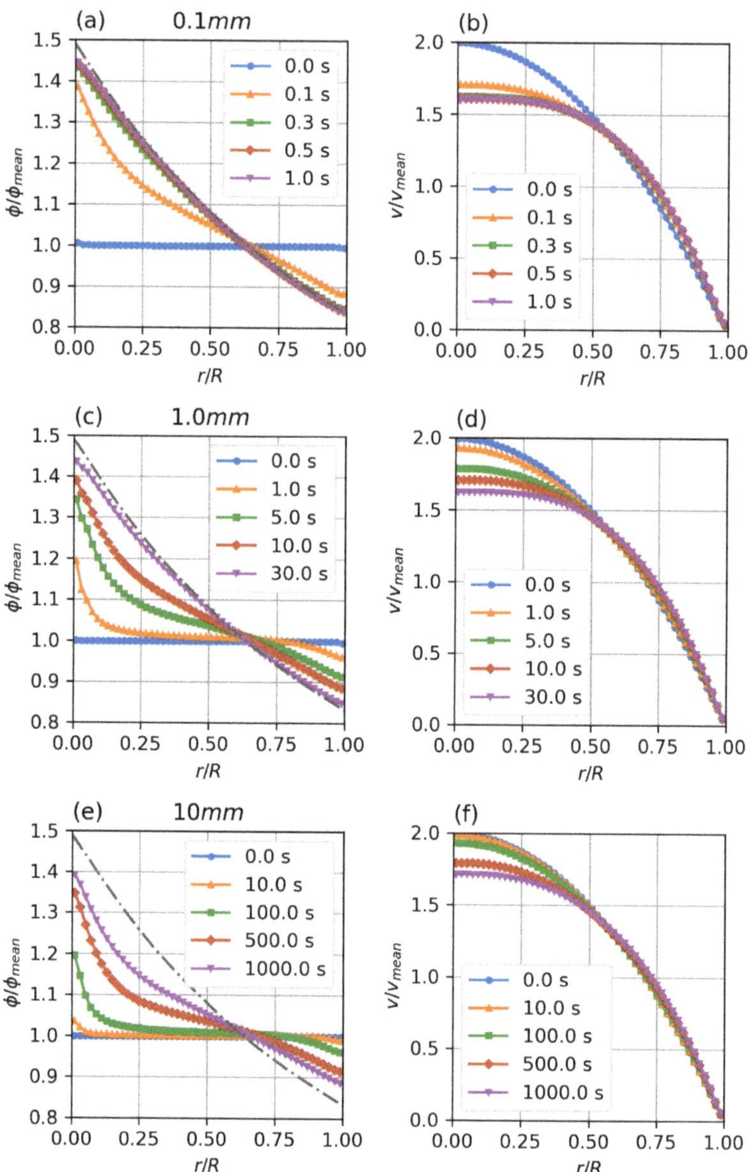

Figure 6. Transient particle distribution (**a**,**c**,**e**) and velocity profiles (**b**,**d**,**f**) for different diameters. Parameters: fully developed pipe flow, $R = 0.1$ (**a**,**b**), 1.0 (**c**,**d**), 10 (**e**,**f**) mm, $V = 0.0065$ (**a**,**b**), 0.065 (**c**,**d**), 0.65 (**e**,**f**) m s^{-1}, $K_c = 0.41$, $K_\mu = 0.62$, $n = 1.82$, $\phi^* = 0.68$, Standard Krieger–Dougherty Model.

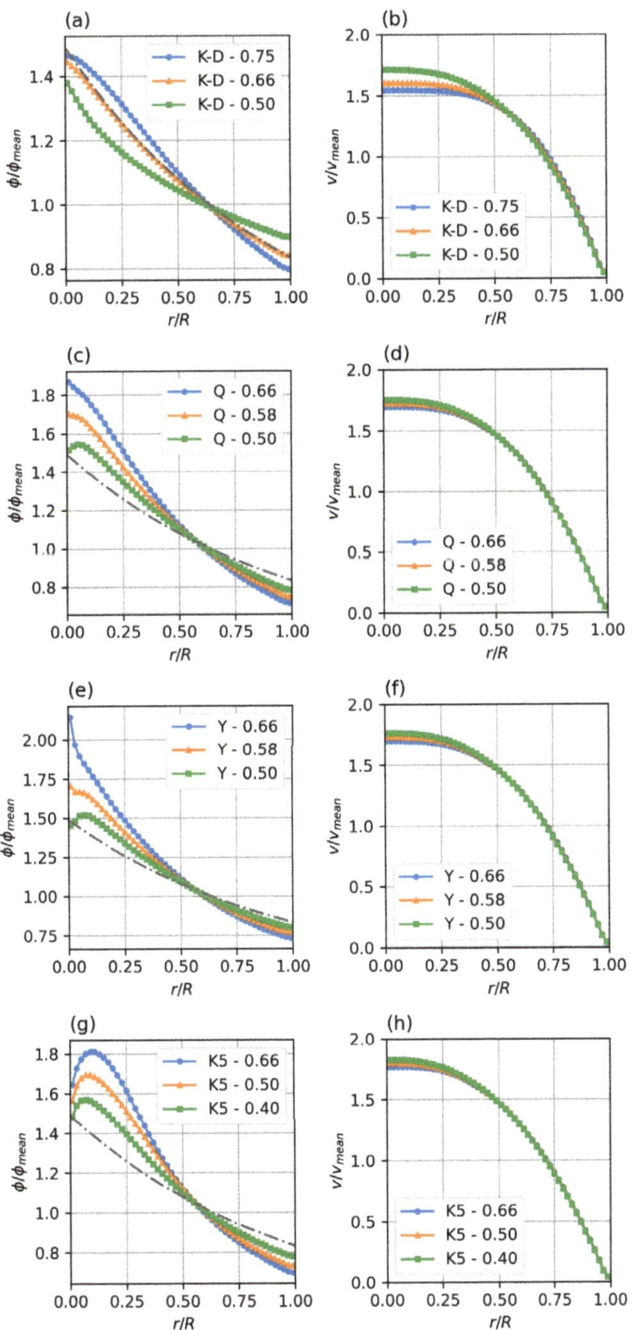

Figure 7. Steady state particle distribution (**a**,**c**,**e**,**g**) and velocity profiles (**b**,**d**,**f**,**h**) for different viscosity models and collision parameter ratios (K_c/K_μ). Parameters: fully developed pipe flow, $R = 0.05$ mm, $V = 0.0065$ m s^{-1}, $K_c/K_\mu = 0.4$–0.75 (given in legends). Standard Krieger–Dougherty (**a**,**b**), Quemada (**c**,**d**), Yeleswarapu (**e**,**f**), and modified 5-parameter Krieger (**g**,**h**) model.

3.4. Application to Realistic Vessel Model

The hybrid implicit–explicit implementation presented here allows for transparent change of viscosity model and does not use any artificial stabilisation, allowing for low residual solutions on realistic complex geometries.

Figure 8 shows the time-averaged haematocrit concentration near the wall and the time-averaged WSS for a realistic human common carotid artery (CCA).

The variation of haematocrit in this case is around $\pm1\%$, which is in agreement with the expected haematocrit variation for the size ($R_{CCA} = 3$ mm).

While the haematocrit transport is not significant in this case, it shows that the implementation is also stable for physiological vessel shapes (and the reduced grid quality that is typical for clinical applications) and for highly dynamic flow. The haematocrit-dependent viscosity models can be employed, and for larger vessels, where the haematocrit transport can be neglected, the haematocrit equation can be frozen, to reduce simulation overhead.

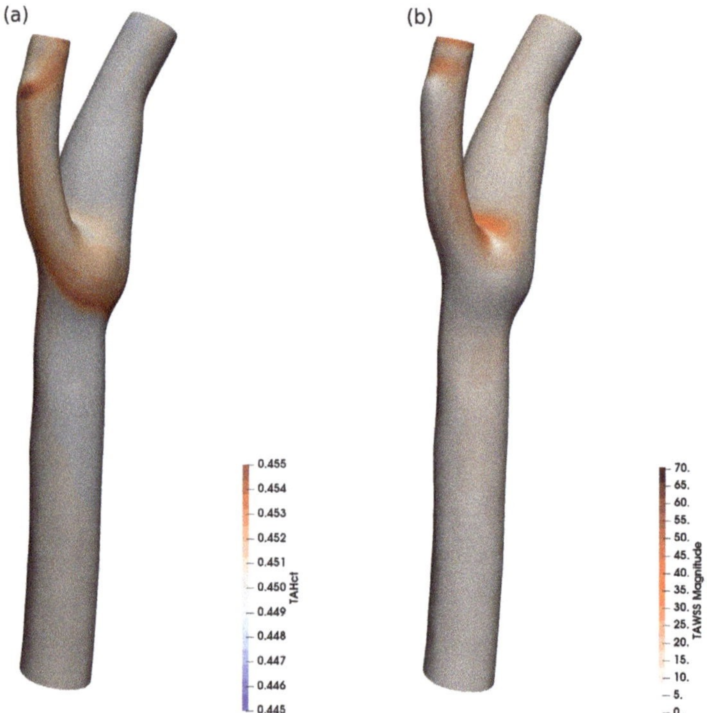

Figure 8. Time averaged haematocrit (**a**) and WSS (**b**) in a realistic human carotid, 900 k cells.

4. Discussion

While previous implementations [17,19] of this class of model are limited by the fact that the viscosity term in Equation (5) is linearised in the viscosity gradient with H, our implementation avoids this by implementing the non-linear term directly, which (i) allows the use of different viscosity models and, crucially, (ii) facilitates the functional forms of transport parameter dependence on field variables we are constructing from the operation of our explicit, meso-scale models. Furthermore, the present implementation avoids the use of artificial stabilisation terms, that lead to underestimation of RBC migration [18].

The particle migration time scales with $(a/R)^2$, where a is the RBC collision radius. This means that the particle migration is most relevant for small vessels of a diameter of 1 mm or lower, where the migration occurs on physiologically relevant timescales. For larger vessels, minor effects are caused by a synergy of particle migration and secondary flows [18] (Figure 8).

Limitations of the Model and Future Work

It has to be noted that the implementation presented here uses the magnitude of the shear in the particle flux formulation. As noted by Phillips [14], this assumes an essentially one-dimensional shear state, and an isotropic response. This limits the application of the model macroscopic to flow situations where the shear tensor is aligned with the flow and the main shear in radial direction, though it also provides a convenient initial simplification to our microscopic modelling. As with isotropic turbulence modelling the isotropic migration model will over-predict migration pressure in regions with high anisotropy, e.g., stagnation points, strong acceleration, or rotational shear. The authors are currently looking to implement an explicit formulation for a localised, anisotropic shear and migration pressure tensor, similar to approaches proposed by Miller [37] or Fang et al. [15].

Clearly, the parameters of the particle migration model we have used in this work are intentional targets for assignment using data from the operation of our explicit meso-scale simulations. Of course, for immediate impact, one might calibrate them to suitable experimental data, using established phenomenology. (Though note that the improved quantitative understanding we hope to obtain from our direct, meso-scale studies might produce the further benefit of insight into the functional form of transport parameter dependence upon field variables). While such experimental data are available, albeit scarce, for rigid particles in suspension, e.g., based on nuclear magnetic resonance measurements of particle profiles, the authors are not aware of any such data for soft vesicles, in particular RBCs. Here again, we hope eventually to use our recent meso-scale models (MCLBM) which account for the cellular-scale interactions in full detail [11] to refine and parameterise a constitutive equation for concentrated suspensions of deformable vesicles and hence to inform the present continuum model parameters. The results presented here provide us with the reassurance that, the final discussions below notwithstanding, a suitably sensitive, macro-scale model exists for this undertaking.

5. Software

The continuum-scale haemorheology framework was implemented in OpenFoam, version 1912 and 2012.

haemoFoam is a modelling framework for vascular flow simulation based on FOAM, that is intended to cater for the particular requirements of haemodynamics, in particular with respect to WSS related phenomena like atherosclerosis. At the time of writing it includes:

- Haematocrit transport model, modelling the shear driven transport of red blood cells in direction of the shear gradient.
- Blood specific non-Newtonian rheology models including haematocrit dependency and shear thinning behaviour:
    - Krieger Dougherty (non shear-thinning);
    - Modified K-D [25] (shear-thinning);
    - Quemada;
    - Yeleswarapu;
    - Casson–Merrill;
    - Carreau model (not concentration dependent, Fluent implementation).
- Windkessel boundary conditions for outlets.
- Fluid-Structure-Interaction (FSI) for flexible vessel walls.
- Post-processing for WSS and established WSS derived parameters:
    - TAWSS, TAWSSMag;
    - OSI;

-   Transverse WSS;
-   Relative Residence Time;
-   Temporal and spatial WSS gradients.

Planned future features are:

-   Viscoelastic rheology models (e.g., Oldroyd B);
-   Platelet transport;
-   Low density lipoprotein (LDL) transport.

haemoFoam is open-sourced under GPL3 and will be made available to interested parties upon request.

Author Contributions: Conceptualization, T.S. and I.H.; methodology, T.S. and I.H.; software, T.S.; validation, T.S.; formal analysis, T.S.; writing—original draft preparation, T.S.; writing—review and editing, T.S. and I.H.; visualization, T.S.; funding acquisition, I.H. and T.S. All authors have read and agreed to the published version of the manuscript.

Funding: This research was funded by NHS Heart of England Trust grant number REC 15/WM/0164.

Institutional Review Board Statement: Not applicable.

Informed Consent Statement: Not appicable.

Data Availability Statement: The software (haemoFoam) is freely available to interested parties on github (TS-CUBED/haemoFoam). Please contact the author for testing and developer access.

Conflicts of Interest: The authors declare no conflict of interest.

Abbreviations

The following abbreviations are used in this manuscript:

LBM	Lattice Boltzmann Method
CFD	Computational Fluid Mechanics
ADC	Antibody Drug Conjugates
MKM5	Modified 5-parameter form of the Krieger model
FVM	Finite Volume Method
SIMPLE	Semi-Implicit Method for Pressure Linked Equations
PISO	Pressure-Implicit with Splitting of Operators
PIMPLE	compination of PISO and SIMPLE
CCA	Common Carotid Artery
WSS	Wall Shear Stress
OSI	Oscillatory Shear Index
TAWSS	Time Averaged WSS
TAWSSMag	Time Averaged WSS magnitude
RRS	Relative Residence Time

1. Holton, J.R. *An Introduction to Dynamic Meteorology*; Academic Press: Cambridge, MA, USA, 2004.
2. Secomb, T.W. Blood Flow in the Microcirculation. *Annu. Rev. Fluid Mech.* **2017**, *49*, 443–461. [CrossRef]
3. Bessonov, N.; Sequeira, A.; Simakov, S.; Vassilevskii, Y.; Volpert, V. Methods of Blood Flow Modelling. *Math. Model. Nat. Phenom.* **2016**, *11*, 1–25. [CrossRef]
4. Tosenberger, A.; Salnikov, V.; Bessonov, N.; Babushkina, E.; Volpert, V. Particle Dynamics Methods of Blood Flow Simulations. *Math. Model. Nat. Phenom.* **2011**, *6*, 320–332. [CrossRef]
5. Aidun, C.K.; Clausen, J.R. Lattice-Boltzmann Method for Complex Flows. *Annu. Rev. Fluid Mech.* **2010**, *42*, 439–472. [CrossRef]
6. Clausen, J.R.; Reasor, D.A.; Aidun, C.K. Parallel Performance of a Lattice-Boltzmann/Finite Element Cellular Blood Flow Solver on the IBM Blue Gene/P Architecture. *Comput. Phys. Commun.* **2010**, *181*, 1013–1020. [CrossRef]
7. Ladd, A.J.C. Numerical Simulations of Particulate Suspensions via a Discretized Boltzmann Equation. Part 1. Theoretical Foundation. *J. Fluid Mech.* **1994**, *271*, 285–309. [CrossRef]
8. Ladd, A.J.C. Numerical Simulations of Particulate Suspensions via a Discretized Boltzmann Equation. Part 2. Numerical Results. *J. Fluid Mech.* **1994**, *271*, 311–339. [CrossRef]

9. MacMeccan, R.M., III. Mechanistic Effects of Erythrocytes on Platelet Deposition in Coronary Thrombosis. Ph.D. Thesis, Georgia Institute of Technology, Atlanta, GA, USA, 2007.
10. Dupin, M.M.; Halliday, I.; Care, C.M.; Alboul, L.; Munn, L.L. Modeling the Flow of Dense Suspensions of Deformable Particles in Three Dimensions. *Phys. Rev. E Stat. Nonlinear Soft Matter Phys.* **2007**, *75*, 066707. [CrossRef]
11. Spendlove, J.; Xu, X.; Schenkel, T.; Seaton, M.A.; Halliday, I.; Gunn, J.P. Three-Dimensional Single Framework Multicomponent Lattice Boltzmann Equation Method for Vesicle Hydrodynamics. *Phys. Fluids* **2021**, *33*, 077110. [CrossRef]
12. Burgin, K. Development of Explicit and Constitutive Lattice-Boltzmann Models for Food Product Rheology. Ph.D. Thesis, Sheffield Hallam University, Sheffield, UK, 2018.
13. Leighton, D.; Acrivos, A. The Shear-Induced Migration of Particles in Concentrated Suspensions. *J. Fluid Mech.* **1987**, *181*, 415–439. [CrossRef]
14. Phillips, R.J.; Armstrong, R.C.; Brown, R.A.; Graham, A.L.; Abbott, J.R. A Constitutive Equation for Concentrated Suspensions That Accounts for Shear-Induced Particle Migration. *Phys. Fluids A Fluid Dyn.* **1992**, *4*, 30–40. [CrossRef]
15. Fang, Z.; Mammoli, A.A.; Brady, J.F.; Ingber, M.S.; Mondy, L.A.; Graham, A.L. Flow-Aligned Tensor Models for Suspension Flows. *Int. J. Multiph. Flow* **2002**, *28*, 137–166. [CrossRef]
16. Nott, P.R.; Brady, J.F. Pressure-Driven Flow of Suspensions: Simulation and Theory. *J. Fluid Mech.* **1994**, *275*, 157–199. [CrossRef]
17. Mansour, M.H.; Bressloff, N.W.; Shearman, C.P. Red Blood Cell Migration in Microvessels. *Biorheology* **2010**, *47*, 73–93. [CrossRef]
18. Biasetti, J.; Spazzini, P.G.; Hedin, U.; Gasser, T.C. Synergy between Shear-Induced Migration and Secondary Flows on Red Blood Cells Transport in Arteries: Considerations on Oxygen Transport. *J. R. Soc. Interface* **2014**, *11*, 20140403. [CrossRef]
19. Chebbi, R. Dynamics of Blood Flow: Modeling of Fåhraeus and Fåhraeus–Lindqvist Effects Using a Shear-Induced Red Blood Cell Migration Model. *J. Biol. Phys.* **2018**, *44*, 591–603. [CrossRef] [PubMed]
20. Krieger, I.M.; Dougherty, T.J. A Mechanism for Non-Newtonian Flow in Suspensions of Rigid Spheres. *Trans. Soc. Rheol.* **1959**, *3*, 137–152. [CrossRef]
21. Quemada, D. Rheology of Concentrated Disperse Systems and Minimum Energy Dissipation Principle—I. Viscosity-Concentration Relationship. *Rheol. Acta* **1977**, *16*, 82–94. [CrossRef]
22. Quemada, D. Rheology of Concentrated Disperse Systems II. a Model for Non-Newtonian Shear Viscosity in Steady Flows. *Rheol. Acta* **1978**, *17*, 632–642. [CrossRef]
23. Quemada, D. Rheology of Concentrated Disperse Systems III. General Features of the Proposed Non-Newtonian model. Comparison with Experimental Data. *Rheol. Acta* **1978**, *17*, 643–653. [CrossRef]
24. Das, B.; Johnson, P.C.; Popel, A.S. Effect of Nonaxisymmetric Hematocrit Distribution on Non-Newtonian Blood Flow in Small Tubes. *Biorheology* **1998**, *35*, 69–87. [CrossRef]
25. Hund, S.; Kameneva, M.; Antaki, J. A Quasi-Mechanistic Mathematical Representation for Blood Viscosity. *Fluids* **2017**, *2*, 10. [CrossRef]
26. Merrill, E.W.; Gilliland, E.R.; Cokelet, G.; Shin, H.; Britten, A.; Wells, R.E. Rheology of Human Blood, near and at Zero Flow: Effects of Temperature and Hematocrit Level. *Biophys. J.* **1963**, *3*, 199–213. [CrossRef]
27. Yeleswarapu, K.K.; Kameneva, M.V.; Rajagopal, K.R.; Antaki, J.F. The Flow of Blood in Tubes: Theory and Experiment. *Mech. Res. Commun.* **1998**, *25*, 257–262. [CrossRef]
28. Brooks, D.E.; Goodwin, J.W.; Seaman, G.V. Interactions among Erythrocytes under Shear. *J. Appl. Physiol.* **1970**, *28*, 172–177. [CrossRef]
29. Papir, Y.S.; Krieger, I.M. Rheological Studies on Dispersions of Uniform Colloidal Spheres: II. Dispersions in Nonaqueous Media. *J. Colloid Interface Sci.* **1970**, *34*, 126–130. [CrossRef]
30. Cokelet, G.R.; Merrill, E.W.; Gilliland, E.R.; Shin, H.; Britten, A.; Wells, R.E. The Rheology of Human Blood—Measurement near and at Zero Shear Rate. *Trans. Soc. Rheol.* **1963**, *7*, 303–317. [CrossRef]
31. Sequeira, A.; Janela, J. An Overview of Some Mathematical Models of Blood Rheology. In *A Portrait of State-of-the-Art Research at the Technical University of Lisbon*; Seabra Pereira, M., Ed.; Springer: Dordrecht, The Netherlands, 2007; pp. 65–87. [CrossRef]
32. Wu, W.; Pott, D.; Mazza, B.; Sironi, T.; Dordoni, E.; Chiastra, C.; Petrini, L.; Pennati, G.; Dubini, G.; Steinseifer, U.; et al. Fluid–Structure Interaction Model of a Percutaneous Aortic Valve: Comparison with an In Vitro Test and Feasibility Study in a Patient-Specific Case. *Ann. Biomed. Eng.* **2015**, *44*, 590–603. [CrossRef]
33. Jung, J.; Hassanein, A. Three-Phase CFD Analytical Modeling of Blood Flow. *Med. Eng. Phys.* **2008**, *30*, 91–103. [CrossRef]
34. Casson, M. A Flow Equation for Pigment-Oil Suspensions of the Printing Ink Type. In *Rheology of Disperse Systems*; Pergamon Press: Oxford, UK, 1959; pp. 84–104.
35. Patankar, S.V.; Spalding, D.B. A Calculation Procedure for Heat, Mass and Momentum Transfer in Three-Dimensional Parabolic Flows. *Int. J. Heat Mass Transf.* **1972**, *15*, 1787–1806. [CrossRef]
36. Issa, R.I.; Gosman, A.D.; Watkins, A.P. The Computation of Compressible and Incompressible Recirculating Flows by a Non-Iterative Implicit Scheme. *J. Comput. Phys.* **1986**, *62*, 66–82. [CrossRef]
37. Miller, R.M.; Singh, J.P.; Morris, J.F. Suspension Flow Modeling for General Geometries. *Chem. Eng. Sci.* **2009**, *64*, 4597–4610. [CrossRef]

mathematics

Article

A Parametric Tool for Studying a New Tracheobronchial Silicone Stent Prototype: Toward a Customized 3D Printable Prosthesis

Jesús Zurita-Gabasa [1], Carmen Sánchez-Matás [2], Cristina Díaz-Jiménez [3], José Luis López-Villalobos [2] and Mauro Malvè [1,4,*]

[1] Department of Engineering, Public University of Navarra (UPNA), Campus Arrosadía s/n, E-31006 Pamplona, Spain; jesus.zurita@unavarra.es
[2] Department of Thoracic Surgery, University Hospital Virgen del Rocío, Avenida Manuel Siurot s/n, E-41013 Sevilla, Spain; nem.csm@gmail.com (C.S.-M.); jllopezvillalobos@gmail.com (J.L.L.-V.)
[3] AIN-Asociación de la Industria Navarra, Ctra. Pamplona, 1. Edif. AIN, E-31191 Cordovilla, Spain; cdiaz@ain.es
[4] CIBER-BBN, Research Networking in Bioengineering, Biomaterials & Nanomedicine, C/Mariano Esquillor s/n, E-50018 Zaragoza, Spain
* Correspondence: mauro.malve@unavarra.es

Abstract: The management of complex airway disorders is challenging, as the airway stent placement usually results in several complications. Tissue reaction to the foreign body, poor mechanical properties and inadequate fit of the stent in the airway are some of the reported problems. For this reason, the design of customized biomedical devices to improve the accuracy of the clinical results has recently gained interest. The aim of the present study is to introduce a parametric tool for the design of a new tracheo-bronchial stent that could be capable of improving some of the performances of the commercial devices. The proposed methodology is based on the computer aided design software and on the finite element modeling. The computational results are validated by a parallel experimental work that includes the production of selected stent configurations using the 3D printing technology and their compressive test.

Keywords: tracheobronchial stent; finite element method; parametric model; 3D printing; customized prosthesis

Citation: Zurita-Gabasa, J.; Sánchez-Matás, C.; Díaz-Jiménez, C.; López-Villalobos, J.L.; Malvè, M. A Parametric Tool for Studying a New Tracheobronchial Silicone Stent Prototype: Toward a Customized 3D Printable Prosthesis. *Mathematics* 2021, 9, 2118. https://doi.org/10.3390/math9172118

Academic Editor: Rafael Sebastian

Received: 22 July 2021
Accepted: 24 August 2021
Published: 1 September 2021

Publisher's Note: MDPI stays neutral with regard to jurisdictional claims in published maps and institutional affiliations.

1. Introduction

Tracheobronchial or airway stents are tubular scaffolds used for enlarging a constricted airway or supporting the trachea and/or bronchi from collapse. This can be due to airways obstructions caused by prolonged intubation, or benignant or malignant carcinoma and tracheomalacia that may lead to several morbidities. Normally, tracheobronchial prosthesis is necessary as a last chance if the stenosis cannot be treated with surgical means [1]. Airway stents are available in different materials and shapes and can be classified into four categories: silicone, balloon-expandable metal, uncovered and covered self-expanding metal [2]. Solid silicone tubes have been designed for avoiding in-stent restenosis, but are affected by migration and obstruction as main side effects [3]. Metallic devices, usually made of steel or nitinol, are similar to those used in the cardiovascular field. They promote re-epithalization that avoids migration, but cannot avoid restenosis, becoming less efficient as soon as the airway tissue grows and cause frequent granulations [4]. Partially, the covered metallic stents or hybrid stents have solved the problem of the restenosis [5]. However, covered self-expanding metal devices are associated to necrosis of mucosa and fistula formation due to the radial force applied by the stent [6].

The choice of a specific stent is determined by the type of lesion [7]. Silicone prostheses are usually used for both benign and malign pathologies. Metallic devices are indicated for malignant obstructions, but their use has been recently not recommended if not fully covered [8]. Silicone prostheses are commonly used in Europe and Japan [9,10] while metallic or hybrid stents are more frequently implanted in U.S.A. so that the use of these scaffolds are also driven by the clinicians school. In this sense, each stent has advantage and contraindications, but the response of the tissue to their implantation is always problematic due to unavoidable inflammations and reaction to the foreign body. Additionally, all stent categories are, in general, very rigid and tend to impede the physiological maneuvers of coughing, swallowing and forced breathing [2,11–13]. For these reasons, the design of new prostheses capable of addressing the flaws of the existing stents is necessary. Several absorbable polymeric stents have been experimentally tested for their possible use in the airways. The associated inflammatory reaction has been less extensive with respect to permanent stents [14]. Unfortunately, the degradation time of these type of stents in the trachea or bronchi remains difficult to control, or is even undefined [15]. In the last three decades, the clinical experience has demonstrated that there is a large number of situations in which commercial stents could not solve the clinical problem [16].

A tracheobronchial prosthesis should satisfy different requirements. It should be easy to be inserted and eventually removed. It should ideally avoid migration and be biocompatible, limiting the tissue reaction [10]. It should adapt to the airway and possibly be patient customizable [17]. Airway stent implantation could in fact result in inefficient clinical results due to the poor fit of the stent in the airway [18]. Personalized prosthesis can already be ordered to commercial factories. The personalization regards size, diameters and angles measured by computerized tomography (CT) scan and bronchoscopy. Nevertheless, the baseline design remains unchanged despite the personalization, as the stents are still straight tubes [16]. In this context, clearly, a customization that only provides changes of main geometrical characteristics is too simple and not efficient. The three dimensional (3D) printing offers a new opportunity that is rapidly entering in the clinics and allows rapid prototyping and fabrication of patient-specific anatomical shapes [10]. This technique has been rapidly entered in the clinics and it has been used recently for surgical treatment [19]. In the literature, it is stated that it is already capable of manufacturing optimized devices made of silicone or elastic thermoplastics for a particular patient. Stents can be designed for matching a particular patient specific anatomy and for exerting the necessary radial force [16]. The FDA has recently released guidelines on the 3D printing of medical devices [20].

The 3D-dimensional printing technique has already been reported in many clinical applications, including the thoracic surgery [21,22]. The possibility to convert anatomic images into 3D objects using this technique helps the surgeon to overcome specific problems and prepare the surgery [23]. Miyazaki et al. [24] used a 3D printed airway model to manage a post-transplant airway complications of bronchial anastomosis. Guilbert et al. [10] utilized a 3D printed models of corrected airways to select and customize airway stents. Although promising results have been obtained, other clinical complications such as mucus plugging and migration have been not solved. In the same line, Gildea et al. [18] treated complex stenoses due to granulomatosis with polyangiitis in two patients while Cheng et al. [25] treated a patient with a tracheal dehiscence. Morrison et al. [26] successfully applied the 3D printing technology to produce a personalized medical device for treating the tracheobronchomalacia. Debiane et al. [27] use the stereology to quantify the granulation. They have analyzed which type of tissue contributes more to the pathology and used this information for designing a customized drug eluting tracheo-bronchial stent.

Furthermore, a few studies have presented engineering tools for the design of new tracheobronchial parametric and/or customizable stents. Melgoza et al. [28,29] presented an integrated tool for the design of an innovative customized tracheal stent with aim of meeting the most critical requirements of a prosthesis. For this, the experience collected by clinicians and patients during the interviews and hospital visits has been used. Vearick and coworkers proposed a modification of the commercial Dumon stent [30] and introduced a fiber reinforced silicone prosthesis [31]. Schopf et al. [14] introduced a new polymer absorbable stent with a spiral shape evaluating the clinical signs and histological reaction in an experimental rabbit model.

Taking into account these aspects, in the present study we focus on the silicone Dumon stent. We aimed to design, simulate, produce and test a new tracheobronchial prosthesis that is customized to the patient, printable with 3D technology and overtakes some of the limitations of the commercially available devices. In particular, we proposed a parametric tool capable of simulating the influence of several variations in the geometry of the prosthesis, yet assessing the importance of each single parameter. Additionally, the tool was validated against an experimental test aimed to prove the reliability of the computational results.

2. Materials and Methods

2.1. Parametric Geometry of the Tracheobronchial Prosthesis

The new stent was designed to be tubular. The baseline CAD model, created with the in-house code, is represented in Figure 1. The outside of the tube was designed with an upward reinforcing structure that is similar to the typical X-pattern metallic stents (see Figure 2). The device geometry has been parametrized in order to study the effect of each single feature on the mechanical properties. Through modulating the different parameters in fact, the flexibility, the radial stiffness and the mechanical strength of the stent can be manipulated. In Figure 2, the parameters considered in this work are depicted on the unwrapped stent (Figure 2a), on the frontal view of the stent (Figure 2b) and on a detail representing the fibered stent wall Figure 2c). The baseline tube resembles the widely known Dumon prosthesis [32]. The outer skewed fibers that reinforce the baseline tube are located only in a percentage of the total outer surface area and they progressively reduce their thickness (see Figure 2). Differently from the commercial Dumon prosthesis, but similarly to the natural stent [33], the new prototype presents a novel design in which the radial stiffness varies. The reason is the particular behavior of the trachea during the physiological maneuvers of forced breathing, coughing and swallowing. The trachea is composed by a transversal muscular membrane and stiffer cartilage rings. During breathing and coughing, for example, the trachea dilates and collapses, enlarging and reducing its diameter. Especially during coughing, it is the muscular membrane that considerably deforms. It has been reported that the Dumon stent is rigid, as it has a constant thickness. For this reason, the side of the prosthesis that corresponds to the transversal membrane has been designed without fibers. Hence, as visible in Figures 1 and 2 the fibers cover only a portion of the outer prosthesis surface that could be also varied as it is the parameter p as explained below.

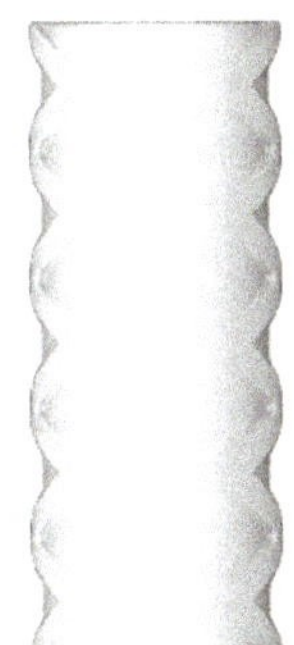

Figure 1. Baseline geometry of the stent prototype.

The shape of the cells that reinforces the outer prototype surface is governed by their number in longitudinal (n_L) and radial (n_R) direction. An increase of the cells number in longitudinal direction promotes smaller pitch angle α (also called braiding angle in metallic stents, see Figure 2) while an increase of the cells number in circumferential direction promotes higher pitch angle α. As visible in Figure 2, with smaller pitch angles, the cell shape is a rhombus circumferentially oriented around the outer prosthesis surface. With higher pitch angles, the cell shape is a rhombus longitudinally oriented. As visible in Equation (1), the pitch angle α can be computed as a function of the number of circumferential and longitudinal cells given to the prosthesis:

$$\alpha = atan\left(\frac{L/n_L}{p(2\pi R)/n_R}\right) \tag{1}$$

where L is the length of the prosthesis, n_L is the number of cells in longitudinal direction, p is the percentage of the external prosthesis surface that is fiber reinforced, r is the inner radius of the prosthesis and n_R is the number of cells in radial direction. The parameter p governs the extension of the fibers around the external surface of the tube and, as a consequence, the width of the region of the prosthesis without fibers.

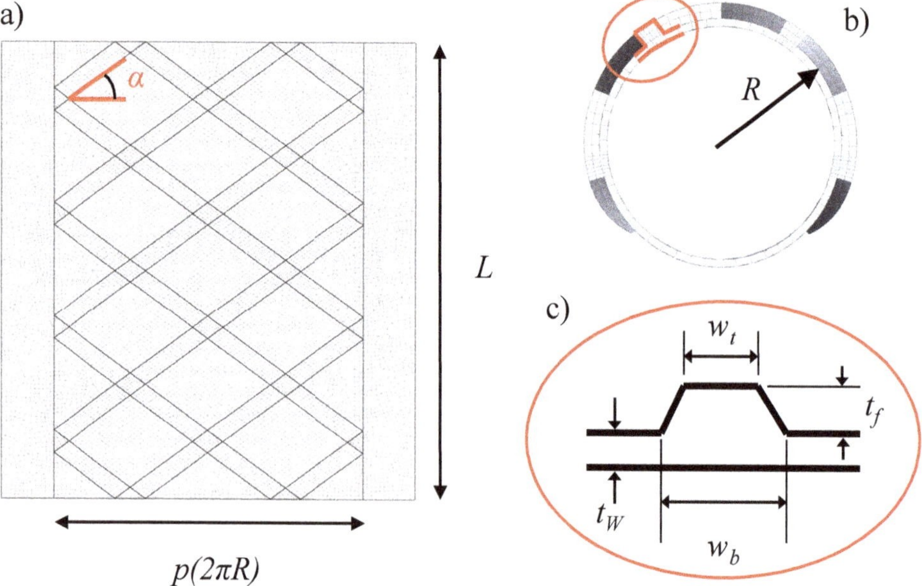

Figure 2. Parametrization of the tracheobronchial stent prototype: (**a**) unwrapped geometry, (**b**) top section of the tubular configuration and (**c**) detail of the stent fiber.

The dimensions of the fibers are parametrized. In particular, fiber bottom and top width are represented with w_b and w_t in Figure 2c). Finally, the thickness of the stent is represented by t_w and that of the fibers by t_f. Of course, stent radius R and length L are also parameters and they can be adapted to a specific patient. In this way, the device is customizable.

In Figure 2b), it is visible that the fiber thickness gradually decreases (see also Figure 1. This thickness has a biomechanical relevance, as it is known that a roundoff of the edges of the whole geometry of the prosthesis reduces tissue damages. For these reasons, the thickness was gradually reduced (from 2 to 0.8 mm in Figure 2b) for the prosthesis fabrication and for computational models of the experimental validation. On the contrary, as the mechanical response of the prosthesis is not affected by this parameter, the overall numerical simulations of the computational study were generated with a constant thickness.

In Table 1, the parameters and corresponding values are summarized. The number of the necessary computations that consider all the variations would imply a number $n = 972$ simulations. In this work, only $n = 86$ simulations were run, considering the most important parameter variations and taking into account the computational costs. In fact, the goal of the study is twofold: for one side, a parametric analysis of the new prototype is carried out. On the other side, the final aim of the study is to present a computational tool capable of performing all the necessary simulations for the customization of a prosthesis for a specific patient. For this reason, not all the possible parameter combinations are given or simulated.

Table 1. Summary of the parameters and their corresponding values.

t_w [mm]	t_f [mm]	n_R	n_L	L [mm]	p [%]	R [mm]	w_b	w_t/w_b
0.5–1	1–0.5	3–5	3–5	50	0.75	9	1–4	0.5–1

2.2. Stent Fabrication

Three tracheobronchial stent prototypes were fabricated by ACEO - 3D PRINTING WITH SILICONES (Wacker Chemie AG, ACEO Campus, Burghausen, Germany) starting from the STL (Stereolithography) files of the prostheses. The stents were 3D-printed using medical silicone technology. They were printed using medical silicone of modulus of elasticity of 15 ± 0.4 MPa and hardness 3.5 ± 0.2 MPa that were obtained after a material property analysis at AIN (Asociación de la Industria Navarra, Pamplona, Spain). In particular, the stress–strain curve was obtained by means of a traction test. The obtained curve evidenced a linear elastic region for deformation until about 37%. The three samples were built adding single layers composed by thin longitudinal slices using a manufacturing process. In Figure 3, the three prosthesis are shown after the printing process. The stents' height and outer diameter were 50 mm and 18 mm, respectively. The stent wall and fiber thickness was 1 mm each and the fibered reinforced surface covered the 75% of the device perimeter. The pitch angle α was changed within the three prototypes. Its values, summarized in Table 2, produce different cell configurations longitudinally and transversally. The number of cells in the two direction changes: prototype #1, later called $A3$, has 3×5 cells, prototype #2 or $A4$ has 4×5 cells, and prototype #3 or $A5$ has 5×3 cells. As can be seen in the Figure 2, the difference between the three fibered stents is represented by the pitch angle that promotes different cell shapes as explained in the previous section.

Table 2. Summary of the geometrical feature of the printed prototypes.

	t_w [mm]	t_f [mm]	n_R	n_L	α [°]	L [mm]	p [%]	R [mm]	w_t/w_b
#1	1	1	3	5	35.27	50	75	0.9	1
#2	1	1	4	5	43.35	50	75	0.9	1
#3	1	1	5	3	63.03	50	75	0.9	1

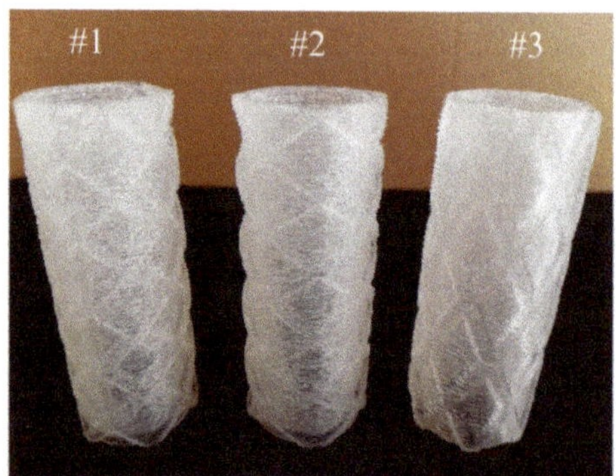

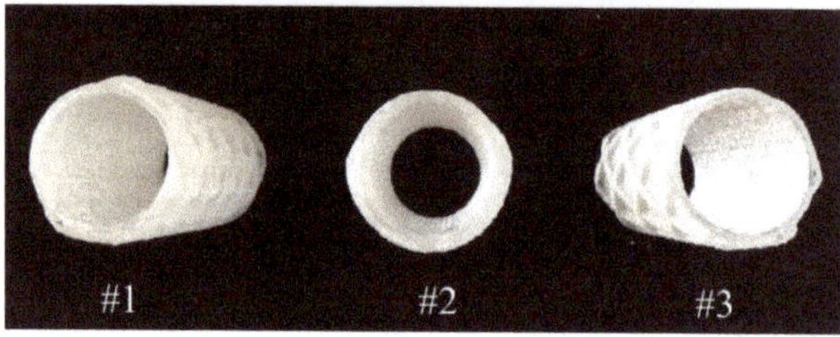

Figure 3. Stent prototypes printed by ACEO and used for the experimental compressive tests: #1 prototype, #2 prototype, #3 prototype.

2.3. Prosthesis Experimental Testing

As the prototypes are tubular, it was not possible to perform a traditional radial compressive test for assessing their radial stiffness as in the case of bare metallic stent for cardiovascular applications. With the aim of assessing the resistance of the prosthesis to compressive loads and to determine the effect of the fibers on the mechanical response of the prosthesis, a flat plate test have been carried out in NAITEC (Navarre Technology Center of Automotive and Mechatronics, Pamplona, Spain). The printed prototypes were placed between two flat plates (Figure 4a), compressed under displacements control until the inner prosthesis diameter was reduced to zero (Figure 4b) and unloaded with a compression rate of 0.1 mm/s. The force was measured by means of an Advanced Digital Force Gauge (AFG250N) LE01/50 (Mecmesin, Slinfold, West Sussex, UK).

Figure 4. Flat plate experimental testing: (**a**) unloaded and (**b**) crushed configuration.

2.4. Prosthesis Computational Modelling

The geometry was parametrized using an in-house software that automatically generates the computational grid. The aim of the parametrization is the analysis of the effect of each geometrical feature and of the customization of the stent to different patients. For facilitating the automatization of the simulations, the geometries and grids were generated contemporaneously. The generation of the mesh has been carried out using prismatic elements selecting the desired element size directly on a specific geometrical configuration hence fixing first the desired values of each parameter described in the previous section. In details, the in-house code was developed in C language and consists in five steps.

In the first step, the basic planar geometry has been generated. The corresponding parameters are the prosthesis length L, radius R, number of longitudinal fibers n_L and fiber thickness t_f. According to Figure 5a six areas are firstly defined: rectangular surfaces corresponding to the fibers (a), rhomboidal or quadratic surfaces corresponding to the cell unit (b), triangular surfaces corresponding to regions between the unit cells and the non-fibered region (c), triangular surfaces between fibers (d) or unit cells (e) and boundary of the fibered region of the prosthesis sections and rectangular non-fibered surfaces (f).

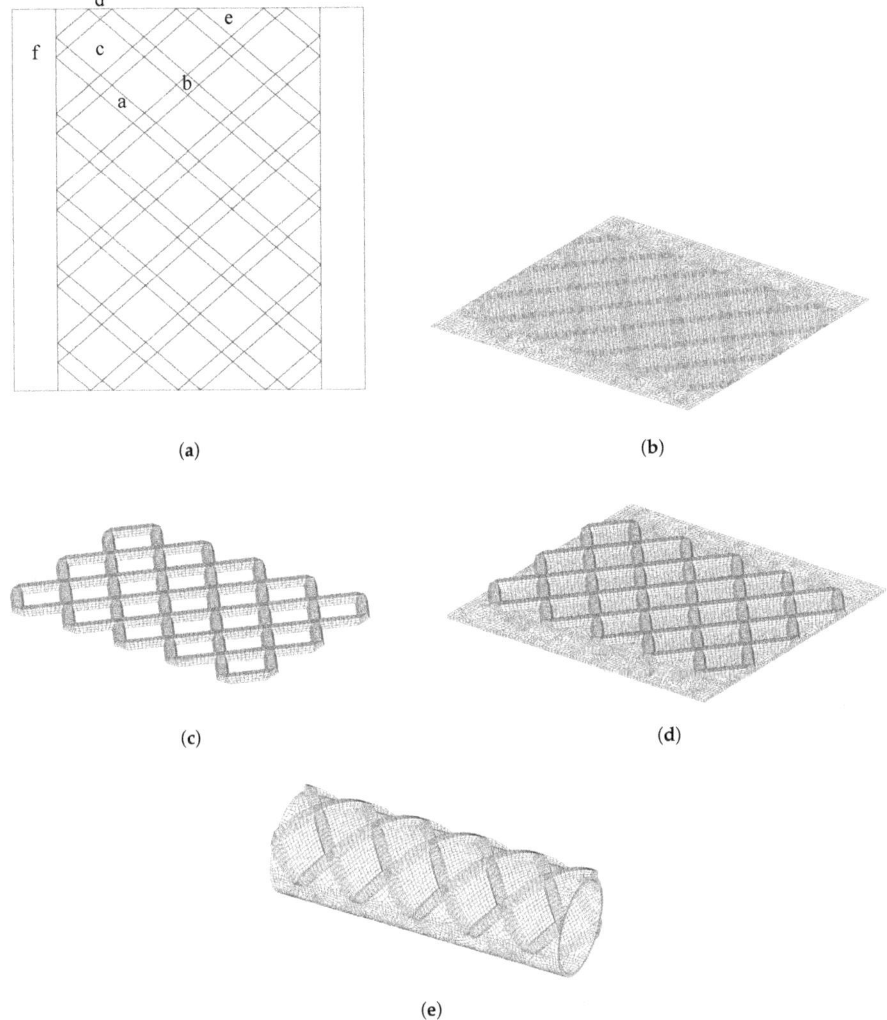

Figure 5. Generation of the prosthesis using the in-house code. (**a**) Planar geometry: generated surfaces a, b, c, d e, f; (**b**) Planar mesh; (**c**) Fiber extrusion; (**d**) Planar 3D mesh; (**e**) Final 3D mesh of the prosthesis.

In the second step, the surfaces a, b, c are meshed with structured quadrilateral elements as sketched in Figure 5b. In contrast, surfaces d, e and f are meshed with unstructured quadrilateral elements. The quadrilateral mesh is obtained starting from a previous triangularization of the surfaces performed by means of a Delaunay algorithm. The grid obtained after step 2 is represented in Figure 5b.

In step 3, the planar mesh of the surfaces a, b and c is extruded in normal direction obtaining the tridimensional brick mesh sketched in Figure 5c. Here, the considered parameters are the fiber thickness t_f and the top and bottom width w_t and w_b, respectively, (see Figure 2c).

Then, in step 4, the rest of the surfaces mesh are extruded in normal direction, obtaining the grid sketched in Figure 5d.

In the last step, the mesh is bent, obtaining a cylindrical solid meshed with hexahedral elements obtaining the tubular prosthesis depicted in Figure 2e.

In the Figure 6, the mesh topology is shown for the unwrapped geometries used for the experimental validation. The total number of elements used for the computations varies depending on the particular configuration to be studied. In particular, it ranges from 8198 to 61,480 prismatic elements, while the number of nodes ranges correspondently between 14,218 to 81,387. The average element size of the grid is 0.7 mm. The mesh element size was selected after an appropriate mesh size sensitivity analysis. The Figure 7 shows the force-displacements curve for grids with different maximum element sizes from 1 mm to 2 mm (that correspond to average element sizes of 0.7 mm and 1.6 mm, respectively). For this analysis, we used a prosthesis with $n_R = 5$, $n_L = 5$, $t_w = 0.5$ mm, $t_f = 2$ mm, $w_b = 2$ and $w_t = 2$. A comparison of the presented force-displacement curves proves the independency of the results on the discretization. The finest mesh refinement (black line with circles) converged adequately within 2% of the densest evaluated mesh (yellow line with circles).

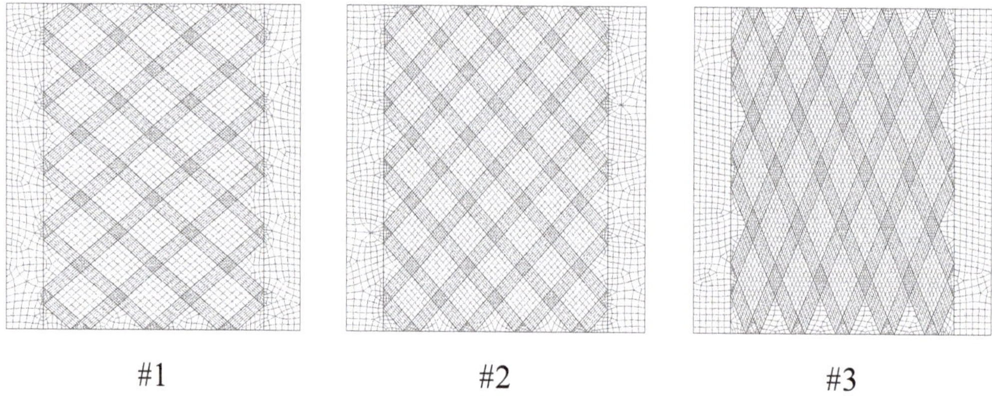

Figure 6. 2D grid topology of the unwrapped model #1, #2 and #3.

The computational analysis was performed using ANSYS Mechanical APDL Release 18 (ANSYS Inc., Canonsburg, PA, USA). In this commercial software, the simulation of compression test of biodegradable stents was performed by a static structural analysis with the aim of determining the load-displacements diagram and eventual additional variables such as the principal stresses and strains. The grids generated with the in-house code described above were imported in Ansys where the set-up of the models and the corresponding boundary conditions were applied. Steady loading was assumed. The order of element shape function was selected as linear and of the first order. The material properties of the medical silicone estimated during the experimental test were specified in Ansys for defining the material. In particular, the modulus of elasticity and the Poisson coefficient used for the simulations are 15.2 MPa and 0.29, respectively. The linear elastic behavior of the material used in the presented simulations is reasonable as the strains are achieved up to 0.2. This value is overtaken only in high compressive states and specific flexible models. Furthermore, as mentioned before, the linear elastic behavior can be correctly assumed until a strain value of about 0.37.

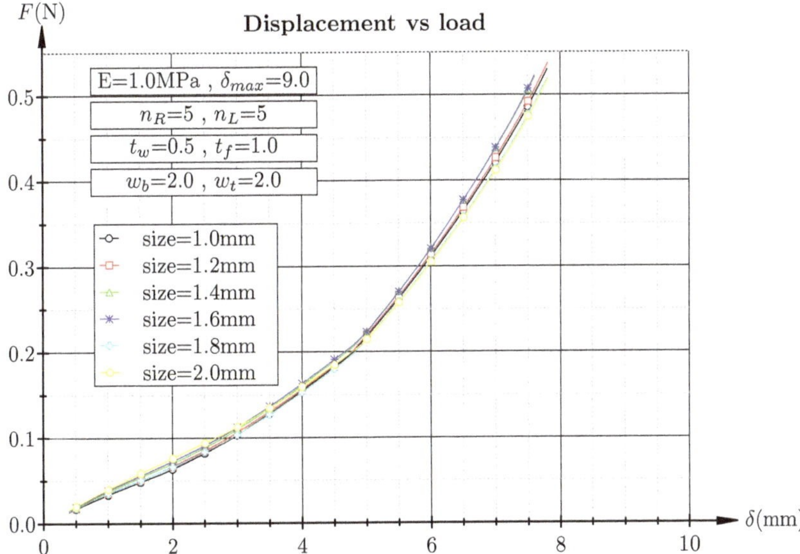

Figure 7. Mesh size sensitivity analysis: for a specific geometrical configuration, it is visible that the force-displacement curves obtaining compressing the stent varies only reasonably slightly within different grids.

Based on the experimental compression test, the lower plate was fixed, and the upper plate was moved until the desired radial displacements are reached (14 mm). The contact between the plates and the prosthesis has been carried out frictionlessly. Large displacements have been switched on in the Ansys solver. The loads and displacements were registered and plotted as shown in the next section. It has to be noted that the radial stiffness was measured only for a single orientation of the prosthesis. In particular, the prosthesis was compressed in the direction in which the trachea usually experiments with the larger displacements [11].

3. Results

3.1. Flat Plate Test Simulation

In Figure 8, the radial force is plotted versus the radial displacement for different values of the fiber bottom width w_b and fiber thickness t_f. The comparison is shown for a specific configuration in which the ratio w_t/w_b is fixed and equal to unity and the same number of cells in radial and longitudinal direction have been considered ($n_L = n_R = 5$). The plot shows an increase of radial force caused by an increase of the fiber bottom width w_b and fiber thickness t_f. For a fixed fiber thickness t_f the increase of the radial force becomes more marked for increasing fiber bottom widths w_b, being the increase of the last case ($t_f = 1, w_b = 1$ and 2) the highest (blue lines with circles).

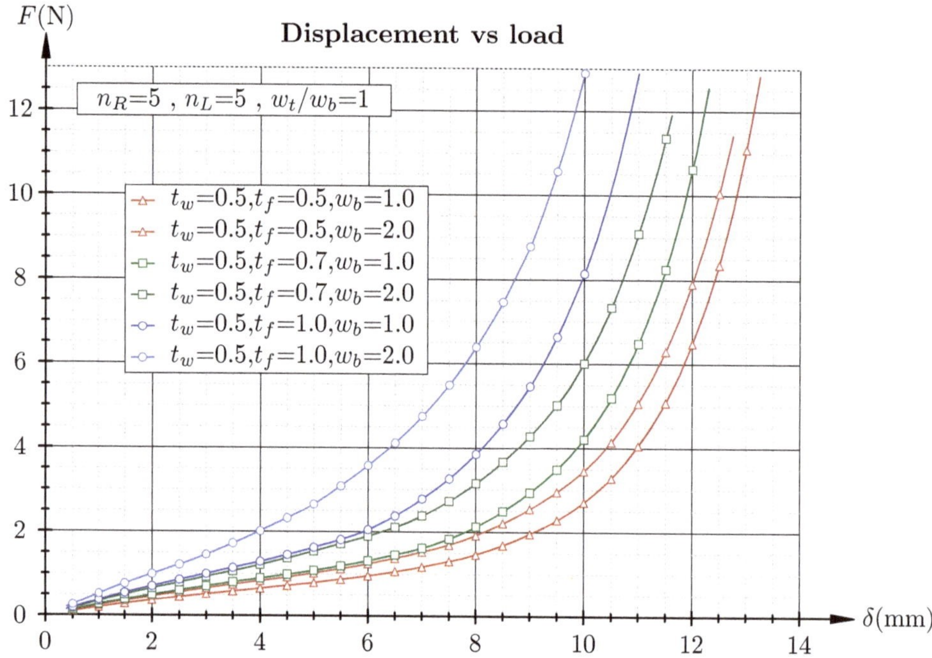

Figure 8. Variation of the radial force for selected values of fiber bottom width w_b and fiber thickness t_f.

In Figure 9, the influence of the number of cells located in circumferential (n_R) and longitudinal direction (n_L) on the radial force is analyzed. Figure 9 shows selected prosthesis configurations with a combination of $n_R = 3, 4, 5$ and $n_L = 3, 4, 5$. However, the presented results can be generalized for different values of n_R and n_L. The comparison is shown for fixed width ratio $w_t/w_b = 1$ and fiber and wall thickness $t_w = t_f = 0.5$ mm. The comparison allows clarifying the role of the cell shape on the radial stiffness. It is clearly visible in fact that an increase of the stiffness is promoted when $n_R < n_L$, i.e., for smaller pitch angle α (see Section 2.1). On the contrary, if $n_R > n_L$, i.e., for higher pitch angle α, the radial stiffness decreases. Summarizing, fixing the fiber and wall thickness and the top and bottom fiber width, a prosthesis with $\alpha < 45°$ is stiffer than a prosthesis with $\alpha > 45°$. For $\alpha = 45°$ the shape of the stent cell is squared and this configuration corresponds to a change of the trend. Maintaining $\alpha = 45°$, the plot of Figure 9 highlights that if the number of cells increases, the radial stiffness also increases as demonstrated by a comparison between the black line with triangle ($n_L = n_R = 3$), the red line with squares ($n_L = n_R = 4$) and the green line with circles ($n_L = n_R = 5$). In any case, the increase is not extremely marked, as demonstrated by other cases such as $n_L = n_R = 3$ and $n_L = n_R = 4$.

In Figure 10, the total thickness of the prosthesis fixed and equal to 1.5 mm is distributed to the wall thickness t_w and to the fiber thickness t_f. Different percentages are given to these two parameters, keeping their sum constant. The figure shows the case of number of radial and longitudinal cells $n_R = 3$ and $n_L = 4$, respectively. As for the previous plots, the results can be generalized for other configurations with different numbers of cells (with $n_R < n_L$). Additionally, the bottom and top width are $w_b = 2$ mm and $w_t = 1.2$ mm. The figure shows that, as expected, a higher radial stiffness can be obtained if $t_w > t_f$. Notwithstanding, the radial stiffness can be increased when $t_w < t_f$, changing the fiber dimensions, increasing w_b and eventually w_t.

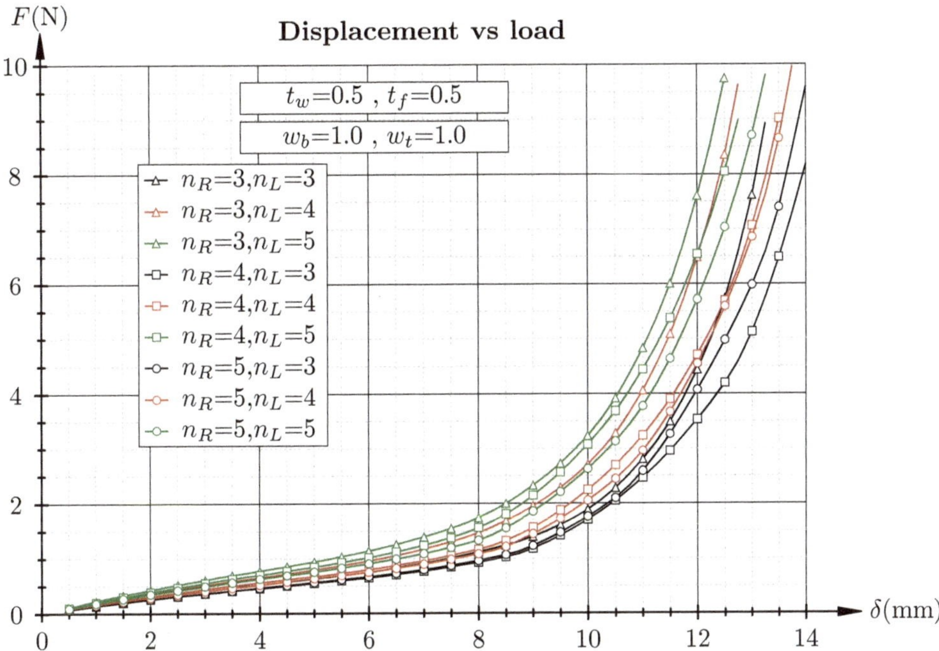

Figure 9. Variation of the circumferential and longitudinal cells number.

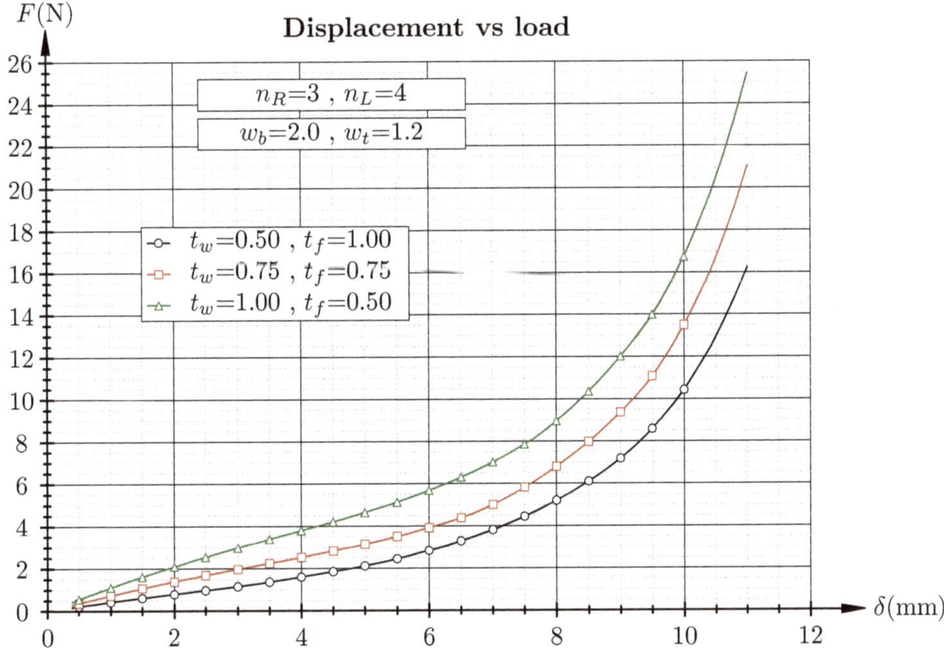

Figure 10. Variation of the thickness distribution within the prosthesis.

In Figure 11, the variation of the fiber shape is studied varying the bottom width w_b and the bottom and top width ratio w_t/w_b. The figure clearly highlights that an increase of the fiber width promotes a increase of the radial stiffness for a fixed wall thickness t_w and fiber thickness t_f. In the figure, it is shown the specific case in which again $n_R = 3$ and $n_L = 4$. Furthermore, in this case, similar curves have been obtained for different configurations (with $n_R < n_L$).

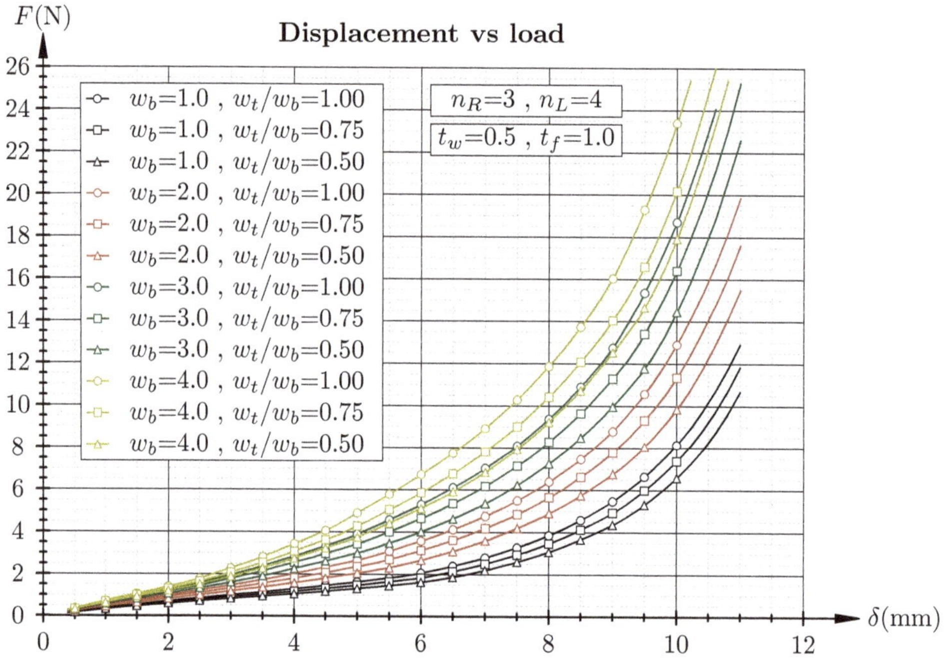

Figure 11. Variation of the fiber bottom width within the prosthesis.

In Figure 12, the stiffness of selected prosthesis configurations is compared. The figure illustrates that different stiffness can be obtained, modulating the different parameters in which the prosthesis has been designed. In the figure, we can see the specific configuration $n_R = 3$ and $n_L = 4$ as in the Figures 10 and 11. The figure shows the case of prosthesis with the same total thickness $t_w + t_f = 1.5$ mm. As visible, the stiffness of the prosthesis with the wall thickness $t_w = 0.75$ mm (black line with squares, Figure 12 is very similar to that obtainable with a reduced wall thickness $t_w = 0.5$ mm and a higher fiber bottom width w_b = 2 mm, with a constant fiber thickness $w_b/w_f = 1$ (red line with circles). Additionally, the same stiffness could be also obtained using a fiber bottom width w_b = 3 mm and a ratio w_b/w_f = 0.5 (green line with triangles). In the same way, the prosthesis with t_w = 1 mm, with a fiber bottom width of w_b = 2 mm and a ratio $w_b/w_f = 0.6$ (black line with triangles, Figure 12) has a similar stiffness of a prosthesis with wall thickness $t_w = 0.5$ mm, with a fiber bottom width of w_b = 4 mm and a ratio $w_b/w_f = 0.5$ (yellow line with triangles in the Figure 12) and of a prosthesis with wall thickness $t_w = 0.5$ mm, with a fiber bottom width of w_b = 3 mm and a ratio $w_b/w_f = 0.5$ (green line with circles). In this last case, the stiffness is different among the three configurations at the beginning of the deformation (for a compressive force in the range 0–7 N). Here, a thicker tube promotes a stiffer prosthesis of course. For higher deformations (after a displacement $\delta > 7.5$ mm), the stiffness of the three prostheses tends to be the similar.

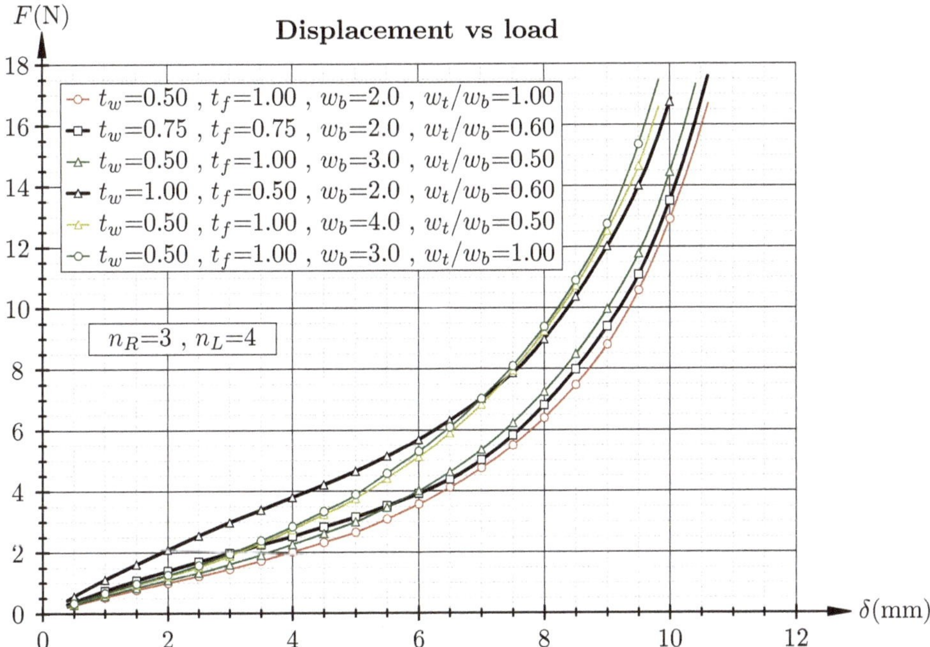

Figure 12. Comparison of different radial compression force within different prostheses.

The equivalent Von Mises stress and strain evolution during the compression test is shown in Figure 13a,b. The figure refers to specific configurations indicated in a and b. Nevertheless, this behavior is the same in all the configurations. Von Mises stresses and strains are initially located at the top of the prosthesis surface (Figure 13c) and tend to increase. Furthermore, at the prosthesis sides, the stresses and strains tends to increase, until the values at both locations meet (close to the 68% of the total compression, Figure 13a,d. Then, the maximum stress changes location from the top to the side of the prosthesis surface (Figure 13e).

3.2. Flat Plate Experimental Testing

The comparison of the simulations results with the experimental test performed on the three 3D-printed prostheses allows the validation of the computational parametric framework. In Figure 14, the compressive force is represented as a function of the radial displacements for both in silico and experimental model for the prototype #1 (a), #2 (b) and #3 (c). The results offer a very good match. In particular, an excellent match can be seen in the range of radial displacements between zero and 8 mm for all the prototypes while between 8 mm and 15 mm the curves tend to separate. Prototypes #1 and #2 seem to moderately overestimate the force in the displacement range 8–14 mm while prototype #3 seems to slightly underestimate the necessary force in the same interval. The maximum difference between curves in this range is 1 mm for prototype #1, 0.5 mm for prototype #2 and 0.4 mm for prototype #3. However, the agreement between curves for prototypes #1 and #2 is very good until a a radial displacement of 10 mm while the prototype #3 shows a separation between experimental and computational curve that starts already at a radial displacement of 2 mm. From this displacement, the separation between the curves increases moderately, it stays almost constant between 8 and 10 mm then further increases, even very slightly, between 10 and 14 mm. The reason of these differences at higher displacements may be due to the different value of the friction in the experimental and computational study and to the complex deformation of the prototypes at crushing. In

Figure 4b, the prototype is compressed between plates. Unfortunately, in the computational analysis, the complete buckling cannot be obtained.

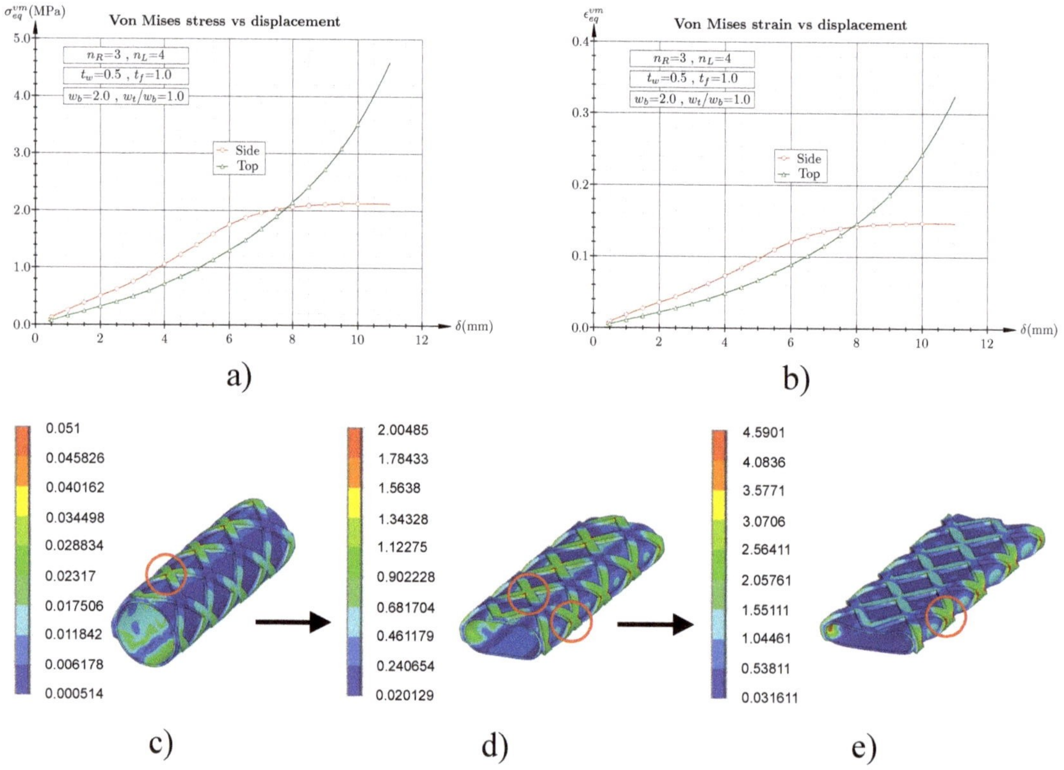

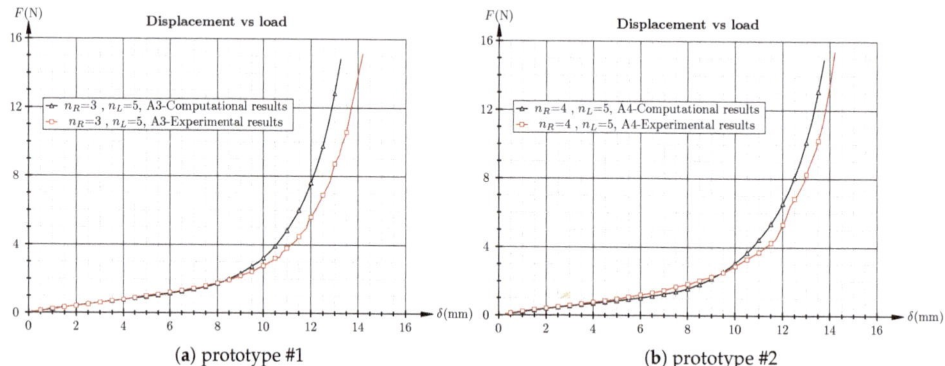

Figure 13. Evolution of the equivalent Von Mises stress (**a**), and strain (**b**) during the compressive load simulation for a specific node on the top and side of the prosthesis. The stress distribution on the prosthesis (in [MPa]) during the compression is sketched at different instants during the compression in (**c**–**e**). The red circle highlights the location of the maximum equivalent Von Mises stress.

Figure 14. *Cont.*

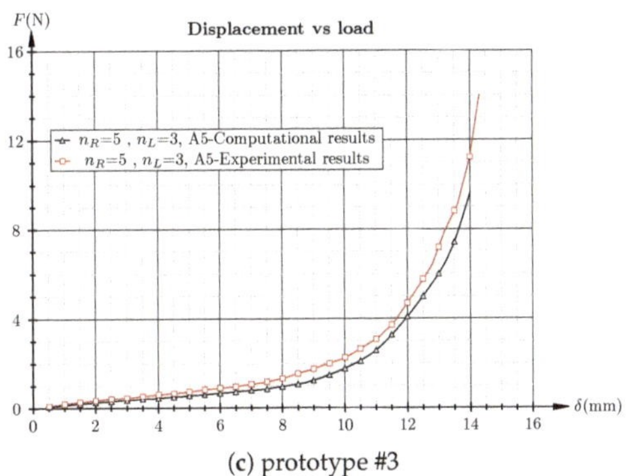

(c) prototype #3

Figure 14. Experimental and computational results: flat plate test of the printed prototypes.

4. Discussion

The goal of this study was the evaluation of the mechanical properties of a new tracheobronchial stent prototype, through the use of computational and experimental modeling. For this reasons, we have presented a parametric in-house tool capable of designing and meshing several geometrical configurations of the prosthesis. Then, the effect of each single parameter variation on the stent radial stiffness was analyzed by means of numerical simulations. In the presented tool, there are additional parameters that could be changed and even they do not contribute to the mechanical performance of the prosthesis, they are useful for creating a patient specific device. For example, the typical main variables such as the inner radius and the length of the prosthesis can be adapted to the patient necessities and a customized device can be rapidly created. In the literature, it is stated that one of the important requirements of a medical device should be patient-customizable [17]. Progress towards customization of commercial airway stents has been made in the recent years. In fact, many manufacturers already produce personalized prostheses. Parameters such as prosthesis shape, size, diameter and angles can be directly measured using CT scan and bronchoscopy [10]. Notwithstanding, the personalization of the device is based on the commercial design of the prosthesis of each fabricant and only a few dimensions can be changed. As a result, almost all stents are still straight and round-shaped [16]. Clinical studies have reported complications [10] so that the use of such stents has been demonstrated to not be definitive. In many cases, of course the use of customized commercial prosthesis could be a good compromise, but frequently this is not sufficient [16]. Thus, the proposed methodology offers the possibility of addressing a part of these limitations. In addition to the personalization of the main typical dimensions, the creation of an individual prosthesis using the degree and the type of the lesion and its exact location, for instance, could be possible.

Several different polymers can be adopted in medicine, taking advantage of their specific properties [16]. In general, polymeric stents are made from silicone, and only a few contain additional copolymers and additives. The material used to produce the stent must exhibit a good resistance to the deformation as it is placed into the bronchoscope and then, once open inside the trachea or the bronchi, it must be capable of adapting to the physiological activities of the respiratory system [31]. The radial strength of the prosthesis depends mainly on its thickness and, as stated in the literature, unfortunately the ratio between inner and outer diameter is unfavorable for silicone stents respect to that of the metallic stents [30]. Due to their much lower modulus of elasticity, silicone prostheses have in fact normally a thicker walls respect to metallic stents. Despite this fact, the Food and Drug Administration (FDA) recommends the insertion of a metallic stent only when the pathology cannot be treated by other means such as surgery or insertion of silicone stents [34,35]. For this reason, silicone prostheses is still widely used and their design needs to be improved for reducing their post placement complications [32].

With the introduction of the additive manufacturing in the medical field, patient-specific implants made from 3D-printing technique represents a new opportunity for the airway surgical treatment and stent design [26]. Differently from the patient-adapted commercial devices, the 3D-printing based on the patient CT-images offers a new opportunity for a more accurate stent choice and on-site customization [10,18]. In the very recent years, the 3D printing and prosthesis customization has been entered in the clinics and several studies have reported the associated experience [18,23–26,36]. Nevertheless, in these studies, the radial force, that is the most important mechanical property of the prosthesis is only estimated, using the difficulty to pass over the stenotic portion [10]. Thus, the presented computational tool could be helpful as the customization is performed with the computation of the radial stiffness of the device. The prosthesis could be tailored to the patient, and the necessary prosthesis radial stiffness could be computed as a function of the degree of stenosis and the type of lesion. With this information, among others, the parameters of the stent design could be optimized, retaining the estimated radial stiffness, even though, as discussed, this information is difficult to be obtained [10]. In the literature, a few optimization methodologies have been proposed, especially for cardiovascular stent [37,38]. Nevertheless, the necessary radial stiffness of a prosthesis remains a variable depending on several aspects. In any case, with the estimated values, an optimization of the prosthesis with the stent radial stiffness as goal may improve the actual devices and the clinical results.

The presented work demonstrates that the thickness of the medical device can be reduced by adequately increasing the external fibers thickness, for instance. This could have important applications to the stent design, as the obstruction and mucus plugging caused by these devices is one of the more frequent problem after the surgery and it is caused by the thickness of the prosthesis [16]. The outer fibers add in fact radial strength without the necessity of increasing the entire device thickness, yet limiting the use of the material in the sites where necessary. Then, similar radial stiffnesses can be obtained accurately modifying the design. The compression tests indicated that at high deformation and at buckling, the prosthesis shows a high compressive strength. The latter can be obtained varying the fibers pitch angle, once fixed all the other parameters. In particular, the study shows that a reduction of the pitch angle promotes an increase of the radial stiffness. Additionally, as discussed above, the increase of stiffness can be obtained increasing other parameters, such as the fiber bottom width without increasing the fiber thickness. In this context, it is clear that the entire design of the prosthesis can be adapted to the patient and its clinical situation.

5. Limitations

Limitations of the work include the use of the type of loading applied to the tracheobronchial stent. These are difficult to be reproduced in experimental settings even though extremely important. The loading conditions have been reported to influence the way how to evaluate the stent design [39]. In the present study we have performed a flat compressive tests, but a nonuniform compression is probably more adequate as it is clear looking at the tracheal physiology and the loading conditions of the human trachea [39]. Furthermore, a comprehensive computational analysis that takes into account the interaction between prosthesis and biological tissue would also be interesting and necessary. The latter could in fact assess the regions where higher stresses are located during the physiological maneuvers. These regions, that normally are located in the proximity of the device have been found to produce tissue reaction and inflammation [40]. In addition, an experimental study needs to be carried out for assessing the foreign body reaction after implantation and observe in situ possible re-epithalization, inflammation and granulation among other biological responses. While this study is pursued in parallel work, in the present study we have focused on the mechanical properties and on the capability to design customizable devices and in the generation of a useful tool for prosthesis design, customization and analysis. In a next step, the interaction of the device with the human patient specific trachea could be also taken into account. Because of the patient variability, of course this aspect is challenging. In the literature, it is suggested for example a categorization of patients and of pathologies [37]. Besides trying to overcome the aforementioned limitations, it is of course possible to extend the proposed parametric strategy by implementing additional design parameters, such as the material of the prosthesis, and the composition of the fibers. The presented study is limited to the use of one specific silicone type. A composition of different polymers or silicones may add useful information for the optimal design or patient customization of the tracheo-bronchial prosthesis. Of course, the analysis of different polymers including, in the future, biodegradable materials, could be interesting. Finally, even though the presented method is aimed to overtake some of the existing stenting technique limitations, it has to be accepted that a prosthesis is a foreign body, and as such, it will always affect the tissue and promote obstruction [16].

6. Conclusions

This study proposed a computational tool for designing and analyzing a new tracheobronchial stent prototype. By means of an in-house code, a baseline model has been parametrized and meshed, allowing a comprehensive finite element analysis. The computational simulations consider several variations of the presented main parameters for elucidating their effect on the stent radial stiffness. Furthermore, in the code, additional parameters such as the inner diameter and the length of the prosthesis can be changed so that the presented parametric tool allows a customization of the device to the patient necessities. The results of the computational study show that the radial stiffness of the prosthesis increases for decreasing pitch angle. Additionally, the radial stiffness also increases if the fiber width or the fiber thickness increases, i.e., the cells inner dimensions reduce. The presented tool allows a manipulation of the fibers geometry for obtaining different prostheses with equivalent radial stiffnesses by reducing for example the tube thickness, the value of which is normally one of the most important problems of the silicone stent. Lastly, the computational study was validated by means of an experimental study performed on three selected geometries. The latter proves that the in-house code and subsequent simulations are reliable and the presented computational tool could be used for the design of printable patient specific tracheobronchial stents.

Author Contributions: Study design, M.M., J.Z.-G, C.S.-M., J.L.L.-V.; conceptualization, J.Z-G, J.L.L.-V., M.M.; stent geometries, M.M., J.L.L.-V.; numerical model and simulations, J.Z.-G; data curation experimental model, C.D.-J., M.M.; writing—original draft preparation, M.M., C.S.-M., J.Z.-G, C.D.-J., J.L.L.-V.; writing—review and editing, M.M., C.S.-M., J.L.L.-V., J.Z.-G, C.D.-J.; supervision, M.M and J.L.L.-V.; funding acquisition, M.M. All authors have read and agreed to the published version of the manuscript.

Funding: The research is supported by the Spanish Ministry of Economy, Industry and Competitiveness through research project DPI2017-83259-R (AEI/FEDER,UE).

Institutional Review Board Statement: Not applicable.

Informed Consent Statement: Not applicable.

Data Availability Statement: Not applicable.

Acknowledgments: The support of the Instituto de Salud Carlos III (ISCIII) through the CIBER-BBN initiative is gratefully acknowledged.

Conflicts of Interest: The authors declare no conflict of interest.

References

1. Folch, E.; Keyes, C. Airway stents. *Ann. Cardiothorac. Surg.* **2018**, *7*, 273–283. [CrossRef]
2. Ratnovsky, A.; Regeva, N.; Walda, S.; Naftali, M.K.S. Mechanical properties of different airway stents. *Med. Eng. Phys.* **2015**, *37*, 408–415. [CrossRef] [PubMed]

3. Dumon, J.; Cavaliere, S.; Díaz-Jimenez, J.P.; Vergnon, J.; Venuta, F.; Dumon, M.; Kovitz, K. Seven-Year Experience with the Dumon Prosthesis. *J. Bronchol.* **1996**, *3*, 6–10. [CrossRef]
4. Gaafar, A.H.; Shaaban, A.Y.; Elhadidi, M.S. The use of metallic expandable tracheal stents in the management of inoperable malignant tracheal obstruction. *Eur. Arch. Otorhinolaryngol.* **2012**, *269*, 247–253. [CrossRef] [PubMed]
5. Sun, F.; Uson, J.; Ezquerra, V.; Crisostomo, V.; Luis, L.; Maynar, M. Endotracheal stenting therapy in dogs with tracheal collapse. *Vet. J.* **2008**, *175*, 186–193. [CrossRef]
6. Charokopos, N.; Foroulis, C.N.; Rouska, E.; Sileli, M.N.; Papadopoulos, N.; Papakonstantinou, C. The management of post-intubation tracheal stenoses with self-expandable stents: Early and long-term results in 11 cases. *Eur. J. Cardiothorac. Surg.* **2011**, *40*, 919–925. [CrossRef]
7. Freitag, L.; Eicker, K.; Donovan, T.J.; Dimov, D. Mechanical properties of airway stents. *J. Bronchol. Interv. Pulmonol.* **1995**, *2*, 270–278. [CrossRef]
8. Food and Drug Administration. FDA Public Health Notification: FDA Recommends Avoiding Use of Metallic Tracheal Stents. *Food Drug Adm.* 2019, Available online: https://www.medscape.org/viewarticle/510184 (accessed on 15 March 2021) .
9. Dutau, H.; Breen, D.; Bugalho, A.; Dalar, L.; Daniels, J.; Dooms, C.; Eberhardt, R.; Ek, L.; Encheva, M.; Current practice of airway stenting in the adult population in Europe: A survey of the European Association of Bronchology and Interventional Pulmonology (EABIP). *Respiration* **2018**, *95*, 44–54. [CrossRef]
10. Guilbert, N.; Saka, H.; Dutau, H. Airway stenting: Technological advancements and its role in interventional pulmonology. *Respirology* **2020**, *25*, 953–962. [CrossRef]
11. Malvè, M.; Pérez del Palomar, A.; López-Villalobos, J.L.; Ginel, A.; Doblaré, M. FSI Analysis of the Coughing Mechanism in a Human Trachea. *Ann. Biomed. Eng.* **2010**, *38*, 1556–1565. [CrossRef]
12. Malvè, M.; Pérez del Palomar, A.; Mena, A.; Trabelsi, O.; López-Villalobos, J.L.; Ginel, A.; Panadero, F.; Doblaré, M. Numerical modeling of a human stented trachea under different stent designs. *Int. Commun. Heat Mass Transf.* **2011**, *38*, 855–862. [CrossRef]
13. Malvè, M.; Pérez del Palomar, A.; Chandra, S.; López-Villalobos, J.L.; Ginel, E.F.A.; Doblaré, M. FSI Analysis of a Human Trachea Before and After Prosthesis Implantation. *J. Biomech. Eng.* **2011**, *133*, 071003. [CrossRef] [PubMed]
14. Schopf, L.F.; Fraga, J.C.; Porto, R.; Santos, L.A.; Marques, D.R.; Sanchez, P.R.; Meyer, F.S.; Ulbrich, J.M. Experimental use of new absorbable tracheal stent. *J. Pediatr. Surg.* **2018**, *53*, 1305–1309. [CrossRef] [PubMed]
15. Novotny, L.; Crha, M.; Rauser, P.; Hep, A.; Misik, J.; Necas, A.; Vondrys, D. Novel biodegradable polydioxanone stents in a rabbit airway model. *J. Thorac. Cardiovasc. Surg.* **2012**, *143*, 437–444. [CrossRef] [PubMed]
16. Freitag, L.; Gordes, M.; Zarogoulidis, P.; Darwiche, K.; Franzen, D.; Funke, F.; Hohenforst-Schmidt, W.; Dutau, H. Towards individualized tracheobronchial stents: technical, practical and legal considerations. *Respiration* **2017**, *94*, 442–456. [CrossRef]
17. Xu, J.; Ong, H.X.; Traini, D.; Byrom, M.; Williamson, J.; Young, P.M. The Utility of 3D-Printed Airway Stents to Improve Treatment Strategies for Central Airway Obstructions. *Drug Dev. Ind. Pharm.* **2019**, *45*, 1–10. [CrossRef]
18. Gildea, T.R.; Young, B.P.; Machuzak, M.S. Application of 3D printing for patient-specific silicone stents: 1-year follow-up on 2 patients. *Respiration* **2018**, *96*, 488–494. [CrossRef]
19. Ghosh, S.; Burks, A.C.; Akulian, J.A. Customizable airway stents- personalized medicine reaches the airways. *J. Thorac. Dis.* **2019**, *11*, S1129–S1131. [CrossRef]
20. Food and Drug Administration. 3D Printing of Medical Devices. *Food Drug Adm..* 2018. Available online: https://www.fda.gov/medical-devices/products-and-medical-procedures/3d-printing-medical-devices (accessed on 15 March 2021).
21. Nakada, T.; Akiba, T.; Inagaki, T.; Morikawa, T. Thoracoscopic anatomical subsegmentectomy of the right S2b + S3 using a 3D printing model with rapid prototyping. *Interact. Cardiovasc. Thorac. Surg.* **2014**, *19*, 696–698. [CrossRef]
22. Malik, H.H.; Darwood, A.R.; Shaunak, S.; Kulatilake, P.; El-Hilly, A.A.; Mulki, O.; Baskaradas, A. Three-dimensional printing in surgery: A review of current surgical applications. *J. Surg. Res.* **2015**, *199*, 512–522. [CrossRef]
23. Guibert, N.; Moreno, B.; Plat, G.; Didier, A.; Mazieres, J.; Hermant, C. Stenting of complex malignant central-airway obstruction guided by a three-dimensional printed model of the airways. *Ann. Thorac. Surg.* **2017**, *103*, e357–e359. [CrossRef]
24. Miyazaki, T.; Yamasaki, N.; Tsuchiya, T.; Matsumoto, K.; Takagi, K.; Nagayasu, T. Airway stent insertion simulated with a three-dimensional printed airway model. *Ann. Thorac. Surg.* **2015**, *99*, e21–e23. [CrossRef]
25. Cheng, G.Z.; San Jose Estepar, R.; Folch, E.; Onieva, J.; Gangadharan, S.; Majid, A. Three-dimensional printing and 3D slicer: powerful tools in understanding and treating structural lung disease. *Chest* **2016**, *149*, 1136–1142. [CrossRef] [PubMed]
26. Morrison, R.J.; Hollister, S.J.; Niedner, M.F.; Ghadimi Mahani, M.; Park, A.H.; Mehta, D.K.; Ohye, R.G.; Green, G.E. Mitigation of tracheobronchomalacia with 3D-printed personalized medical devices in pediatric patients. *Sci. Transl. Med.* **2015**, *7*, 285ra64. [CrossRef]
27. Debiane, L.; Reitzel, R.; Rosenblatt, J.; Gagea, M.; Chavez, M.A.; Adachi, R.; Grosu, H.B.; Sheshadri, A.; Hill, L.R.; Raad, I.; et al. A Design-Based Stereologic Method to Quantify the Tissue Changes Associated with a Novel Drug- Eluting Tracheobronchial Stent. *Respiration* **2019**, *98*, 60–69. [CrossRef]
28. Melgoza, E.L.; Serenó, L.; Rosell, A.; Ciurana, J. An integrated parameterized tool for designing a customized tracheal stent. *Comput.-Aided Des.* **2012**, *44*, 1173–1181. [CrossRef]
29. Melgoza, E.L.; Vallicrosa, G.; Sereno, L.; Rosell, A.; Rodríguez, C.; Elias, A.; Ciurana, J. Rapid Tooling Using 3D Printing System For Manufacturing Of Customized Tracheal Stent. *Rapid Prototyp. J.* **2014**, *20*, 2–12. [CrossRef]

30. Gastal Xavier, R.; Stefani Sanches, P.R.; Viera de Macedo Neto, A.; Kuhl, G.; Bianchi Vearick, S.; Dall'Orden Michelon, M. Development of a modified Dumon stent for tracheal applications: an experimental study in dogs. *J. Bras. Pneumol.* **2008**, *34*, 21–26.

31. Bianchi Vearick, S.; Bendo Demfrio, K.; Gastal Xavier, R.; Moreschi, A.H.; Frotta Muller, A.; Stefani Sanches, P.R.; Loureiro dos Santos, L.A. Fiber-reinforced silicone for tracheobronchial stents: An experimental study. *J. Mech. Behav. Biomed. Mater.* **2018**, *77*, 494–500. [CrossRef]

32. Dumon, F. A dedicated tracheobronchial stent. *Applied Sciences* **1990**, *97*, 328–332. [CrossRef] [PubMed]

33. Hye Yun Park.; Hojoong Kim.; Won-Jung Koh.; Gee Young Suh.; Man Pyo Chung.; O Jung Kwon. Natural stent in the management of post-intubation tracheal stenosis. *Respirology* **2009**, *14*, 583–588. [CrossRef]

34. Lund, M.E.; Force, S. Airway Stenting for Patients With Benign Airway Disease and the Food and Drug Administration Advisory. *Chest J.* **2007**, *132*, 1107–1108. [CrossRef]

35. Food.; Administration, D. Airway stenting for benign tracheal stenosis: what is really behind the choice of the stent? *Eur. J. Cardiothorac. Surg.* **2011**, *40*, 924–925.

36. Guibert, N.; Mhanna, L.; Didier, A.; Moreno, B.; Leyx, P.; Plat, G.; Mazieres, J.; Hermant, C. Integration of 3D printing and additive manufacturing in the interventional pulmonologist's toolbox. *Respir. Med.* **2018**, *134*, 139–142. [CrossRef] [PubMed]

37. De Beule, M.; Van Cauter, S.; Mortier, P.; Van Loob, D.; Van Impe, R.; Verdonck, P.; Verhegghe, B. Virtual optimization of self-expandable braided wire stents. *Med. Eng. Phys.* **2009**, *31*, 448–453. [CrossRef] [PubMed]

38. Gundert, T.J.; Marsden, A.L.; Yang, W.; LaDisa Jr., J.F. Optimization of Cardiovascular Stent Design Using Computational Fluid Dynamics. *J. Biomech. Eng.* **2012**, *134*, 011002. [CrossRef] [PubMed]

39. McGrath, D.J.; O'Brien, B.; Bruzzi, M.; Kelly, N.; Clauser, J.; Steinseifer, U.; McHugh, P.E. Evaluation of cover effects on bare stent mechanical response. *J. Mech. Behav. Biomed. Mater.* **2016**, *61*, 567–580. [CrossRef]

40. Chaure, J.; Serrano, C.; Fernández-Parra, R.; na, E.P.; Lostalé, F.; De Gregorio, M.A.; Martínez, M.A.; Malvè, M. On Studying the Interaction Between Different Stent Models and Rabbit Tracheal Tissue: Numerical, Endoscopic and Histological Comparison. *Ann. Biomed. Eng.* **2016**, *44*, 368–381. [CrossRef]

MDPI
St. Alban-Anlage 66
4052 Basel
Switzerland
Tel. +41 61 683 77 34
Fax +41 61 302 89 18
www.mdpi.com

Mathematics Editorial Office
E-mail: mathematics@mdpi.com
www.mdpi.com/journal/mathematics